STRUCTURE AND EVOLUTION
OF THE GALAXY

ASTROPHYSICS AND SPACE SCIENCE LIBRARY

A SERIES OF BOOKS ON THE RECENT DEVELOPMENTS

OF SPACE SCIENCE AND OF GENERAL GEOPHYSICS AND ASTROPHYSICS

PUBLISHED IN CONNECTION WITH THE JOURNAL

SPACE SCIENCE REVIEWS

VOLUME 22

STRUCTURE AND EVOLUTION

OF THE GALAXY

PROCEEDINGS OF THE NATO ADVANCED STUDY INSTITUTE
HELD IN ATHENS, SEPTEMBER 8–19, 1969

Edited by

L. N. MAVRIDIS

University of Thessaloniki

D. REIDEL PUBLISHING COMPANY

DORDRECHT-HOLLAND

Library of Congress Catalog Card Number 77–135107

ISBN 90 277 0177 6

Printed in The Netherlands by D. Reidel, Dordrecht

PREFACE

The present volume contains papers presented during the Advanced Study Institute on the 'Structure and Evolution of the Galaxy' held under the auspices of the Science Committee, North Atlantic Treaty Organization (NATO) in Lagonissi near Athens, Greece between 8–19 September 1969.

Seventeen astronomers from eight countries acted as lecturers in the Institute, which was attended by sixty-three students from twelve countries. All the lectures given in the Institute were included in the volume with the exception of two lectures given by Professor P. van de Kamp on 'Stellar Parallaxes' and 'Objects of Very Low Mass', the manuscripts of which were not submitted in time in order to be considered for publication.

It is obviously impossible in a two-weeks course to cover the vast field of the Structure and Evolution of the Galaxy in all its aspects. Effort has been made, however, to give the students a general outline of the field. In addition to the lectures ample time was spent in discussions and personal contacts. In order to simplify the publication it was felt reasonable, however, not to include the discussions in the present volume.

It is a great pleasure for the Editor in his capacity as Scientific Director and Local Organizer of the Institute to thank all those who contributed to the success of the Institute. Especially he would like to express his sincere thanks to the following organizations and persons:

(a) To the members of the NATO Science Committee and the staff of the NATO Scientific Affairs Division, who approved the necessary funds and supported in many ways the organization of the Institute. Special thanks are due to the representative of the NATO Science Committee in the Institute Professor P. Bourgeois, who not only delivered during the Inaugural Ceremony of the Institute an invited lecture on the 'New Trends in the Activities of the NATO Science Committee' but also attended the lectures and took active part in the discussions.

(b) To the Greek Government and the Greek National Tourism Organization, which approved supplementary funds for the social events program of the Institute. Many thanks are also due to the Greek National Committee for Astronomy and especially to its President Professor J. Xanthakis for his support for the organization of the Institute.

(c) To the lecturers, who kindly accepted the invitation to teach in the Institute. In particular cordial thanks are due to Professor A. Blaauw for his helpful advice and suggestions during the preparation of the scientific program of the Institute.

(d) To the students for their active contribution to the discussions of the Institute.

Finally, the Editor wishes to express his gratitude to the officials and the staff of the D. Reidel Publishing Company for the excellent care they have taken in printing this book.

THE EDITOR

TABLE OF CONTENTS

HISTORICAL DEVELOPMENT OF OUR IDEAS
CONCERNING THE STRUCTURE OF THE GALAXY

H. KIENLE

Landessternwarte Heidelberg, Germany, and Ege University Observatory, Izmir, Turkey

1. Prelude

The first ideas concerning the structure of the universe were only speculations based on the visual aspect of the sky. Demokritos (460–370 B.C), it is true, had already expressed the view that the stars would be suns of the same nature as our sun, and that the glittering band of the Milky Way consisted of small particles, but these views were not generally adopted. Copernicus and Kepler spoke only of a 'sphere of fixed stars' – *sphaera immobilis stellarum fixarum* – whose diameter would be so large compared to the diameter of the Earth's orbit around the Sun, that no parallactic displacement of the stars caused by the yearly motion of the Earth could be observed.

G. Galilei (1564–1642) from observations with his telescopes sketched maps of the Pleiades and of Orion (1610), and was able to resolve parts of the Milky Way into clouds of stars. Chr. Huyghens (1629–1695) was the first to give a rough idea of the true distances to the stars by determining what we now call the 'distance-modulus' of Sirius, adopting the luminosity of Sirius being equal to the luminosity of the Sun and its apparent brightness 800 million times smaller. The distance derived from these figures – 28000 times the distance from Earth to Sun or in modern units 0.14 parsec – is no doubt too small by a factor of at least 20, but nevertheless it gave a first idea of the order of size of stellar distances.

Another important observation is due to Halley (1656–1742), who, by comparing the apparent positions of some bright stars with the positions in the Almagest of Hipparchos, found (1718) that Arcturus, Sirius and Aldebaran were displaced by rsp. 60, 45 and 6 min of arc within 1850 years. Thus the first proper motions of 'fixed' stars became known. In 1760 Tobias Mayer (1723–62) published a catalogue of 80 proper motions which he had derived from comparisons of his observations with those of Olaus Roemer (1644–1710).

About the middle of the 18th century it was generally accepted that there exists a 'stellar system', an arrangement of the stars in space somehow symmetrical to the Milky Way. Moreover Th. Wright in a book entitled *Theory of the Universe* (1750), with wonderful engravings, uttered the opinion that the whole universe would be filled up with stellar systems in a hierarchical manner. I. Kant took over these ideas regarding the Andromeda nebula as a remote galaxy similar to our Milky Way system, and J. H. Lambert (1728–1777) published extensive speculations on the hierarchy of cosmical systems of different order which he believed to be constructed according to the same principles as the solar system, with a big mass in the centre, around which all other members would move in nearly circular orbits. The search

L. N. Mavridis (ed.), Structure and Evolution of the Galaxy, 1–16. *All Rights Reserved*
Copyright © 1971 by D. Reidel Publishing Company, Dordrecht - Holland

for a 'central sun' of the Milky Way system became one of the problems of the 19th century.

2. Dawn of Stellar Astronomy (1780–1850)

The foundations of stellar astronomy have been laid by W. Herschel (1738–1822). He introduced the method of star counts and sketched the first picture of the Galaxy (1785) from 'star gauges' with his 20-ft telescope. From the proper motions of Tobias Mayer he derived the direction of the motion of the Sun (1783), fixing the coordinates of the apex at $A = 262°$, $D = +26°$. Some time later (1805) he rediscussed the material and found $A = 247°$, $D = +49°$.

The first picture of the stellar system derived from star counts was based on the assumption that the system would be finite and that within the limits of the system the stars were distributed uniformly. Assuming further that the penetrating power of his telescope would be so large as to reach in every direction the borders of the system Herschel concluded that the number N of stars per unit area from the apparently brightest ones to the faintest would be proportional to the third power of distance r. Thus, he could outline the borders of the system according to the formula

$$r \sim N^{1/3}$$

and draw the cross section perpendicular to the plane of the Milky Way which is reproduced in Figure 1.

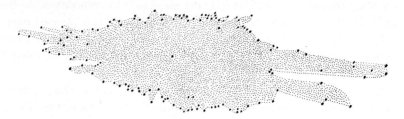

Fig. 1.   Section of the galactic system according to Sir William Herschel's first theory (1785).

But when continuing his sweeps of the sky with still larger telescopes, Herschel became more and more impressed by the patchy structure of the Milky Way and by the detection of such different features as star clusters and clouds of stars, bright and dark nebulae of different shapes, regular and chaotic, that near the end of his life (1817) he abandoned the picture of a nearly lens-shaped stellar system with uniform space distribution and replaced it eventually by a wholly different picture which is strikingly similar to most modern views of the structure of the Galaxy (Figure 2).

Herschel based his determinations of distances on photometric principles assuming the luminosities of the stars all equal to the luminosity of the Sun. He concluded from the fact that with larger and larger telescopes he found fainter and fainter stars in directions along the plane of the Milky Way, whilst in directions perpendicular to this plane the number of faint stars fades out rapidly, that the Milky Way was a stratum of small thickness but of nearly infinite extension in all directions parallel to the plane. If we transform the distance unit used by Herschel – distance of first magnitude stars – into parsecs we get for the diameter of the circle which contains all naked eye stars about 48 pc whilst the thickness of the layer amounts to about 156 pc.

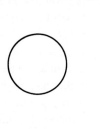

Fig. 2.   William Herschel's diagram illustrating his second theory (1817).

In this way Herschel anticipated a picture of the Galaxy which resembles that proposed by H. Shapley a century later: a flat disk containing all stars down to the faintest ones, whilst the globular clusters would be found 'in the two comparatively vacant spaces on each side adjoining the Milky Way'.

There is no remarkable progress concerning the structure of the stellar system during the first half of the 19th century. It is true F. G. W. Struve tried to derive the space density from the apparent surface distribution of the stars all over the sky, adding to Herschel's star counts other counts of stars of different magnitude in the catalogues of F. W. Argelander and of Bessel which then became available. In his analysis he assumed the stars to be arranged with constant density in planes parallel to the Milky Way, but the density decreasing towards the poles of the Milky Way being a function of the distance z from the galactic plane only. Retaining the assumption of Herschel that the luminosities of the stars are constant, and deriving the distances therefrom Struve found contradictions which he tried to remove by assuming that there is interstellar absorption which falsifies the distances calculated from apparent magnitudes. It is interesting to note that the coefficient of absorption found by Struve – 1% of the light over the distance to first magnitude stars – is just of the order of magnitude which is generally adopted today.

The problem of interstellar absorption and its influence on the determination of the structure of the universe was treated by W. Olbers (1826). His statement that the whole sky would appear in all directions "as bright as the Sun" if the infinite universe were filled with stars of equal luminosity and uniform density, is well known as 'Olbers' paradox' in classical literature. Olbers believed absorption of light in space to be the only way to escape the inference of infinite surface brightness of the sky. But as was first pointed out by H. Seeliger (1898) the paradox does not cause any difficulties if besides the luminous stars the many non-luminous masses which certainly exist in space are taken into account, the surface brightness of the sky then being equal to the mean surface brightness of all cosmical objects together.

As to the kinematics of the stellar system important work was done by F. W. Arge-lander (1799–1875), the initiator of the famous BD (Bonner Durchmusterung). He derived the coordinates of the apex from 390 proper motions larger than 0″.2 known at his time (1837) and confirmed within narrow limits the direction of the solar motion found by Herschel: $A = 261°$, $D = +32°$.

If the Sun moves relative to the neighbouring stars then it must be accepted that these stars too move relative to one another. The observed proper motions therefore may be split up into two parts: the 'parallactic motion' (μ_{par}) arising from the motion of the Sun, and the 'peculiar motion' (μ_{pec}) inherent to the stars themselves. The paral-lactic motions are of systematic character all pointing to the apex, whilst the peculiar motions were assumed to show random distribution of direction as well as size.

The interpretation of stellar motions could be done in two different ways. If the construction of the stellar system were analogous to that of the planetary system, the Milky Way plane corresponding to the ecliptic, the stars would move in Kepler orbits round a massive central mass. Since all attempts failed to identify any one of the luminous stars (Sirius, Canopus) with this hypothetical central sun, the idea of an invisible large mass in the centre of the system was propagated. It was suspected that perhaps such a large mass could be hidden behind Orion nebulae.

The other interpretation was taken from an application of the laws of kinetic theory of gases to the stellar system, comparing the motions of stars to the random motion of gas molecules. This theory played the dominant role all through the 19th century until there were found traces of systematic deviations from random distribu-tion of peculiar motions for different groups of stars. Some kind of rotation became favored in theories of stellar motions without assuming a central mass.

3. First Highlight: The Milky Way System of Classical Stellar Statistics (1890–1920)

In the last decades of the 19th century two men entered the scene who formed a model of the stellar system which was to dominate the ideas of a whole generation of astron-omers: H. Seeliger (1849–1924) and J. C. Kapteyn (1851–1922). Their contributions to the progress of stellar astronomy were quite different in nature but coincided finally in fairly the same picture of the Milky Way system, which had been generally adopted by early 20th-century astronomers.

When treating the problem of transition from apparent distribution of the stars on the sphere to true distribution in space, Seeliger favoured analytical methods. He set up the fundamental integral equations connecting observed star numbers and mean parallaxes to space density D and distribution of luminosities $\varphi(i)$ which were thought to describe the structure of the stellar system, and discussed the possible solutions of these equations on different assumptions. The observational material was taken from published star catalogues.

Kapteyn, on the other hand, very soon became aware of the fact that the observational material available at these early times was much too scanty to allow reliable solutions. He therefore aimed at stimulating observations and organizing collaboration of astronomers all over the world to procure data on properties of the stars down to faintest magnitudes. In the 'Plan of Selected Areas' (1905) an extensive programme was proposed to observers, and as a centre for treating the empirical material the Groningen Astronomical Laboratory was inaugurated. The 206 areas of the plan, each of approximately 1 degree square, comprise only $\frac{1}{120}$ of the whole surface of the celestial sphere but they are well distributed over the sky in such a manner as to reveal the essential structural features of the stellar system. Kapteyn always preferred numerical methods.

We shall not go into details of these different methods of attacking the problem but shall only sketch the primary assumptions and the final results. You will find an extensive presentation of all researches in question in a big compendium written by E. von der Pahlen (1937).

By the middle of the 19th century it was generally agreed by astronomers that the Milky Way represents a plane of symmetry of the stellar system. Herschel already had proposed the use of 'galactic' coordinates (longitude l, latitude b) referred to the galactic equator as more suited for galactic research than equatorial coordinates.

It was further assumed from the very beginning that the stellar system could be described by two functions only: the space density $D = D(r, l, b)$ and a universal distribution function of luminosities $\Phi(M)$ not varying from point to point in space.

As a first approximation Seeliger treated what he called the 'schematic system', assuming D as function of distance r only. The next approximation took into consideration the obvious variations of numbers of stars per unit surface with galactic latitude: Seeliger's 'typical system' is assumed to have rotational symmetry with respect to the Milky Way plane. Stars were therefore counted in zones between definite limits of galactic latitude.

A third stage of approximation would have been to take also into account the variation of apparent distribution with galactic longitude. This was later performed especially by C. V. L. Charlier (1862–1934) and his school in Lund. Seeliger and Kapteyn considered the fluctuations of density along galactic longitude as of minor importance, though Kapteyn once tried to derive an eccentric position of the Sun from traces of systematic variation of star numbers along the galactic belt.

If we denote the number of stars per unit surface of the sphere between apparent magnitudes m and $m + \mathrm{d}m$ by $A(m)\mathrm{d}m$, the mean parallax by $\bar{\pi}_m$, the fundamental integral equations correlating the unknown functions D and $\Phi(M)$ to the observed quantities $A(m)$ and $\bar{\pi}_m$ may be written in the following form:

$$A(m)\,\mathrm{d}m = \omega\,\mathrm{d}m \int_0^\infty D(r)\,\Phi(m + 5 - 5\log r)\,r^2\,\mathrm{d}r, \qquad (1)$$

$$\bar{\pi}_m \cdot A(m) = \omega \int_0^\infty D(r)\, \Phi(m + 5 - 5\log r)\, r\, dr. \qquad (2)$$

It was found that the $A(m)$ could be represented by

$$\log A(m) = a + bm - cm^2, \qquad (3)$$

where the coefficients a, b, c depend on galactic latitude. For the mean parallaxes of stars of magnitude m and proper motion μ Kapteyn used the formula

$$\bar{\pi}_{m,\mu} = a\varepsilon^{m-5.5}\mu^b, \qquad (4a)$$

a, ε and b are empirical constants. Integrating over μ we can write

$$\log \bar{\pi}_m = p + qm. \qquad (4b)$$

Seeliger started with a density law $D = D_0 r^{-\lambda}$, while Kapteyn at the proposal of K. Schwarzschild used a quadratic exponential

$$\log D(r) = a + b\log r - c(\log r)^2. \qquad (5)$$

Seeliger's system is thought to be finite, the density dropping to zero at distance R; it is nearly lens-shaped with flattening of about 1:5. Kapteyn's system fades out asymptotically to zero at infinity; the surfaces of equal density are approximately spheroids with flattening 1:5. The constants which describe both systems are reproduced in Tables I and II. Figures 3a, 3b and 4 will give an impression of the shapes.

TABLE I

Seeliger's system (1921)

Gal. lat.	λ	R
±80°	0.775	900 pc
60	0.615	1100
40	0.715	1800
20	0.675	2900
5	0.535	3625

TABLE II

Kapteyn's system (1920)

Gal. lat.	b	c	$r\,(D=0.05)$
±90°	3.97	1.54	800 pc
60	2.57	1.06	900
30	1.46	0.65	1750
0	1.65	0.59	4500

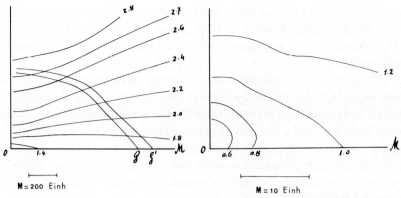

Fig. 3a. Seeliger's 'typical system' (1898). The figures represent cross sections of the surfaces of equal space density for one quadrant of the system. They have to be reflected along OM and rotated around the vertical axis to give an impression of the three-dimensional structure. The unit of distance used by Seeliger ('Siriusweite') corresponds to parallax $0''.2$, i.e. 5 parsec. The figure to the right is a 50 times enlargement of the innermost part of the figure to the left; g resp. g' represent the boundaries of the finite Seeliger system corresponding to different assumptions for the limiting magnitude of Herschel's star gauges: $13^m.5$ resp. $15^m.0$. The numbers marking the curves are the $\log r_1$, r_1 being the distance in direction of galactic latitude $80°$. The transformation of these numbers into space densities D (in units of the density near the Sun) is given by the following table

$\log r_1$	0.6	0.8	1.0	1.2	1.4	1.8	2.0	2.2	2.4	2.6	2.7	2.8
D	0.42	0.31	0.23	0.18	0.13	0.07	0.05	0.04	0.03	0.02	0.02	0.02

The overall dimensions of the system are about 9 kpc in the Milky Way plane and
5 kpc perpendicular to it.

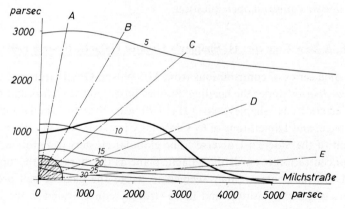

Fig. 3b. The stellar system according to Seeliger (1922).

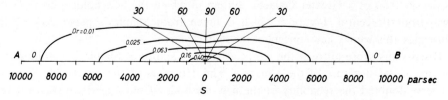

Fig. 4. The stellar system according to Kapteyn-van Rhijn (1920).

A comparison of both models shows much similarity. Only Seeliger puts the borders of the system in distance R where in Kapteyn's system the density has dropped to about 5% of central density. Kapteyn considers the surface $D=0.01$ as the approximate extension of the system.

Kapteyn's luminosity function of 1920 is a normal Gauss curve with dispersion $\sigma = \pm 2^{m}5$ and mean $M_0 = +7^{m}69$. Seeliger's curve does not much differ therefrom.

So far we have only dealt with the internal structure disregarding the motions of the stars. I must now add a few remarks concerning kinematics and dynamics. When deriving the coordinates of the apex from proper motions or radial velocities different values were found from different groups of stars. The galactic latitude indeed always came out nearly the same (a few degrees north), but the longitudes were spread over an arc of about 90° (Kobold, 1906).

In 1905 Kapteyn pointed out that the peculiar motions of the stars were not at all distributed at random as was assumed before but that there exist preferential motions parallel to the Milky Way plane. To account for this Kapteyn created the picture of two starstreams which penetrate another, moving in opposite directions, whilst Schwarzschild proposed a 'unitarian' theory: the mobility of the stars would be greater in directions parallel to the galactic plane than at right angle. The distribution of velocities could be represented by a prolate two-axial ellipsoid.

In any case from now on the kinematics of the stellar system would be characterized by two preferential directions: the old one towards the apex, and a new one towards the vertex. But how to proceed from pure kinematics to dynamics based on Newton's gravitation law was quite an open question.

4. A New Concept: H. Shapley's Greater Galactic System (1918)

From 1915 on a series of contributions from Mt. Wilson Observatory appeared in the *Astrophysical Journal* under the heading 'Studies Based on the Colors and Magnitudes of Stellar Clusters' by H. Shapley. No. VII (1918) bears the title 'The Distances, Distribution in Space, and Dimensions of 69 Globular Clusters'; No. XII 'Remarks on the Arrangement of the Sidereal Universe'. The globular clusters were found to show no apparent concentration towards the galactic plane as the stars and the 'open' galactic clusters did. On the contrary, they seemed to avoid the galactic belt proper within some degrees of galactic latitude and to be strongly concentrated in the direction of Sagittarius Cloud. The space distribution came out nearly uniform within a sphere of about 70 kpc diameter. The system of globular clusters could well be regarded as the skeleton of a Greater Galactic System in which the Sun holds a position far away from the centre. Figure 5 which is taken from Shapley's paper XIV (1918) illustrates this wholly new concept of the stellar system.

Discussions arose all over the world as to the validity of Shapley's revolutionary picture of the structure of the Universe. The defenders of the small-scale Kapteyn universe doubted the reliability of the new methods of photometric distance determination based on the absolute magnitudes of δ Cephei variables. But it was very difficult to attain a reduction of Shapley's distance scale by a factor of 5 to 10 to bring the system of globular clusters within the limits of the Seeliger-Kapteyn system without shaking the foundations of modern astronomy – belief in the physical unity of the world and the overall validity of the laws of physics.

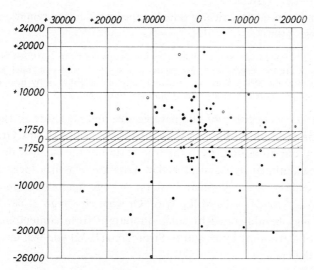

Fig. 5.   The globular clusters and the Milky Way, according to H. Shapley (1918).

In 1920 a discussion took place at the National Academy of Sciences in Washington between H. D. Curtis, defender of the small-scale universe, and H. Shapley, inaugurator of the Greater Galactic System. In the course of this discussion Curtis demonstrated the different aspects by the following confrontation:

Present theory

Our Galaxy is probably not more than 30000 light-years in diameter, and perhaps 5000 light-years in thickness.

The clusters, and all other types of celestial objects except the spirals, are component parts of our own galactic system.

The spirals are a class apart, not intragalactic objects. As island universes, of the same order of size as our Galaxy, they are distant from us 500000 to 10000000, or more, light-years.

Shapley's theory

The Galaxy is approximately 300000 light-years in diameter, and 30000, or more, light-years in thickness.

The globular clusters are remote objects, but a part of our own Galaxy. The most distant cluster is placed about 220000 light-years away.

The spirals are probably of nebulous constitution, and possibly not members of our own Galaxy, driven away in some manner from the regions of greatest star density.

It was not yet possible at that time to come to a unanimously accepted decision though the better arguments already were in favour of Shapley's larger scale universe. The final decision came just a few years later (1924) when E. B. Hubble succeeded in determining the distance of the Andromeda nebula by the same method as has been used by Shapley for globular clusters.

The situation has been very clearly stated in the introductory remarks to the Hubble Atlas of Galaxies (1961):

What are galaxies? No one knew before 1900. Very few people knew in 1920. All astronomers knew after 1924. Galaxies are the largest single aggregates of stars in the universe. Each galaxy is a stellar system somewhat like our Milky Way, and isolated from its neighbours by near empty space. In popular terms, each galaxy is a separate 'island universe' unto itself.

If now our galaxy were really of comparable size and structure to the Andromeda galaxy and if the position of our Sun were far off the centre of this Greater Galactic System then the question arose as to the true nature of the small scale stellar system which had been outlined by classical stellar statistics. Was it a local system, a subsystem of the Galaxy as the globular clusters? A huge agglomeration of stars comparable to the star clouds in the Milky Way? Or what else might it be? The very intricate situation may be illustrated by some quotations from contemporary literature. In a textbook of astronomy by E. and B. Strömgren (1932) you will find the following description of the Galaxy.

Open star clusters, globular clusters, loose clusters, Cepheids, red long-period variables, and novae create a spatial impression of the following nature: two great systems at some distance from each other; the globular clusters belong to one system, though some lie far from its centre. These two principal systems are usually referred to as the local system and the Sagittarian system.

It is quite possible, even probable, that the system, which we call local, extends along the plane of the Milky Way. There is therefore a possibility that we belong to a coherent Milky Way system which is arranged along one and the same plane with the Sagittarian system as the centre of the whole. The diameter of the system along the plane of the Milky Way would then be in the nature of 30000 parsec, and the Sun is located fairly eccentrically.

On the other hand, there is also a possibility that the Sagittarian system and the local system are both limited in their extent, which would mean that the Milky Way system is made up of these two systems, and a few globular star clusters, (lying far from the Sagittarius centre), as well as perhaps some other system or systems as yet unknown.

The idea of the entire Milky Way system as one coherent system having rotational symmetry about an axis passing through the Sagittarian centre and perpendicular to the plane of the Milky Way has in its favour the fact that it allows a single explanation of the motions of bodies in the vicinity of the Sun. Against this there is the argument that, apart from the Milky Way system, we know of no extent system whose diameter is in the order of 30000 parsecs, although super-systems, consisting of a number of smaller individual systems, and having the required overall dimensions have been observed.

From the very beginning, I opposed the idea of a local system as a separate unit. In my opinion the typical system of classical statistics could neither be identified with the Galaxy as a whole nor with an isolated local cluster of smaller size. I pretended that it were a pure phantom arising from application of non-adequate methods to statistically insufficient observational material.

One argument for this view could be found in an alarming lecture by G. Malmquist, a member of the Charlier school, at a meeting of the Astronomische Gesellschaft in 1926. He treated a model stellar system, assumed to consist of a central sphere and a concentric spherical shell, both separated by an empty zone. The space density would be constant and the same in core and shell. Adopting Kapteyn's luminosity function, the theoretical numbers $N(m)$ which an observer in the centre of the system would count could be calculated. It appeared that they could be represented well within the wide range of magnitudes from 12^m to 19^m by the quadratic expression (3).

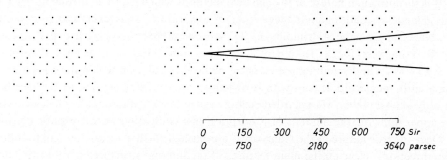

Fig. 6. Malmquist's model stellar system (1926).

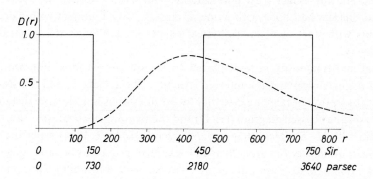

Fig. 7. Malmquist's model space density distribution $D(r)$; —— assumed; --- calculated.

Introducing the calculated constants a, b, c and again Kapteyn's luminosity function $\Phi(M)$ into the integral Equation (1), the space density $D(r)$ could be derived. The result shown in Figure 7 is quite disappointing: the calculated space density does in no respect reproduce the assumed density distribution of the model. There is no intimation of the discontinuities, the core being wholly dropped and the stars of the shell widely spread all over the empty zone.

The reason for this failure of the classical method in this special case is to be found in the large dispersion of Kapteyn's luminosity function; the sharp structural features of the model are wiped out. The same will be true for the Milky Way system if, as we are now certain, the space density does not vary continuously as was anticipated by the adopted form of the density law. Neglecting the irregularities of apparent distribution and smoothing the observed star numbers by applying simple interpolation formulae has evidently distorted the calculated space density to such an extent as to transform the spiral arms of the Galaxy in the vicinity of our Sun into a system with smooth decrease of density in all directions.

The idea that the stellar system could be described at least in a first approximation by two simple functions only, $D(r, b)$ and $\Phi(M)$, proved to be insufficient from the very beginning.

It is this principal failure of the classical methods which led to an erroneous view of the galactic system, and not 'neglection of absorption' as is explained in a recent textbook on Galactic Astronomy by D. Mihalas (1968) following a demonstration by R. Trumpler (1933). Taking into account some kind of a general absorption – exp $(-\kappa r)$ – and retaining the classical method of solution would have resulted in apparently increasing density with increasing distance from the Sun as Seeliger had already shown (1898). He regarded such a system as very improbable and calculated an upper limit for the amount of absorption just compatible with constant density.

Once the true nature of the Galaxy was established further researches had to follow two directions: clear up the main features of the internal structure (spiral arms?) and locate the position of the Sun. To do this methods had to be changed substantially. One could not any longer trust pure statistics but had to turn to the study of single objects, as Shapley had done with globular clusters, and Trumpler with galactic clusters, others with stars of definite physical properties as luminosity, spectral type or space velocity.

Spectral and luminosity criteria played from now on the most important role in outlining the extension and the internal structure of the Galaxy, and the study of interstellar absorption became an essential factor in establishing the true distance scale. The Hertzsprung-Russell diagram (HRD) and the two-dimensional spectrum-luminosity function $\Psi(M, S)$ (Hess diagram, *Seeliger-Festschrift*, 1924) replacing the one-dimensional luminosity function $\Phi(M)$ of Kapteyn, proved to be most effective tools for galactic research. Colours and colour excesses, derived by multicolour photometry, served to analyze interstellar absorption (reddening curve) and to calculate corrections to distance moduli. Study of the space distribution of stars of different spectral classes revealed the fine structure in the neighbourhood of the Sun within about 1 or 2 kpc, whilst picking out blue or red stars of highest luminosity (supergiants) allowed to trace the supposed spiral pattern of internal structure to much larger distances.

When in 1925–26 J. H. Oort and B. Lindblad developed theories of galactic rotation, the very intricate problem of stellar motions became solvable step by step. Apparent solar motion, star streaming and the 'asymmetry of stellar velocities' were explained by differential rotation. From the constants A and B of Oort's rotation formula the distance R_0 to the centre of the Galaxy and the velocity V_0 of the Sun in its orbit around the centre could be derived.

Later on the application of radio telescopes created new powerful means to penetrate to distances far beyond the range of optical telescopes, especially in the direction to the centre, where obscuration by interstellar matter reaches nearly full opacity for optical radiation. Radio astronomers first succeeded in demonstrating the very complex spiral pattern of the Galaxy, which before was only conjectured from analogy of our galaxy with spiral nebulae as M31 or M33.

Further progress in understanding the structure of our Galaxy came from comparable studies of other galaxies, especially M31, where the traces of the spirals could be followed much easier in the apparent distribution of hydrogen emission patches and of dark matter. A last steep rise of our galactic knowledge resulted in the 1950's by the introduction of the notion 'population' (W. Baade, 1952) and considerations about the evolution of stars and stellar systems, by which correlations between 'population' and 'age' could be constructed.

5. Epilogue

It is not the scope of this introductory lecture to trace the picture of the Galaxy as it appears to us today; this will be done in detail by the other lectures of this course. Let me therefore close my report by showing you some pictures which may illustrate the transitional stage as it presented itself in the 1930's from the first conception of the Greater Galactic System by Shapley to a representation of its most probable structure.

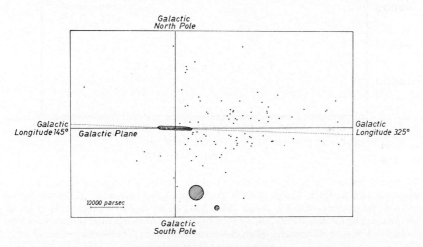

Fig. 8.   Space distribution of open clusters, globular clusters and the
Magellanic Clouds (Trumpler, 1930).

Figure 8 shows the large scale Shapley system of globular clusters (not yet corrected for absorption) and the eccentric location of the supposed local system. Figure 9, drawn to a ten times larger scale, represents the special features of this local system which Trumpler thought to coincide with the system of galactic ('open') clusters. One could imagine groups of clusters to be arranged along vaguely defined spiral arms whilst clouds of dark matter might be responsible for distortions of the apparent distribution.

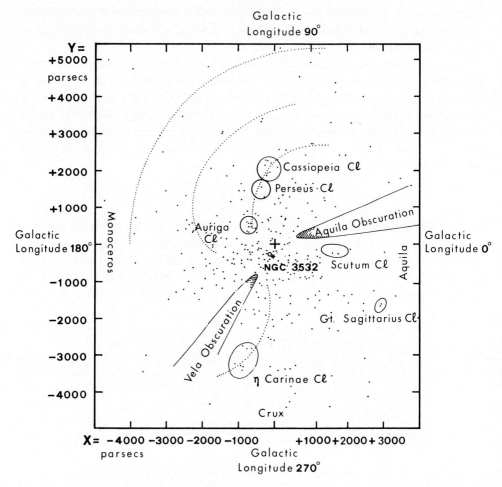

Fig. 9.　Special features of the cluster system (Trumpler, 1930). This figure gives a projection of the clusters on the galactic plane. The position of the Sun is marked by a cross, that of the median point of the system by an open circle, the cluster NGC 3532 by an asterisk. Some traces of spiral structure are indicated by the dotted lines; the large open circles or ovals represent the probable location of cluster groups or star clouds; the shaded areas, dark clouds of absorbing material with their sectors of obscuration.

In Figure 10 Oort (1938) has tried to trace curves of equal space density in a plane through the Sun perpendicular to the galactic plane and in direction to the galactic centre ($x z$-plane). He finds the density decreasing with height above resp. below the galactic plane, but increasing in both directions of x. It looks like a cross-section through two spiral arms, the Sun lying at the outer border of the inner arm.

Finally we reproduce a table (Table III) first shown at a Conference on Stellar Populations (Rome, 1957) which may be considered as a synopsis of the characteristic features of galactic structure and at the same time as a guide to coordination of future galactic research (Blaauw *et al.*, 1959).

TABLE III

Subdivision of objects in the Galaxy into population groups

Population	Halo population II	Intermediate population II	Disk population		Older population I	Extreme population I
			Planet. nebulae, bright red giants, novae	Weak-line stars		
Typical members	Sub-dwarfs, glob. clusters, RR Lyrae var. with period $> 0^d\!.4$	High-vel. FM stars, long-period variables	Planet. nebulae, bright red giants, novae	Weak-line stars	Strong-line stars, A stars Me dwarfs	Gas, super-giants, T Tauri stars
z (pc)	2000	700	450	300	160	120
Z (km/sec)	75	25	18	15	10	8
Axial ratio of sub-system	2	5	$\simeq 25$	–	–	100
Concentration toward centre	Strong	Strong	Strong	?	Little	Little
Distribution	Smooth	Smooth	Smooth	?	Patchy, spiral arms	Extremely patchy, spiral arms
Heavy element content (Schwarzschild)	0.003	0.01	–	0.02	0.03	0.04
Age (10^9 years)	6	6.0–5.0	5	1.5–5	0.1–1.5	< 0.1
Total mass (10^9 ⊙)	16	47			5	2

16 H. KIENLE

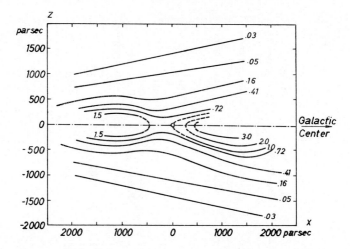

Fig. 10. General density distribution of all types of stars together according to Oort (1938), in the intersection of the galactic system with a plane perpendicular to that of the Galaxy, passing through the Sun and the centre of the system. The dotted parts of the equidensity lines are extrapolated. Unit of density is the density near the Sun.

References

Blaauw, A., Larsson-Leander, G., Roman, N. G., Sandage, A., Thackeray, A. D., and Weaver, H. F. (ed.): 1959, *Second Conference on Co-ordination of Galactic Research*, IAU Symposium No. 7, Cambridge University Press.
Bok, B. J.: 1937, *The Distribution of the Stars in Space*, Chicago.
Charlier, C. V. L.: 1926, *The Motion and Distribution of the Stars*, California Lectures.
de Sitter, W.: 1932, *Kosmos*, Cambridge, Mass.
Eddington, A. S.: 1914, *Stellar Movements and the Structure of the Universe*.
Kapteyn, J. C. and van Rhijn, P. J.: 1920, *Astrophys. J.* **52**, 23 and 300.
Kienle, H.: 1922, *Naturwiss.* **10**, 679.
Kienle, H.: 1969, *Modern Astronomy*, Thomas Y. Crowell Co., New York.
Kobold, H.: 1906, *Der Bau des Fixsternsystems*, Braunschweig.
Malmquist, G.: 1926, *VJS Astron. Ges.* **61**, 244.
Mihalas, D.: 1968, *Galactic Astronomy*, Freeman and Co., San Francisco.
O'Connell, J. K. (ed.): 1958, *Stellar Populations*, Ric. Astron. Specola Astron. Vatic. **5**, North-Holland Publishing Co., Amsterdam.
Oort, J. H.: 1938, *Bull. Astron. Inst. Neth.* **8**, 308.
von der Pahlen, E.: 1937, *Lehrbuch der Stellarstatistik*, Barth, Leipzig.
Schwarzschild, K.: 1916, *Über das System der Fixsterne*, Leipzig.
Seeliger, H.: 1909, *Sitz.-Ber. kgl. Bayr. Akad. Wiss., math.-phys. Kl.*, **4**. Abh.
Seeliger, H.: 1924, *Festschrift zum 75. Geburtstag*, Berlin.
Shapley, H.: 1930, *Star Clusters*, Harvard Monographs No. 2.
Shapley, H. and Curtis, H. D.: 1921, *Bull. Natl. Res. Council*, **2**, Part 3, Washington.
Trumpler, R. J.: 1930, *Lick Observ. Bull.* **14**, No. 420.

SYSTEMS AND CATALOGUES OF PROPER MOTIONS*

W. FRICKE

Astronomisches Rechen-Institut, Heidelberg, Germany

1. Introduction

Up to the second half of the 19th century astronomical research was almost exclusively based on the measurement of positions of celestial objects. F. W. Bessel (1784–1846), who considered the study of the motions of celestial bodies as the most essential part of astronomy, suggested "that observations should be made and repeated as many times as necessary to secure the orbits of the stars in space with such precision that their positions can be predicted for any arbitrary instant of time". The most valuable measurements of stellar positions available to Bessel were those made by Bradley and Maskelyne at Greenwich. Bessel's first great work (1818) was the reduction and publication of Bradley's observations from 1750–1762; the catalogue of 3222 stars is titled *Fundamenta Astronomiae*. Then Bessel made many thousands of measurements of stellar positions with a transit circle at Koenigsberg. The measurements carried out by Bradley, Maskelyne, and by Bessel form pioneer work in the derivation of accurate proper motions.

Whenever positions of objects are to be determined at the celestial sphere, so that motions can be derived from them, the question of a suitable reference frame arises. There are first of all practical requirements which influence the choice of the reference system. Since we do not know any real 'fixed' points in space, the reference system has to be defined in relation to celestial objects which can be observed with available instruments. The system must be accessible for all times, and if possible, at any arbitrary instant of time. It must be invariable for all times. Bright stars have to be selected whose positions at the sphere serve as representatives of the reference system. If the positions are determined at various epochs without reference to other stellar positions, then in principle, a fundamental system of coordinates and proper motions can be set up. The constant of precession must be known to enable the reduction from one equinox to any other.

The conventional reference system is that of the equator as the fundamental plane from which declinations are counted, and the vernal equinox as the origin for counting right ascensions. At least for the determination of the vernal equinox, measurements of positions of the Sun and other bodies of the planetary system are needed. In practice, however, the observed positions of the Sun and planets also serve to adjust the determination of the position of the equator. In other words, they contribute to the determination of the declination system.

The fundamental proper motions form an integral and indispensable part of a reference system that is based on stars as representatives of the system. Moreover, they can serve as primary standards for the derivation of so-called absolute proper motions

* *Mitteilungen Astron. Rechen-Inst. Serie A No. 41.*

from relative displacements of stars, if fundamental stars are among the observed stars.

An understanding of the methods of deriving fundamental proper motions implies that the principles of the measurement of fundamental declinations and right ascensions (Section 2) are clearly understood. In Section 3, the construction of a fundamental system of coordinates and proper motions will be described using the example of the Fourth Fundamental Catalogue (FK4), published in 1963. In Section 4, some characteristics of the proper motion systems of FK4, FK3, N30 and GC will be compared; in Section 5, we describe catalogues of relative proper motions, and in Sections 6 and 7 we report the results of recent determinations of precessional corrections and of the parameters of galactic rotation.

2. The Principles of the Measurement of Fundamental Declinations and Right Ascensions

For the definition of the reference system, declinations and right ascensions of stars have to be determined without reference to other stellar positions. Observations of such kind are called fundamental or absolute of the 1st order. Let us suppose that the observations shall be carried out at a transit circle, which is the classical instrument for this purpose.

The declination δ of a star is obtained by the measurements of an angle, namely the zenith distance z, when the star is in the meridian. Then, in upper culmination,

$$\delta = \varphi \mp z_{uc},$$

where φ is the latitude of the instrument. If circumpolar stars are selected, and the zenith distances z_{UC} and z_{LC} in upper and lower culmination are measured, then one obtains

$$90° - \varphi = \tfrac{1}{2}(z_{UC} + z_{LC}).$$

It is easily seen that the measurements of zenith distances in upper and lower culmination also lead to the knowledge of φ and of the azimuth of the instrument.

Now let us suppose that the instrument is an ideal one, free of errors. The most essential errors, in practice, are those of azimuth and pivot, errors in collimation and level, division errors of the circle and of the nadir reading, and errors caused by flexure of the tube and other parts of the instrument. We also assume that refraction is properly taken into account.

Then observations of zenith distances of several hundred stars, well distributed over the whole sky, and measured from the north pole to some angular distance south of the equator and from the south pole to some distance north of the equator, will provide a system of declinations through which the position of the equator is defined. We will see later that this method needs an important refinement because of the difficulties arising from the various instrumental errors. In practice, the declination system has to be adjusted by the measurement of positions of members of the planetary system within every observational program and by comparison of the observed planetary declinations with those computed from planetary theories.

The right ascension α of a star is obtained in measuring the time interval between the passages of the vernal equinox and the star through the meridian. To determine the time of passage of the vernal equinox through the meridian, a clock reading has to be found for the instant when the declination $\delta_\odot$ of the Sun is zero (the vernal equinox is defined by $\delta_\odot = 0$, $d\delta_\odot/dt > 0$). The instant when $\delta_\odot$ is zero can be found by measuring the declination of the Sun around the equinoxes so that each observation provides a value of the continuously changing right ascension $\alpha_\odot$ of the Sun according to

$$\sin\alpha_\odot \tan\varepsilon = \tan\delta_\odot .$$

(The obliquity ε of the ecliptic can also be determined independently by means of absolute measurements of declinations of the Sun around the solstices.) Theoretically, it would be sufficient to measure the position of one star with respect to the Sun in the vicinity of the vernal equinox. In fact, the first extensive observations of remarkable accuracy for determining the zero point of right ascensions were carried out by Maskelyne who measured the position of α Aquilae with respect to the Sun. He selected 35 stars near the equator for measuring right ascensions differentially with respect to α Aquilae. At present, a few hundred stars well distributed in right ascension and lying near the equator are used as 'clock stars'. Direct observations of stars with respect to the Sun, however, are no longer the only source for the determination of the zero point of right ascensions. Series of observations of major and minor planets within the fundamental programs provide both the equinox correction and the equator point with higher accuracy. Most suitable planetary objects for observations with the conventional transit circle are Ceres and Vesta because they appear with star-like images, an advantage that distinguishes all minor planets compared to the disk-shaped major planets.

Absolute observations of right ascensions and declinations carried out over a period of a few years show clearly that equator, ecliptic and equinox are in continuous motion as a consequence of precession and nutation. There is further the periodic variation of the apparent places of stars caused by the annual aberration, and a more or less periodic latitude variation. While the constants of nutation and aberration are determined by other than fundamental observations, the constant of precession has to be derived from observations of fundamental character.

The product of the observations of several years at an instrument is a star catalogue whose positions can be considered to represent the reference system at the mean epoch of the observations and at the nearest conventional equinox (at present either 1950.0 or 1975.0). It may be that not both coordinates α and δ were observed in fundamental manner, so that the catalogue provides either the α- or the δ-system alone.

Observational catalogues can, of course, only be considered as fundamental as far as they obey the rules which lead to an independent system. In particular, the instrumental errors must have been eliminated as far as possible without reference to stellar positions, and the local refraction must have been known.

For the reduction of observations, the same conventional constants of precession, nutation and aberration should be used by all observers. For the reduction to the mean epoch, some observers use proper motions of the observed stars from other sources. Such proper motions have to be eliminated later by the compiler of the fundamental system.

Two articles by Clemence (1961, 1963) may be consulted for more details on the principles underlying the construction of the astronomical reference system.

3. The System of Positions and Proper Motions in FK4

A number of star catalogues which provide the (α, δ)-system at various epochs can be used to set up a fundamental system of positions and proper motions in the area of the sky they cover. If catalogues, which are absolute of the 1st order, were strictly free of systematic errors, two catalogues of different epochs would be sufficient to provide the fundamental system of proper motions. From the positions x_1 and x_2 at the epochs t_1 and t_2 (both reduced to the same equinox) a straight line $x = x_0 + \mu(t - t_0)$ would give the position x_0 at an initial epoch and the proper motion μ. Two hundred years of praxis have shown that a large number of absolute catalogues with epochs well distributed over many decades are necessary to achieve a system that fulfils best the requirements of accuracy at the time when it shall serve as reference system. One should keep in mind that it is one of the main purposes of the fundamental system to serve as the reference system for measurements in the near future. In other words, it has to fulfil the demands for accurate ephemerides.

In the following we will describe how the system of positions and proper motions of the Fourth Fundamental Catalogue (FK4) was constructed. For a review of previous systems we may refer to the article of Scott (1963). Since 1960, the data of FK4 serve to represent the internationally adopted reference system for astronomical measurements.

The FK4 is the fourth-generation product in a series of fundamental catalogues. The predecessors are

(1) *Fundamental-Katalog für die Zonenbeobachtungen am Nördlichen Himmel.* Compiled by A. Auwers, 1879. (This catalogue, abbreviated FC, contains the positions and proper motions of 539 fundamental stars down to $\delta = -10°$.)

(2) *Neuer Fundamental-Katalog des Berliner Astronomischen Jahrbuchs nach den Grundlagen von A. Auwers.* Compiled by J. Peters, 1907. (The catalogue, abbreviated NFK, contains 925 fundamental stars of the entire sky. These stars were later called 'Auwers' stars'. The system and the individual values of positions and proper motions are based on observations from 1745 to 1900.)

(3) *Dritter Fundamental-Katalog des Berliner Astronomischen Jahrbuchs. Part I. Die Auwers-Sterne.* Compiled by A. Kopff, 1937. *Part II. Die Zusatz-Sterne für die Epoche 1950.* Compiled by A. Kopff, 1938. (The catalogue, abbreviated FK3, contains 1535 fundamental stars of the entire sky. They include 873 Auwers' stars. The system of positions is based on absolute observations carried out after 1900 leading to the mean epoch 1914. The system of proper motions is based on observations from 1845 to about 1930.)

Within this series, the preceding catalogue always served as an 'intermediate system' for the construction of its successor as far as this was possible. In compiling the FK4, the system of FK3 served as the intermediate system. The meaning of this fact has to be explained briefly.

Let us assume that there are n absolute observational catalogues available which shall form the basis of a new fundamental system. A number n_1 of them may already be included in FK3, and the remaining $n-n_1 = n_2$ may be new absolute catalogues. For each of the n catalogues the comparison Cat $-$ FK3 provides differences $\Delta\alpha$, or $\Delta\delta$ for a certain number of fundamental stars at the epoch of observation and equinox 1950.0. By convention, Newcomb's precession is being used for the reduction. By averaging the individual values $\Delta\delta$ over small intervals of δ a set of mean values $\Delta\delta_\delta$ is obtained, and by averaging the residuals $\Delta\delta - \Delta\delta_\delta$ in each declination zone over small intervals of α a set of values $\Delta\delta_\alpha$ results, so that a catalogue provides for each star a correction

$$\Delta\delta = \Delta\delta_\delta + \Delta\delta_\alpha + v_\delta,$$

and, correspondingly,

$$\Delta\alpha = \Delta\alpha_\delta + \Delta\alpha_\alpha + v_\alpha.$$

v_α, v_δ are individual deviations which are taken into account in the improvement of the individual accuracy of positions and proper motions within the new system.

There remain the systematic parts of the differences, namely

$$\text{Cat}_v - \text{FK3} = (\Delta\delta_\delta)_v,$$
$$\vdots$$
$$\text{Cat}_v - \text{FK3} = (\Delta\alpha_\alpha)_v,$$
$$v = 1, ..., n_1, n_1 + 1, ... n,$$

where v indicates the particular catalogue under consideration. By weighted averaging of these differences the definitive systematic corrections $\overline{(\Delta x_y)}_v$ to FK3 ($x = \alpha, \delta$; $y = \alpha, \delta$) are obtained which transform the FK3 system into that of FK4 according to

$$\text{FK4} = \text{FK3} + \overline{(\Delta x_y)}_v.$$

3.1. THE DECLINATION SYSTEM OF FK4

For the derivation of the declination system, comparisons of absolute catalogues from 1846 to 1956 with FK3 were made which provided sets of differences $\Delta\delta_\alpha$ and $\Delta\delta_\delta$ for each catalogue. These differences usually do not include the correction to the equatorial declinations of the observational catalogue from observations of the bodies of the planetary system. We have mentioned before that by the measurement of declinations of the Sun, Moon and planets the position of the equator can be determined.

The first step is therefore the determination of the position of the equator from corrections

$$\Delta\delta = \Delta\delta_\delta - \Delta\delta_0 \quad \text{at} \quad \delta = 0,$$

derived from catalogues which provide the correction $-\Delta\delta_0$ from observations of the Sun, Moon or planets. The results were grouped according to their epochs in successive decades. The group means of $\Delta\delta$ are shown in Table I together with their weights w. The weight of the equatorial correction of each catalogue is based on the number and

TABLE I

Corrections to equatorial declinations of FK3

Group	Number of catalogues	Mean epoch	$\Delta\delta$	w
1	3	1857	$+0''.26$	4
2	3	1869	$-0''.12$	6
3	5	1884	$-0''.09$	10
4	5	1892	$+0''.02$	12
5	5	1906	$+0''.06$	33
6	6	1915	$+0''.01$	47
7	5	1927	$-0''.05$	40
8	4	1937	$-0''.08$	45
9	5	1949	$-0''.01$	50

accuracy of the observations of the Sun and of other bodies of the planetary system. (The weights vary from $w=1$ to $w=15$.) The weight of the group means in Table I is the sum of weights of all catalogues within the corresponding group.

From the data of Table I follows the definitive correction $\Delta\delta(T)$ to the equatorial declinations of FK3

$$\Delta\delta(T) = -0''.017(\pm 0''.021) - 0''.097(\pm 0''.098)(T - 1928.4),$$

where T is counted in units of 100 years. The indicated errors are standard deviations. This solution shows that the position of the equator in FK3 was already well determined, and that the equator of FK4 does not differ significantly from that of FK3.

In computing the definitive corrections

$$\Delta\delta_\delta, \Delta\delta_\alpha, \Delta\mu_\delta, \Delta\mu_\alpha$$

to FK3, weights have been assigned to each catalogue and zone, depending first on the number of fundamental stars within the zone, second on the individual accuracy within the zone, and third on the absolute character of the catalogue as a whole. The first and second partial weight vary from 0.5 to 4 and the third varies from 0.5 to 2. The weighted differences $Cat_{cor} - FK3$ provided the definitive systematic corrections $\Delta\delta_\delta$ and $\Delta(\mu_\delta)_\delta$ to FK3 by a least square solution. The corrections $\Delta\delta_\alpha$ and $\Delta(\mu_\delta)_\alpha$ to FK3 have been determined correspondingly.

While the proper motion system in declination of FK4 (μ_δ-system) has been derived from absolute observations of the period 1846 to 1955, the δ-system has been based on absolute observations carried out after 1900. The mean epoch of the δ-system is about 1925. More details on the procedure and the tabulated catalogue comparisons can be found in a paper by Kopff *et al.* (1964).

The accuracy of the μ_δ-system of FK4 is described by the mean errors (standard deviations) computed from the dispersion of the system of absolute catalogues. The average values of the mean errors of the μ_δ-system decrease from

$$\pm\, 0\!''\!07 \quad \text{per century for} \quad \delta \geqslant +\, 70°$$
$$\text{to} \quad \pm\, 0\!''\!05 \quad \text{per century at} \quad \delta = 0°.$$

and increase in the southern sky with declination

$$\text{to} \quad \pm\, 0\!''\!13 \quad \text{for} \quad \delta \simeq -\, 75°.$$

On the average, the systematic errors of the proper motions in the southern sky are twice the errors in the northern sky as a consequence of fewer absolute observations carried out in the southern hemisphere.

3.2. THE RIGHT ASCENSION SYSTEM OF FK4

The basic material for the derivation of the α- and μ_α-system of FK4 are the catalogue comparisons

$$\text{Cat}_v - \text{FK3} = (\varDelta\alpha_\delta)_v,$$
$$\text{Cat}_v - \text{FK3} = (\varDelta\alpha_\alpha)_v$$

between absolute catalogues ($v = 1, \dots\, n_1,\, n_1 + 1, \dots n$) and FK3. n_1 is the number of investigated catalogues already incorporated in FK3, and $n - n_1$ is the number of new catalogues. The numbers n_1 and n vary for various declination zones as a consequence of the different number of contributing observatories and observations in the northern and the southern sky. It may be noted that in fundamental catalogues the proper motion component in right ascension is denoted by μ instead of μ_α, and in declination by μ' instead of μ_δ.

3.3. THE ZERO POINT OF THE α-SYSTEM (EQUINOX) OF FK4

The observations of the Sun and of other bodies of the planetary system which are included in absolute catalogues permit us to examine whether the position of the equinox N_1 determined by Newcomb needs a correction. The correction $\varDelta N$ has to be applied to all right ascensions. The position of the equinox of FK3 was defined by $N_1 + \varDelta N$, with $\varDelta N = -0\!\overset{s}{.}050$; for comparison, the correction applied in GC and N30 is $\varDelta N = -0\!\overset{s}{.}040$. The derivation of $\varDelta N$ from observations of the past 150 years is particularly beset with difficulties arising from the various changes in observational techniques as was shown by Kahrstedt (1931) and Morgan (1950). The observations carried out after 1900 with impersonal micrometer show values of $(\varDelta N)_{\text{Cat}}$ between $-0\!\overset{s}{.}028$ and $-0\!\overset{s}{.}073$, and no reliable time dependence of the corrections can be deduced. The adopted equinox of FK4 is $N_1 - 0\!\overset{s}{.}050$.

3.4. PROPER MOTION SYSTEM OF FK4 IN RIGHT ASCENSION (μ_α-SYSTEM)

In deriving the μ_α-system, a technique different from that for the μ_δ-system has been applied. The series of catalogues obtained with the same instrument were used for a least square solution of μ_α from the differences

$$\text{Cat}_v - \text{FK3} = (\varDelta\alpha)_v, \quad v = 1, \dots n,$$

thus providing

$$\varDelta\mu_\alpha(\text{instrumental system minus FK3})$$

for each instrumental system.

Large errors of observations during the 19th century, which are at least partly understood by well known deficiencies in the older techniques of observations, suggested that we confine our attention to observations from about 1900 onwards. The following instrumental series were available, where n is the number of catalogues:

Greenwich	1901–1935	$(n=5)$
Pulkovo	1897–1940	$(n=6)$
Washington 6″	1914–1945	$(n=5)$
Washington 9″	1907–1941	$(n=3)$
Cape	1908–1956	$(n=12)$.

Each instrument reaches only part of the sky, and the number of fundamental stars included in each catalogue varies appreciably. Among the Cape catalogues $(n=12)$ are five which contain less than 100 fundamental stars (except clock and azimuth stars).

Among the observations in the northern sky, those of Greenwich deviate remarkably from all others. A detailed study of the Greenwich catalogues revealed some deficiencies in the techniques prior to 1918 which appeared to be sufficiently serious to disregard the Greenwich series for the derivation of proper motions. The Greenwich catalogues from 1918 to 1935, however, contributed to the derivation of the α-system. South of about $\delta = -30°$, the fundamental system of proper motions rests entirely on the Cape series of catalogues.

Essential is the fact that the μ_α-system has been derived from instrumental series alone, thus disregarding isolated observations at some instruments which had led to one catalogue only. In view of the increasing observational activity with improved techniques in the near future, it is unlikely that this procedure can be continued for further improvements of the μ_α-system.

The system of positions in α has been derived without the restriction to instrumental series, and has been based on absolute observations from 1918 to 1956 providing a mean epoch about 1935.

The accuracy of the μ_α-system of FK4 is described by the mean errors (standard deviations) computed from the dispersion of the instrumental systems. In the southern sky (south of $-20°$) the mean errors were obtained from the internal errors of the Cape system itself. The average values of the mean errors of the μ_α-system decrease from

$$\pm 0\!''\!15 \quad \text{per century for} \quad \delta \geqslant + 80°$$
$$\text{to} \quad \pm 0\!''\!09 \quad \text{per century for} \quad \delta = 0°,$$

and increase in the southern sky up

$$\text{to} \quad \pm 0\!''\!36 \quad \text{per century for} \quad \delta \simeq - 75°.$$

On the average the systematic errors of the proper motions south of $-20°$ are twice the errors in the northern sky.

3.5. Differential Observations for the Improvement of Fundamental Positions and Proper Motions

Nothing has so far been said about the use of observations for improving the individual accuracy of positions and proper motions in a fundamental catalogue. Not only the reference system must be as accurate as possible, but also the individual positions and proper motions of the stars that represent the system must be of highest accuracy. There are two sources of very different nature which are being used:

(a) The fundamental observations from which the systems of right ascensions and declinations are compiled provide the residuals v_α and v_δ mentioned before. These residuals do not contain systematic parts, and they are considered as individual corrections to the positions of the stars at the epoch of observation.

(b) In addition to absolute observations yielding v_α, v_δ, also such observations can be used which were made strictly differential to any well defined system. For the construction of FK4 there were many new observations available carried out in a differential manner with respect to FK3. These contributed to the increase of accuracy in FK4. In all cases, however, the relative observations had to be investigated for systematic deviations from FK3. Such deviations originate from instrumental and other errors which have to be eliminated. In some cases, it was possible to disclose the origin of systematic deviations. For these reasons, in the construction of FK4 all available modern differential observations were analysed in the same way as absolute observations, and significant systematic deviations eliminated.

The mean errors of the individual positions and proper motions within the system of FK4 are given for each star in the fundamental catalogue. These errors do not include the errors of the system which we have given above.

4. Comparisons of Various Fundamental Proper Motion Systems

There are important kinematical problems of the Galaxy, the solutions of which must be based on fundamental proper motions. Previous investigations which may still be of interest in the future made use of the systems of GC, FK3 or N30. Some essential characteristics of these systems are summarized briefly in comparison with FK4 in Table II.

TABLE II
Proper motion systems

System	Year of publication	Number of stars	Epochs of absolute observations	Diff. observations yielding individual corrections to μ_α and μ_δ
GC	1937	33 342	1755 to 1930 for μ_α and μ_δ	1755 to 1930
FK3	1937/38	1 535	1845 to 1930 for μ_α and μ_δ	1755 to 1930
N30	1952	5 268	Standard epochs 1900 and 1930 for μ_α and μ_δ	1900 and 1930
FK4	1963	1 535	1845 to 1956 for μ_δ 1897 to 1956 for μ_α	1755 to 1956 1755 to 1956

The GC (Albany General Catalogue) is the result of an enormous enterprise started by Lewis Boss and completed by Benjamin Boss after three decades of work. The catalogue contains all stars brighter than seventh magnitude and some thousands of fainter stars. The large number of stars results from the inclusion of numerous objects which were measured by differential techniques exclusively, many of them at two epochs only. For fundamental observations it appears impracticable even at the present time to have more than about 1000 stars in a program at one observatory. The proper motion system of the GC as a whole can hardly be considered as a primary fundamental system.

Some of the main distinctions between the FK3 and GC are the following:

(1) The number of stars in FK3 observed by differential techniques alone is small compared to that in GC.

(2) Absolute observations from 1755 to 1845 are not included in FK3 because of their low weight and their detrimental influence on the proper motion system.

(3) The techniques of compiling were different for GC and FK3. An important example is the determination of the declination system of GC without regard to observations of bodies of the planetary system.

The direct way to judge the quality of a system consists in a comparison of modern observations with the positions of the stars computed on the basis of different systems. Such comparisons have revealed very drastically the superiority of FK3 over GC. They also disclosed the existence of an appreciable magnitude equation in the GC-system, which means that the proper motion system of stars fainter than seventh visual magnitude is different from that of the brighter stars. An accurate determination of this difference is one of the future tasks. Since FK3, N30 and FK4 contain only stars brighter than 7.5th magnitude, it is not possible, at present, to transfer the proper motions of faint GC stars to any of these systems of better quality.

The N30 (Catalogue of 5268 Standard Stars, Based on the Normal System N30) is the result of a partial but successful revision of the GC carried out by H. R. Morgan. The system of proper motions has been derived from two normal systems of positions at mean epochs around 1900 and 1930. The system at 1900 is essentially that of the positions in GC with some corrections applied. The system of positions at 1930 was obtained from modern absolute observations. The main uncertainty in the proper motion results from the corrections applied to the GC positions for epoch 1900, which have been questioned later.

The systematic differences FK4−FK3 in the proper motions have been listed in FK4 for the whole sky; the systematic differences FK4−GC and FK4−N30 were determined by Brosche *et al.* (1964). A common feature of all comparisons is the regional structure of the deviations of the systems from each other. There are parts of the sky where new observations contributed appreciably to the systematical improvement, thus showing remarkable differences between FK4 and the older systems. In other parts of the sky the number and weight of old observations had already led to a satisfactory approximation to a reliable system.

If we disregard the GC proper motion system because of proved deficiencies, the comparisons between FK3, N30 and FK4 show several highly interesting features.

(1) In the northern sky (down to $-20°$) the μ_δ-systems are in fair agreement with each other; the deviations have amounts of the order of the systematic errors computed for the μ_δ-system of FK4. The conclusion appears to be permitted that the present μ_δ-system is already a good approximation to a reliable system in this part of the sky. South of $\delta = -20°$ the deviations between the μ_δ-systems of FK3, N30 and FK4 are greater than the systematic errors. This demonstrates that each new Cape catalogue caused a noticeable change of the μ_δ-system.

(2) The deviations of the μ_α-systems from each other are greater than the computed systematic errors of FK4 in large parts of the northern sky as well as in the southern sky. This leads to the conclusion that, besides the difficulties in the determination of the equinox, it has been much more difficult to measure right ascensions correctly than declinations. For the determination of the latter, however, the observations of the bodies of the planetary system should have favorably influenced the results.

5. Absolute Proper Motions of the 2nd Order

Fundamental proper motions play an important part in investigations on motions of the stars in the Galaxy. They permit the determination of the constant of precession, and they contribute to the establishment of a distance scale in the Galaxy. However, the number of fundamental proper motions is much too small to provide directly the material for all desired purposes. Their main purpose is to provide the reference system for the determination of proper motions of many more objects. Concerning the methods applied in the past for the derivation of absolute proper motions of the 2nd order (and also 3rd order) we refer to papers by Heckmann and Dieckvoss (1958) and by Dieckvoss (1963). They shall not be reviewed here.

At the present time, the photographic method yields proper motions on the system of FK4 of stars down to about 9th visual magnitude in the most economical way. In principle, plates are taken at two epochs, and the positions of all stars down to a certain magnitude limit are measured with respect to several reference stars on the plates.

The positions of the reference stars have to be measured with respect to the fundamental stars with transit circles at each epoch.

An international program which has been carried out recently consists in the repetition of the AGK2-catalogue of the Astronomische Gesellschaft. In AGK2, the positions of 183 600 stars of the northern sky down to about 9th visual magnitude are given for an epoch about 1930. They were obtained by the measurement of photographic plates taken at astrographs about 1930. The positions of the stars on the plates covering a $5° \times 5°$ field each were measured with respect to 'reference stars' whose positions were tied to the fundamental system by transit circle observations. About 13 750 stars were observed from 1928 to 1930 at transit circles with respect to the fundamental stars in NFK, and later transferred to the system of FK3. These stars form the reference stars (published in AGK2A by Kopff) for the AGK2 positions at 1930.

During the past few years, the second-epoch plates were taken at an astrograph of the Hamburg Observatory which had already contributed most of the first-epoch plates. The positions of about 21 000 reference stars for the second-epoch (AGK3R) have been determined on the system of FK4 by transit circle observations taken at a number of observatories in the northern hemisphere.

The result of the large undertaking will be the AGK3, a catalogue containing proper motions of 183 600 stars on the system of FK4, which may be considered as absolute of the 2nd order. Every possible provision has been made to prevent systematic errors depending on magnitude and color of the stars.

Another international program similar to that of AGK3 has been started to determine proper motions in the southern sky down to 9th magnitude as strictly as possible on the fundamental system. Several observatories located in the southern hemisphere and some northern observatories which have sent their instruments to the south are carrying out transit circle observations of agreed reference stars (SRS-program). These stars will be observed differentially with respect to FK4. Also fundamental observations will be carried out which will contribute to the improvement of the FK4 system. The Cape Observatory – and possibly some others – will take the photographic plates at this epoch.

6. Catalogues of Relative Proper Motions

In the preceding section the photographic method of deriving proper motions has been described. It has been shown how by measuring positions of stars with respect to reference stars absolute proper motions may be obtained. The photographic method has also been used with great success for measuring relative proper motions of special objects, where a reduction to a fundamental frame of reference would not have presented any advantages. The method consists in taking two or more plates of the same area of the sky with the same telescope at different epochs. Then either plate coordinates or their differences are measured, or the plates are surveyed for stars of large proper motions by means of a blink microscope.

Relative proper motions may be sufficient and extremely useful, if nearby stars are searched. Pioneers in this field were M. Wolf, Heidelberg, and R. T. A. Innes, Johannesburg. Wolf (1919) made searches in scattered regions in the northern hemisphere, occasionally down to magnitude 17, and published a catalogue containing about 1000 stars of large proper motions. Innes blinked a large number of pairs of plates taken at various telescopes in the southern hemisphere. He published lists of proper motions from 1915 to 1927 in the Union Observatory Circulars, Johannesburg. We owe Wolf and Innes the discovery of a great number of nearby stars, among them many interesting objects, like 'van Maanen's star', a white dwarf of abnormal high density (found by Wolf), and 'Proxima Centauri', the nearest star known (found by Innes).

The most extensive and successful surveys for stars of large proper motions were made by Luyten (1963) who used pairs of plates taken at various telescopes. Luyten published a great number of catalogues, the most recent ones present the results of surveying with the Palomar 48-inch Schmidt telescope. We owe him the discovery of many white dwarfs, of faint blue objects in high galactic latitudes, and of many peculiar objects of great interest.

7. Determination of the Constants of Precession

By international convention Newcomb's values of precession are used in the compiling of star catalogues. All determinations of precession carried out after Newcomb (1905) have shown that the adopted value of lunisolar precession needs an appreciable correction Δp_1. The results of the various determinations lie in the interval of about

$$+ 0\overset{''}{.}75 \leqslant \Delta p_1 \leqslant + 1\overset{''}{.}50 \quad \text{per century}.$$

Moreover it turned out that the right ascensions require a constant correction Δe as a consequence of an error of Newcomb's motion of the equinox. If one takes further into account that the adopted value of planetary precession may require a correction $\Delta \lambda$, the proper motions contain fictitious parts $\Delta \mu_\alpha$ and $\Delta \mu_\delta$ caused by Δp_1, Δe, and $\Delta \lambda$. Thus, all proper motions μ_α, μ_δ given in catalogues have to be corrected in the sense

$$(\mu_\alpha)_{\text{cor}} = \mu_\alpha - \Delta \mu_\alpha,$$
$$(\mu_\delta)_{\text{cor}} = \mu_\delta - \Delta \mu_\delta,$$

where, in the usual notations,

$$\Delta \mu_\alpha = \Delta k + \Delta n \sin \alpha \tan \delta,$$
$$\Delta \mu_\delta = \Delta n \cos \alpha,$$
$$\Delta k = \Delta m - \Delta e,$$
$$\Delta m = \Delta p_1 \cos \varepsilon - \Delta \lambda,$$
$$\Delta n = \Delta p_1 \sin \varepsilon,$$

ε denotes the obliquity of the ecliptic.

Fricke (1967a) determined the values of the precessional corrections Δp_1 and $\Delta \lambda + \Delta e$ from the large material of McCormick and Cape proper motions which were reduced to the systems of FK3, N30 and FK4. It turned out that all three systems give the same values for the precessional corrections but it was also found that the Cape proper motions cannot be reduced with sufficient accuracy to anyone of the three systems. The results for Δp_1 are

$$+ 1\overset{''}{.}08 \pm 0\overset{''}{.}12 \text{ from McCormick motions alone},$$

$$+ 1\overset{''}{.}38 \pm 0\overset{''}{.}08 \text{ from McCormick and Cape motions with equal weight}.$$

(These are centennial values with their probable errors.)

The suspicion against the Cape proper motions was fully confirmed by an extensive investigation based on proper motions of fundamental stars only. Fricke (1967b) made use of FK4/FK4 Sup stars with distances greater than 100 parsec and carried out solutions with proper motions of these stars as they are given in the four different catalogues FK4, N30, FK3 and GC. The most essential result is that the correction to Newcomb's lunisolar precession has the same value, within $0\overset{''}{.}01$ per century, in the systems FK4, N30 and FK3,

$$\Delta p_1 = + 1\overset{''}{.}10 \pm 0\overset{''}{.}10 \quad \text{(p.e.) per century},$$

while the value in the GC system is slightly different. The value $\Delta \lambda + \Delta e$ describing the combined correction due to incorrect planetary precession and a non-precessional motion of Newcomb's equinox is

$$\Delta \lambda + \Delta e = + 1\overset{''}{.}20 \pm 0\overset{''}{.}11 \quad \text{(p.e.) per century}$$

in both FK4 and N30, while the value is slightly different in the older systems FK3 and GC. On the basis of these results Fricke (1968) recommended values of the precessional corrections to be used according to the equations given above. The recommendation is the following: for investigations to be based on proper motions freed from the effects of incorrect precession and of the zero point error in μ_α it is recommended that the corrections

$$\Delta n = + 0\overset{''}{.}44 \quad \text{corresponding to } \Delta p_1 = + 1\overset{''}{.}10 \text{ per century},$$

$$\Delta \lambda + \Delta e = + 1\overset{''}{.}20 \quad \text{per century}$$

should be applied to motions in the systems FK4 and N30. Proper motions in the systems GC or FK3 should in all cases be reduced to the system of FK4, and then the precessional corrections applied. It is not recommended that, for investigations of stellar motions, averages be taken of proper motions in N30 and FK4. In case of doubt it is preferable to base investigations on each system, separately considered, in order to make apparent the variational width of results due to the differences between the best proper motion systems presently available.

8. Determination of Oort's Constant B from Proper Motions

The determination of galactic rotation from fundamental (or absolute) proper motions is possible, if it is true that the proper motions do not contain rotational effects other than those arising from the precessional corrections described in the preceding section and those arising from galactic rotation. Let us consider fundamental proper motions freed from precessional errors.

For an understanding of the principle of determinations of Oort's constant B let us assume that the Galaxy would be in rigid rotation with an angular velocity ω and that all stars are at rest with respect to the Sun. Then, by fundamental observations of the stars at various epochs the rotation of the reference frame would be noticed. The common fundamental proper motion would be

$$\mu_l = B = -\omega, \quad \mu_b = 0,$$

where μ_l is measured in counterclockwise direction (the direction of increasing galactic longitudes), and B is Oort's constant

$$B = -\omega - \frac{1}{2}\frac{d\omega}{dR}R,$$

where R is the distance of the Sun from the galactic center. In rigid rotation $d\omega/dR=0$, and Oort's constant A, which describes the deformation of the velocity field due to differential rotation, is

$$A = -\frac{1}{2}\frac{d\omega}{dR}R \equiv 0.$$

In practice, the stars are not at rest with respect to the Sun, and the angular velocity is not constant but a function of R. Then the constant A can be determined from the proper motions in addition to B, and it is important to notice that the determination does not require the knowledge of stellar distances.

The parameters of galactic rotation as determined by Fricke (1967b) from fundamental proper motions in the systems of FK4 and N30 are (in units of km/sec per kpc)

$$\text{FK4: } A = +14.2 \pm 1.9, \quad B = -11.8 \pm 1.9 \quad \text{(p.e.)},$$
$$\text{N30: } A = +17.1 \pm 1.9, \quad B = -10.0 \pm 1.9 \quad \text{(p.e.)}.$$

These values clearly confirm that a value for the angular velocity of the Galaxy near the Sun of about

$$\omega = A - B = 25 \text{ km/sec per kpc}$$

appears to be a good approximation. If the distance of the Sun from the galactic center is $R=10$ kpc, then the circular velocity near the Sun would be 250 km/sec.

The validity of the results for B have recently been questioned by Aoki (1967) who gave arguments in favor of the existence of a kind of rotation of the equatorial plane which was unknown so far and not taken into account in determinations of B. Aoki suggested that there is a rotation of the equatorial plane about an axis through the equinoxes which decreases the obliquity of the ecliptic at a rate of about $0\overset{''}{.}3$ per century. The justification for this suggestion was derived from the fact that the observations of the Sun and some planets appear to have indicated a decrease of the obliquity of the ecliptic which cannot be explained by the action of planetary perturbations. According to Aoki a rotation of the equatorial plane by $0\overset{''}{.}3$ per century may be caused by friction between mantle and core of the Earth.

Aoki's suggestion has given strong impetus for an investigation of the unexplained change of obliquity. If this change would be caused by a motion of the equator, the value of Oort's constant B would be about $B = -30$ km/sec per kpc instead of -10 km/sec per kpc. The circular velocity of the Galaxy near the Sun would be by about 450 km/sec instead of 250 km/sec. In a joint effort by J. L. Lieske, Pasadena, and the author it was found that there is, in fact, no good reason for the assumption of a significant motion of the ecliptic or the equator. Lieske (1970), has re-examined the observations of the minor planet Eros for the determination of a secular error in the obliquity, and he found that the data do not indicate the existence of a secular error in obliquity which is significantly different from zero. There remains the fact that the observations of Sun, Moon, Mercury, and Venus have all given negative corrections to the change of obliquity of the order of $-0\overset{''}{.}3$ per century. Even if there should be inaccuracies in the determinations, there remains some justification for suggesting that either Newcomb's values for the planetary masses are erroneous (in particular the mass of Venus), or that the observations of the Sun, Moon and planets are affected by an error of systematic nature which escaped detection. In the first case, which means a motion of the ecliptic, there would be no influence on the determination of the constant B and of precessional corrections. In the second case the equator would perform a fictitious rotation which erroneously would have been interpreted as an error in the change of obliquity.

An investigation carried out by the author of this article has quite recently resulted in the discovery of a systematic error in the declination systems of old fundamental observations that must have affected the observations of the Sun and planets. If the dynamical inertial system and the system of FK4/N30 (averaged) are identical, the analysis of the motions of the Sun, Moon and planets from about 1750 to the present time must reveal a fictitious rotation of the equator of about $-0\overset{''}{.}2$ per century due to the observational error. This error does not affect the determination of Oort's constant B from proper motions in the system of FK4 or N30.

These results, which will be published elsewhere, invalidate largely Aoki's arguments against previous determinations of galactic rotation. This does not mean that there would be no room anymore for a correction to the mass value of Venus or for a motion of the axis of rotation of the Earth due to friction between mantle and core of the Earth, but it is certain that the effects are smaller than the one presented by the unraveled mystery of the change of obliquity.

References

Aoki, S.: 1967, *Publ. Astron. Soc. Japan* **19**, 585.

Bessel, F. W.: 1818, *Fundamenta Astronomiae pro Anno MDCCLV*, Königsberg.

Brosche, P., Nowacki, H., and Strobel, W.: 1964, *Veröffentl. Astron. Rechen-Inst. Heidelberg*, No. 15.

Clemence, G. M.: 1961, *Proc. Astron. Observ. La Plata*, p. 17.

Clemence, G. M.: 1963, in *Basic Astronomical Data* (ed. by K. Aa. Strand), Univ. of Chicago Press, Chicago, p. 1.

Dieckvoss, W.: 1963, in *Basic Astronomical Data* (ed. by K. Aa. Strand), Univ. of Chicago Press, Chicago, p. 40.

Fricke, W.: 1967a, *Astron. J.* **72**, 642.

Fricke, W.: 1967b, *Astron. J.* **72**, 1368.

Fricke, W.: 1968, in *Highlights of Astronomy* (ed. by L. Perek), D. Reidel, Dordrecht-Holland, p. 306.

Fricke, W. and Kopff, A.: 1963, *Fourth Fundamental Catalogue* (FK4) (compiled in collaboration with W. Gliese, F. Gondolatsch, T. Lederle, H. Nowacki, W. Strobel, and P. Stumpff), *Veröffentl. Astron. Rechen-Inst. Heidelberg*, No. 10.

Gliese, W.: 1963, *Veröffentl. Astron. Rechen-Inst. Heidelberg*, No. 12.

Heckmann, O. and Dieckvoss, W.: 1958, *Astron. J.* **63**, 156.

Kahrstedt, A.: 1931, *Astron. Nachr.* **244**, 33.

Kopff, A., Nowacki, H., and Strobel, W.: 1964, *Veröffentl. Astron. Rechen-Inst. Heidelberg*, No. 14.

Lieske, J. L.: 1970, *Astron. Astrophys.* **5**, 90.

Luyten, W. J.: 1963, in *Basic Astronomical Data* (ed. by K. Aa. Strand), Univ. of Chicago Press, Chicago, p. 46.

Morgan, H. R.: 1950, in *Constantes Fondamentales de l'Astronomie*, Colloques Internat. du C.N.R.S., Paris, p. 37.

Newcomb, S.: 1905, *Astron. Pap. Washington* **8**, Part 1.

Scott, F. P.: 1963, in *Basic Astronomical Data* (ed. by K. Aa. Strand), Univ. of Chicago Press, Chicago, p. 11.

Wolf, M.: 1919, *Veröffentl. Sternw. Heidelberg* **7**, 195.

PHOTOMETRY AND PHOTOMETRIC SYSTEMS

M. GOLAY

Observatoire de Genève, Genève, Switzerland

1. Introduction and Definitions

The parts of Astronomy and Astrophysics which do not require photometric information are certainly not numerous. The photometric measurement is almost instinctive, and the relevant information seems to be very easy to obtain. This apparent simplicity hides numerous pitfalls which it is not always easy to avoid. We will not discuss here the methods used for reducing the measurement to its value outside the atmosphere, this having been done very well by Hardie (1962), nor will we dwell on the subjects of photoelectric and photographic techniques handled in the same volume by Baum, Lallemand, Johnson, Stock and Williams (1962). Let us first point out three fundamental articles. First, the article by Lamla (1965) which describes a large number of photometric systems, the expressions allowing to pass from one system to another as well as an extensive bibliography. Next, the one by Strömgren (1963) in which he presented the principal methods of spectral classification based on photometric systems. Finally, the same author has discussed, in another article (1966), the case of narrow- and intermediate-band photometries with special attention to the u, v, b, y, β system. We propose to present an introduction to these three fundamental articles in the following pages.

After correcting for atmospheric absorption and instrumental effects, the information given by the measuring instrument is, in magnitudes:

$$m_j = -2.5 \log \int_{\lambda_a}^{\lambda_b} E(\lambda)\, S_j(\lambda)\, d\lambda + \text{const.}, \tag{1.1}$$

where $E(\lambda)$ is proportional to the spectral energy distribution at the entry of the atmosphere, $S_j(\lambda)$ is the response curve of the photometer (filter + photocells + various reflections) for the pass-band j, λ_a, λ_b extreme limits of the pass-band.

The width of the pass-band is not generally given by $\lambda_b - \lambda_a$. We can define it with the following expression:

$$\mu_j^2 = \frac{\int_{\lambda_a}^{\lambda_b} (\lambda - \lambda_0)^2\, S_j(\lambda)\, d\lambda}{\int_{\lambda_a}^{\lambda_b} S_j(\lambda)\, d\lambda} \quad \text{with} \quad \lambda_0 = \frac{\int_{\lambda_a}^{\lambda_b} \lambda S_j(\lambda)\, d\lambda}{\int_{\lambda_a}^{\lambda_b} S_j(\lambda)\, d\lambda}. \tag{1.2}$$

In the case where $S_j(\lambda)$ has a rectangular form with the limits λ_a, λ_b, we have

$$\mu = (\lambda_b - \lambda_a)/2\sqrt{3}.$$

L. N. Mavridis (ed.), Structure and Evolution of the Galaxy, 34–69. *All Rights Reserved*
Copyright © 1971 by D. Reidel Publishing Company, Dordrecht-Holland

If the distribution $S_j(\lambda)$ has the shape of a Gaussian, the width at half-maximum $\Delta\lambda$ equals about 2.36 μ. It is usual to distinguish between three types of photometry, depending upon the value of $\Delta\lambda$:

Wide-band photometry, if $\Delta\lambda > 300$ Å.

Intermediate-band photometry, if 100 Å $< \Delta\lambda \leqslant 300$ Å.

Narrow-band photometry, if 10 Å $\leqslant \Delta\lambda \leqslant 100$ Å.

When a photometric system consists of pass-bands with very different $\Delta\lambda$, it is the narrowest of these which defines the type of photometry.

A photometric system consists of a set of n pass-bands ($j = 1$ to n), each having a width of $\Delta\lambda_j$ (equal or different to each other), which cover a series of spectral intervals distributed, uniformly or not, in the whole spectral interval accessible to the technical means at the moment. Thus, photometric systems were restricted to the visible region of stellar spectra until the end of the first half of the 20th century. Since then, this interval has been extended into the infrared and submillimeter wavelengths and, on the side of the shorter wavelengths, as far as the X-rays by use of artificial satellites. By increasing the number n of pass-bands and by limiting $\Delta\lambda$, we finally obtain the actual description of $E(\lambda)$, which is the aim of spectrophotometry.

Photometric measurement techniques have undergone considerable progress during these last 5 years. This fact, added to the increasing number of available large telescopes and to the possibility of using artificial satellites, has made possible the even more frequent use of the direct recording of the spectrum. These spectro-photometric recordings give us $E(\lambda)$ of expression (1.1), which allows us to calculate m_j for any pass-band S_j we may desire. Nevertheless, in spite of the advantages inherent in the direct recording, the integral expression (1.1) shows us that by measuring through a very limited number of pass-bands of adequate $\Delta\lambda$, it is possible to acquire information, which is of course more limited, on a much greater number of stars. The wider the pass-bands are, the more the number of bands of the system are limited, the less information is received from the star, the greater is the number of measurable stars, the more there is information to be had about populations of stars. The extraordinary multiplicity of photometric systems shows that each one of them has attempted to establish a subtle balance between sometimes conflicting requirements. It is, therefore, useless to look for a unique system, as it is also useless to try and describe all the existing systems. We, therefore, intend in the following lines to present some rudiments which should enable the reader, for whom astronomical photometry is not a speciality, to understand better that which he may expect of a given photometry.

2. Description of the Observed Energy Distribution

Let us reconsider expression (1.1). The energy distribution $E(\lambda)$ of the star can be put in the form

$$E(\lambda) = E_\lambda^0 e^{-K_\lambda M}, \tag{2.1}$$

where E_λ^0 is the energy distribution in the absence of interstellar extinction, K_λ is the law of interstellar extinction, M the mass of interstellar matter traversed.

Whence the expression for the magnitude

$$m_j = -2.5 \log \int_{\lambda_a}^{\lambda_b} E_\lambda^0 e^{-K_\lambda M} S_j(\lambda) \, d\lambda + \text{const}. \qquad (2.2)$$

m_j follows directly from observation. Starting from a series of measurements m_j ($j = 1$ to n) we can deduce some properties of E_λ^0, if K_λ is known, or of K_λ, if E_λ^0 is known. Thus, we must now examine how a series of photoelectric measurements can describe $E(\lambda)$, E_λ^0, K_λ and which are the photometric parameters which contribute to the description of $E(\lambda)$, E_λ^0, K_λ. Assuming that the discontinuities of the energy distribution $E(\lambda)$ are not too great (which has to be examined in each case), it is possible to calculate a point $E(\lambda_{ej})$ of $E(\lambda)$ for a wavelength λ which is generally close to the

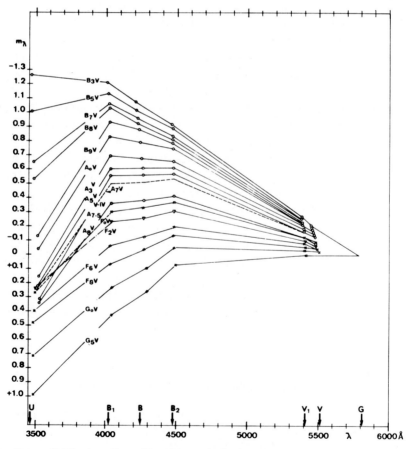

Fig. 1. Energy distribution of unreddened stars obtained with a photometry having 7 intermediate and wide pass-bands.

effective wavelength λ_{ej} derived for the pass-band j. We have

$$E(\lambda_{ej}) \cong \frac{10^{-(m_j + Cte)/2.5}}{\displaystyle\int_{\lambda_a}^{\lambda_b} S_j(\lambda)\, d\lambda},$$ (2.3)

with

$$\lambda_{ej} = \frac{\displaystyle\int_{\lambda_a}^{\lambda_b} \lambda E(\lambda)\, S_j(\lambda)\, d\lambda}{\displaystyle\int_{\lambda_a}^{\lambda_b} E(\lambda)\, S_j(\lambda)\, d\lambda}.$$ (2.4)

By introducing $E(\lambda_{ej})$ into (2.4) we get a first approximation for λ_{ej}. We thus get a set of monochromatic values. The distribution $E(\lambda)$ is thus given in n points. Figure 1 gives such a description of $E(\lambda)$, for unreddened stars, carried out with the seven-color intermediate and wide-band photometry of Geneva (Golay, 1969 (a)). We call 'pseudo-continuum' the polygonal line which thus describes the energy distribution $E(\lambda)$. In Figure 2, we give the continuum as described by Code (1960). The difference between Code's continuum and the pseudo-continuum comes from line-blocking in the spectral interval covered by the pass-band j. This deviation measures the importance of the discontinuities in the energy distribution $E(\lambda)$. When the number of bands increases and the pass-bands become narrower we arrive at the description of $E(\lambda)$

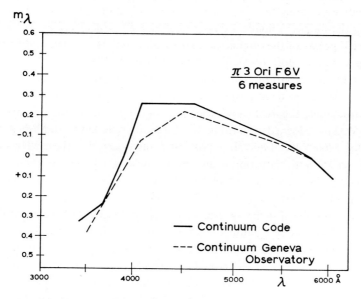

Fig. 2.   Comparison of Code's (1960) continuum with the one obtained with the
U, B_1, B_2, V_1, G system.

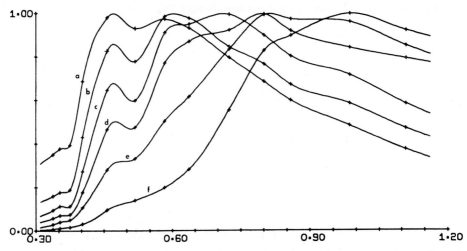

Fig. 3.   Dwarf stars from K0V to M5V, a = K0V, b = K3V, c = K5V, d = K7V, e = M2V, f = M5V
according to the 13-color photometry of Mitchell and Johnson (1969 (d)).

given for example by the 13-color photometry of Mitchell and Johnson (1969 (d))
($\Delta\lambda$ between 150 and 800 Å covering the spectral interval 3300–11000 Å) shown
in Figure 3.

Let us now suppose that a star, reddened by interstellar extinction, of which we
have the monochromatic energy distribution $E(\lambda_{ej})$ in n wavelengths λ_{ej} is sufficiently
well known spectroscopically, so that we may take as intrinsic energy distribution E_λ^0
the one corresponding to a spectroscopically identical but unreddened star. By
comparing the monochromatic values $E(\lambda_{ej})$ and E_λ^0 (calculated at the same λ_{ej}) we
can deduce n points of the interstellar extinction curve $M \cdot K_\lambda$. The wide-band photometries in various colors of Stebbins and Whitford (1945) and of Johnson (1965)
have brought an important contribution to the determination of the interstellar
extinction laws. They have particularly shown that these laws can be different in
different directions. Figure 4 shows the effect of extinction on the pseudo-continuum
of an O9V star reddened according to three different laws established by Nandy
(1964, 1965, 1966, 1967, 1968). It is possible to ascertain that these effects due to the
different laws are sufficiently important and are now easy to detect. These are the laws
that we shall use later on as we attempt to characterize the intrinsic energy distribution
E_λ^0 from the observed distribution $E(\lambda)$.

Let us now examine a stellar spectrum, for instance the one of the B star of Figure 5
taken from an article by Chalonge (1958). This recording clearly shows the Balmer
discontinuity as well as the lines of the hydrogen Balmer series. Let us now compare
the continua (defined by the curved lines which touch the extreme points of the
microphotometric recording) of two different stars and let us plot the differences in
monochromatic magnitude as a function of λ^{-1}. This is shown in Figure 6 (from
Berger and Fringant, 1955). The straight lines allow the determination of the relative

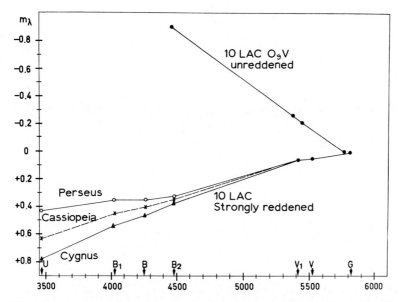

Fig. 4.   Effect of interstellar extinction on the pseudo-continuum of an O9V star, reddened by calculation according to 3 different laws of extinction.

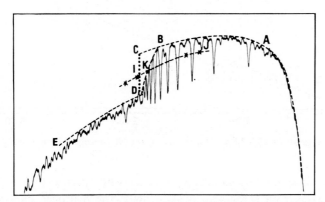

Fig. 5.   Definition of the continuum in the spectrophotometry of Chalonge.

gradient in the wavelength interval where such a straight line can be drawn. Thus, we see that, on either side of the Balmer discontinuity, the energy distribution of one star relative to another can be represented by an array of straight segments. The slopes of these segments being different, several relative gradients are needed to characterize an energy distribution. It is easily shown that two pass-bands placed on either side of the Balmer discontinuity D allow the measurement of this discontinuity. A large number of wide-band as well as intermediate and narrow-band photometries take advantage of this possibility. In Table I more than $\frac{3}{4}$ of the described intermediate and narrow-band systems have bands on either side of D (we find the same proportion for the wide-band photoelectric systems described by Lamla, 1965). The great interest

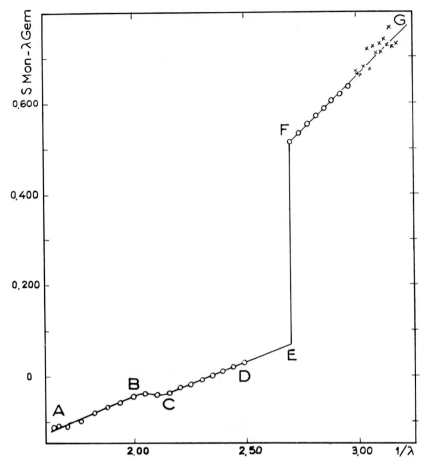

Fig. 6. Relative gradient of S Mon relative to ϱ Gem (according to Berger and Fringant, 1955).

of the Balmer discontinuity D is that it varies with spectral type and luminosity. Figure 7 is taken from Hack and Struve (1969). The hydrogen lines which appear in Figure 5 also carry precious information about the atmosphere of the star. Figure 8, taken from an article by Golay and Goy (1965), gives the variation of the equivalent widths of the Hγ and Hβ lines derived from measurements by Sinnerstad (1961). These properties have been used in narrow-band photometry, for instance by Crawford (1958 (k)) with a parameter β which is a measurement of Hβ as well as by Bappu *et al.* (1962 (v)) with a parameter γ as a measurement of Hγ. These two narrow-band photometric systems also make use of the color index $(U-B)_0$ (this is an $(U-B)$ index of the U, B, V photometry, corrected for interstellar reddening). Now, we have just seen that this index is obtained by the aid of two pass-bands placed on either side of the Balmer discontinuity and that it is a measurement of this discontinuity (for spectral types between B0 and F5). For unreddened spectral types between O and B9, we have the relations $(U-B)=1.20+2.25\ D$. A two-dimensional representation is

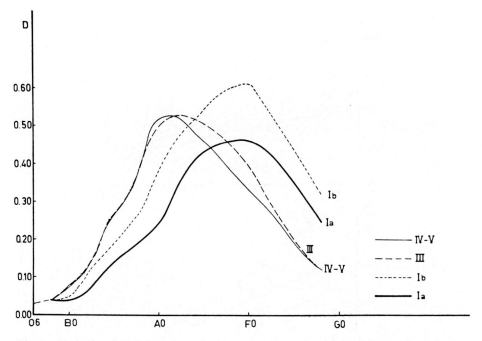

Fig. 7. Variation of the Balmer discontinuity D as a function of spectral type (Figure from Hack and Struve, 1969).

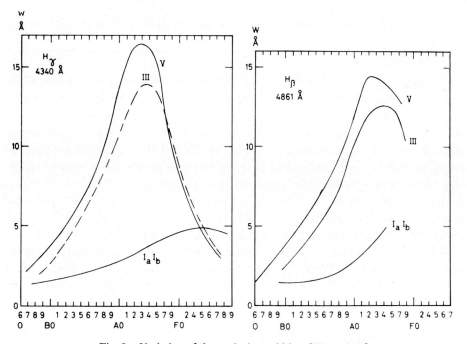

Fig. 8. Variation of the equivalent widths of Hγ and Hβ.

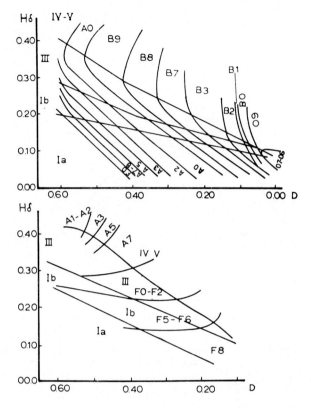

Fig. 9. Two-dimensional representation based on Hδ and D according to Hack. (Figure from Hack and Struve, 1969.)

thus possible. Figure 9, taken from Hack and Struve (1969) gives an example of this representation obtained by the aid of low-resolution spectra. Figure 10, taken from Crawford (1958 (k)), shows this two-dimensional representation obtained photometrically. The set of hydrogen absorption lines varies with spectral type and luminosity more or less in the same manner as do the Hβ, Hγ and Hδ lines which we have just discussed. Table I shows that several photometries have pass-bands in spectral regions very rich in hydrogen line. This is the case for instance for the P filter at 3750 Å in the photometry of Borgman (1960 (b)) as well as for the L filter of Walraven and Walraven (1960 (e)) who use it for establishing an index narrowly correlated with Crawford's β index (Figure 11, taken from Walraven and Walraven, 1960 (e)).

Figure 6 shows that it is possible to establish relative gradients in spectral regions which are poor in lines. We know that the relative gradient $G_{2\cdot1}$ of the continuum of a star 1 relative to that of a star 2, is connected with the monochromatic color indices C_1 and C_2 determined at the wavelengths λ_a, λ_b, by the following relation:

$$G_{2\cdot1} = 0.921 \frac{C_2 - C_1}{\lambda_a^{-1} - \lambda_b^{-1}}.$$

(2.5)

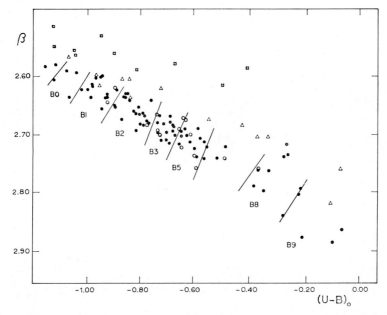

Fig. 10. Parameter β as a function of $(U-B)_0$ according to Crawford (1958 (k)).

This relation becomes approximate when the indices C_1 and C_2 are heterochromatic. Nevertheless, this shows us that properties established by spectrophotometrists in terms of gradients must appear among photometrists in terms of color indices. Divan (1954) has shown that $\Delta\varphi$, the difference between the gradients defined by the continuum which is on the ultraviolet side of D and the one defined in the visible region varies with spectral type but varies very little with interstellar extinction. So we must expect to come across a similar property by establishing a difference of color indices. This has been done by Borgman (1959 (c)) who uses the difference of the color indices $R-Q$ and $N-M$ to define a parameter of spectral type $\delta(\lambda_R=3295$ Å, $\lambda_Q=3560$ Å, $\lambda_N=4056$ Å, $\lambda_M=4580$ Å). The same is applied by Walraven and Walraven (1960 (e)) with the indices $U-W$ and $V-B(\lambda_W=3220$ Å, $\lambda_U=3260$ Å, $\lambda_B=4260$ Å, $\lambda_V=5590$ Å) defining the parameter $[U-W]$. In the same photometry, Walraven and Walraven use a parameter $[B-U]$ which is a measurement of the Balmer discontinuity. Thus, we come back to a photometric diagram which allows a two-dimensional representation based on the difference of the gradients and the Balmer discontinuity. This diagram is particularly useful in the interval between O6 and A3 and is shown in Figure 12.

We have pointed out that the spectrophotometric determination of the difference of the gradients $\Delta\varphi$ is but slightly dependent on interstellar extinction, and this is also the case for the determination of the Balmer discontinuity D. So the photometric measurements must also satisfy this condition. This is approximately the case for the parameters introduced in the photometries of Walraven and Borgman. Let us examine this question further, for it is important in the choice of the widths of the pass-bands.

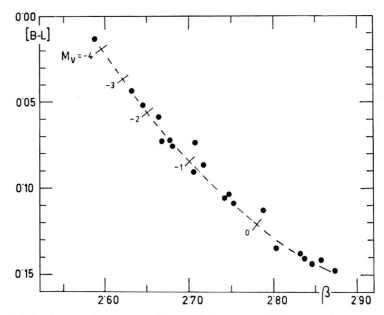

Fig. 11.   Relation between the parameter $[B - L]$ of the system of Walraven and Walraven (1960 (e))
and the parameter β of Crawford's system (1958 (k)).
Figure from Walraven and Walraven (1960 (e)).

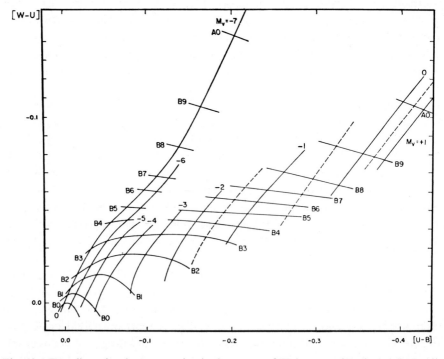

Fig. 12.   Two-dimensional representation in the system of Walraven and Walraven (1960 (e)).

Suppose we have a relative energy distribution similar to the one given in Figure 6. Along the abscissa we plot $I = 2.5 \log(E/E_0)$ (in Figure 6, it is $\log(E/E_0)$ which is plotted, but here we consider magnitudes), where E_0 is relative to the reference star. Let us form the difference between the slopes of segments $\overline{AB}$ and $\overline{CD}$, which comes to establishing a difference of gradients:

$$\delta = \text{slope } \overline{AB} - \text{slope } \overline{CD} = \frac{I_B - I_A}{\lambda_B^{-1} - \lambda_A^{-1}} - \frac{I_D - I_C}{\lambda_D^{-1} - \lambda_C^{-1}}. \tag{2.6}$$

The following quantity

$$\delta' = \delta \cdot (\lambda_B^{-1} - \lambda_A^{-1}) = (I_B - I_A) - \frac{\lambda_B^{-1} - \lambda_A^{-1}}{\lambda_D^{-1} - \lambda_C^{-1}} (I_D - I_C) \tag{2.7}$$

is proportional to the difference of the slopes. We notice monochromatic color indices in $I_B - I_A$ and $I_D - I_C$. Let us suppose one of the stars is reddened by interstellar matter according to a law $K(\lambda)$ proportional to λ^{-1}. The relative intensity I can be written, to a constant, $I = I^0 + M \cdot \lambda^{-1}$ where I^0 is the intrinsic relative intensity and M the mass of the interstellar matter. We can easily ascertain the identity:

$$\delta' = (I_B - I_A) - \frac{\lambda_B^{-1} - \lambda_A^{-1}}{\lambda_D^{-1} - \lambda_C^{-1}} (I_D - I_C) = (I_B^0 - I_A^0) - \frac{\lambda_B^{-1} - \lambda_A^{-1}}{\lambda_D^{-1} - \lambda_C^{-1}} (I_D^0 - I_C^0). \tag{2.8}$$

δ' is a parameter which is independent of the quantity of interstellar matter, if the law of extinction is in λ^{-1} in the spectral interval where the monochromatic color indices are defined.

We can state that $(\lambda_B^{-1} - \lambda_A^{-1})/(\lambda_D^{-1} - \lambda_C^{-1}) = E_{BA}/E_{DC}$, where E_{BA} and E_{DC} are the monochromatic color excesses

$$I_B - I_A = I_B^0 - I_A^0 + E_{BA} \quad \text{and} \quad I_D - I_C = I_D^0 - I_C^0 + E_{DC}.$$

If the law is different from a simple law in λ^{-1}, it is no longer obvious that we should obtain a polygonal line when we plot I as a function of $K(\lambda)$. Under such conditions the following relation:

$$\delta' = (I_B - I_A) - \frac{K(\lambda_B) - K(\lambda_A)}{K(\lambda_D) - K(\lambda_C)} (I_D - I_C)$$

or more generally

$$\delta' = (I_B - I_A) - \frac{E_{BA}}{E_{DC}} (I_D - I_C) \tag{2.9}$$

will be independent of the quantity of interstellar matter traversed, but depends on the type of law of extinction. The linear combination of indices given by expression (2.9) applies to stars poor in lines, where it is easy to define monochromatic indices.

When we consider heterochromatic color indices we can show that it is impossible to make linear combinations which are independent of the quantity of interstellar matter traversed. The coefficients of the linear combination vary strongly not only with the quantity of interstellar matter traversed, but also with the spectral type of the star. The effect of the width of the pass-bands in the linear combinations of indices has been specially studied by Golay (1971). Without the need for a demonstration (this can be found in the book given as reference above) we give the expressions permitting to pass from a monochromatic magnitude to a heterochromatic magnitude and from a monochromatic color excess to a heterochromatic color excess. These relations have been obtained by expanding expression (1.1) in a series

$$m_{\lambda_0}^h = m_{\lambda_0}^0 + 1.086\, M \left[\frac{\alpha}{\lambda_0} + \beta - \left(\frac{\mu}{\lambda_0} \right)^2 \frac{\alpha}{\lambda_0^2} (\Phi - 6\lambda_0) \right]$$

$$- 0.543\, \frac{\alpha^2}{\lambda_0^2} \left(\frac{\mu}{\lambda_0} \right)^2 M^2 , \tag{2.10}$$

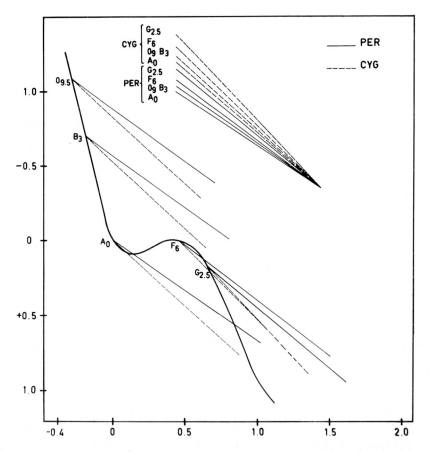

Fig. 13.   The $(U - B)$ vs. $(B - V)$ diagram, $U - B$ in ordinate, $B - V$ in abscissa. The reddening lines are calculated for various spectral types and for two different laws of reddening. The reddening lines have been made to pass through a common point in the upper part of the diagram so as to show the effects due to spectral type and to a difference in law of reddening.

$m_{\lambda_0}^0$ monochromatic magnitude at the wavelength λ_0, $m_{\lambda_0}^h$ heterochromatic magnitude of the pass-band characterized by λ_0 and μ (relation 1.2), $(\alpha/\lambda)+\beta$ law of interstellar extinction, α and β are defined in the spectral interval covered by the pass-band, Φ the absolute gradient in the spectral interval covered by the pass-band in question.

$$E_{\lambda_1\lambda_2}^h = E_{\lambda_1\lambda_2}^0 - 1.086\,M\left[\frac{\alpha_1}{\lambda_1}\left(\frac{\mu_1}{\lambda_1}\right)^2(\Phi - 6\lambda_1) - \frac{\alpha_2}{\lambda_2}\left(\frac{\mu_2}{\lambda_2}\right)^2(\Phi - 6\lambda_2)\right]$$
$$- 0.543\,M^2\left[\frac{\alpha_1^2}{\lambda_1^2}\left(\frac{\mu_1}{\lambda_1}\right)^2 - \frac{\alpha_2^2}{\lambda_2^2}\left(\frac{\mu_2}{\lambda_2}\right)^2\right], \qquad (2.11)$$

$E_{\lambda_1\lambda_2}^h$ is the heterochromatic color excess, $E_{\lambda_1\lambda_2}^0$ is the monochromatic color excess.

As for the ratios $r = E_{\lambda_1\lambda_2}^h / E_{\lambda_3\lambda_4}^h$, we can show that they can be put into the following form:

$$r \cong \frac{E_{\lambda_1\lambda_2}^0}{E_{\lambda_3\lambda_4}^0} + a\cdot\Phi + bE_{\lambda_3\lambda_4}^h, \qquad (2.12)$$

which has been actually confirmed by observation. One must point out that the coefficients a and b are proportional to the $(\mu/\lambda)^2$, which means to the square of the width of the pass-bands. These ratios r are the slopes of the reddening lines in the two-color-index diagrams. Figure 13 illustrates, in the case of wide-band U, B, V photometry, the variation of these slopes as a function of spectral type and in the case of two different laws of extinction with a color excess E_{B-V} of unit value. The term μ^2 also appears in the relations between two photometric systems which are close to each other. By close photometric systems we mean the photometric system copied on a standard system (same photocells, apparently same filters, etc.) or the several ones we believe we have night after night. The wide-band systems are difficult to reproduce, and the deviations from linear relations we will observe in the photometric relations, will be proportional to the differences of the squares of the widths of the pass-bands.

Figure 14 shows all the positions that can be occupied by the locus of the black bodies in the U, B, V diagram. These loci have been calculated with pass-bands slightly different from those of the standard system. In the U, B, V system, the linear combination is the one that gives the parameter Q

$$Q = (U - B) - \frac{\overline{E_{U-B}}}{E_{B-V}}(B - V), \qquad (2.13)$$

introduced by Johnson and Morgan (1953). The coefficient of the linear combination was derived from reddened O and B stars

$$\overline{E_{U-B}/E_{B-V}} = 0.72 \pm 0.03.$$

This parameter is a measurement of the Balmer discontinuity for hot stars. The following relation: $D = +0.525 + 0.525\,Q$ can be used between O and A2V. Applied to the study of globular clusters it is, according to McClure and van den Bergh (1968 (g)), an index of metallicity. Figure 15 illustrates in a general manner the properties of the

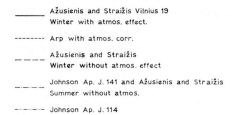

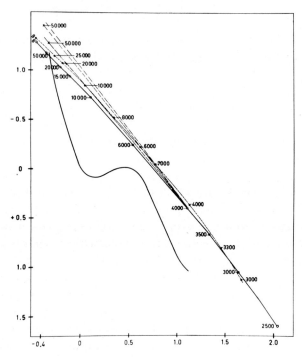

Fig. 14.   Loci of the black bodies calculated with different pass-bands and plotted in the
standard $(U-B)$ vs. $(B-V)$ diagram.

wide-band U, B, V system. The reddening lines for three different known laws have
been plotted. Practically all wide and intermediate-band photometries use linear
combinations to obtain parameters independent of the quantity of interstellar matter
traversed. For instance the two parameters $[U-W]$ and $[B-U]$ of Walraven's
photometry are given by the following linear combinations:

$$[B' - U] = (B - U) - 0.66(V - B)$$
$$[U' - W] = (U - W) - 0.55(V - B).$$

The narrower the pass-bands are, the more limited are the residual effects caused by
the width of the pass-bands. Nevertheless, these combinations will always remain

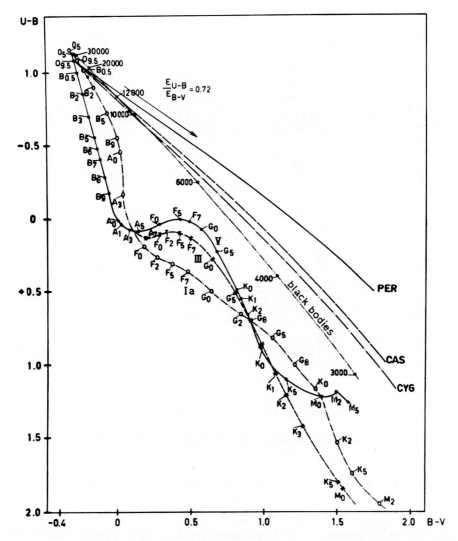

Fig. 15.   Diagram $(U-B)$ versus $(B-V)$ for the classes Ia, III, V, the black bodies and various reddening lines.

dependent on the laws of extinction. If the law is known, it is possible to correct the parameters given by the linear combinations, on condition that one knows an approximate value of the color excess, which is possible with the aid of a $(U-B)$ vs. $(B-V)$ diagram.

3. Introduction of Line-Blocking

Figure 16 shows the fraction of energy $\eta(\lambda)$ lost in the lines (H lines included) in a narrow spectral interval (usually 15 to 25 Å wide). The positions of the bands of the Geneva U, B_1, B_2, V_1, G photometric system are shown schematically.

　　　　　　　　M. GOLAY

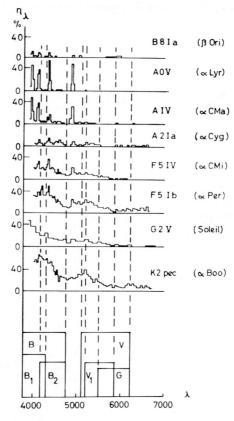

Fig. 16.　Fraction $\eta(\lambda)$ of the energy lost in the lines (H lines included) and schematic representation of the U, B_1, B_2, V_1, G system.

Figure 17, from an article by Wolff (1967), gives the distribution of $\eta(\lambda)$ (H lines not included) for two Ap stars (β CrB and γ Equ) and for the normal star β Ari (A5V). These figures show us that $\eta(\lambda)$ generally decreases towards the infrared, whence the advantage of the infrared for defining indices indicating effective temperature. Besides, we see that we can choose the widths of the pass-band of the filters in such a manner that the relative loss of energy caused by the lines in the interval of one filter be of the same order as the loss in the spectral interval covered by the other filter serving to define a color index. Such an index will be independent of the differences in abundance of lines for a set of spectral types which remains to be defined. This is the case, for instance, for the $B_2 - V_1$ index according to Golay and Goy (1965). This is also the case for a still greater number of cases for the $b-y$ index of Strömgren (1966). Such an independence in regard to abundance of lines applying to a set of stellar types to be defined, can also be arrived at for linear combinations of indices. It is for instance the case for the parameter C_1 ($C_1 = (u-v)-(v-b)$) of Strömgren's photometry (1966) and for $[d]$ ($[d] = (U-B_1)-1.430(B_1-B_2)+\varepsilon_d$) of the U, B_1, B_2, V_1, G photometry. These two parameters measure the Balmer discontinuity in the spectral

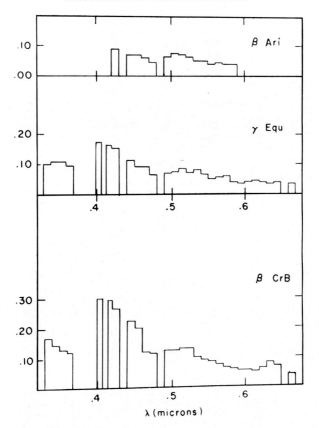

Fig. 17.   Fraction $\eta(\lambda)$ of the energy lost in the lines according to Wolff (1967) for two Ap stars
(β Cr B and γ Equ) and for the normal star β Ari (A5V).

interval B0 to F9. We can on the other hand adopt the contrary attitude and try to
determine a color index which is as sensitive as possible to the difference in abundance
of lines in the different spectral regions. This is the case of the index $B_1 - B_2$ in the
U, B_1, B_2, V_1, G system, the case also of $v-b$ in Strömgren's photometry (1966),
of C_{38-41} of the photometry of McClure and van den Bergh (1968 (g)) and of many
others. Let us point out that the more a photometry is accurate and is homogeneous,
that is to say possesses a good conservation of the standard pass-bands, the better
this photometry allows the detection of slight differences in abundance of lines
between stars which appear identical to spectroscopists. Indices such as $B_1 - B_2$ and
$v-b$ can enter into linear combinations allowing the definition of parameters which
are but slightly dependent on extinction by interstellar matter. In the U, B_1, B_2, V_1,
G photometry it is

$$[g] = (B_1 - B_2) - 1.357(V_1 - G) + \varepsilon_g,$$

in the u, v, b, y photometry, it is

$$m_1 = (v - b) - (b - y).$$

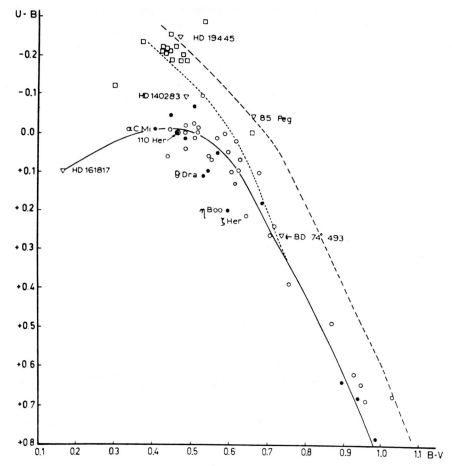

Fig. 18.   Position of 'strong-line' (circles), 'weak-line' (squares) and high-velocity halo-type stars (triangles), in the $(U - B)$ vs. $(B - V)$ diagram, according to Hack and Struve (1969).

Very wide-band photometry also allows the measurement of the abundance of lines in a stellar spectrum. We have already pointed out that the parameter Q can become an indicator of metallicity for globular clusters.

Figure 18, taken from the book by Hack and Struve (1969) shows the position of stars poor and rich in lines in the $(U-B)$ vs. $(B-V)$ diagram. Figure 19 defines the ultraviolet excess $\delta(U-B)$ which is a measurement of the deficiency in lines of the U filter in regard to the B filter. Figure 20, taken from an article by Wallerstein (1962) connects the ultraviolet excess to the degree of metallicity of the stellar atmosphere measured by the ratio [Fe/H].

Figure 2, at the beginning of this article, shows by the aid of the seven-color photometry of the Geneva Observatory, the energy lost in the lines. We can use this representation to illustrate some interesting cases (unreddened stars). Figure 21 compares the pseudo-continua of F7V strong and weak metallic-line stars. It also

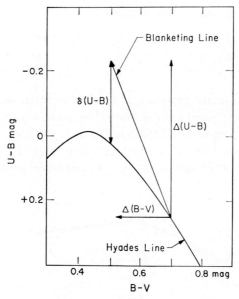

Fig. 19.   Definition of the ultraviolet excess $\delta(U-B)$.

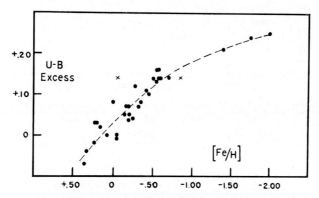

Fig. 20.   Relation between the ultraviolet excess $\delta(U-B)$ and the ratio [Fe/H] according
to Wallerstein (1962).

shows the difference between the pseudo-continua of a weak and a strong metallic-line G0V star. Figure 22 compares metallic-line stars, the metallic characteristics of which are more or less marked. Lastly, Figure 23 compares the pseudo-continua of normal and peculiar stars. These various illustrations allow one to get an idea of the order of magnitude of the effects.

4. Description of a Photometric System

We have often used in this article information supplied by the seven-color photometry described by Golay (1969 (a)). This system has been in use at the Geneva Observatory since 1960. Its characteristics are the following:

	U	B	V	B_1	B_2	V_1	G
λ_0	3458	4248	5508	4022	4480	5408	5814
$\mu(\text{Å})$	170	283	298	171	164	202	206

which make it a system on the boundary between intermediate and wide-band systems.
Thus, it is of a certain interest for it allows the measurement of weak objects. It
possesses most of the properties of Strömgren's u, v, b, y system. Nevertheless, because
of the width of the pass-bands, we must take great care in the reductions and especially
in the conservation of the response curves. The results obtained can be given either

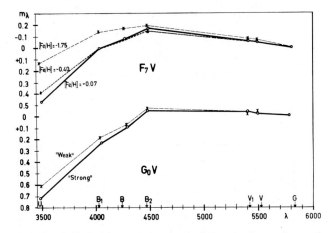

Fig. 21.   Comparison, in 7 colors, of pseudo-continua of stars having various values of [Fe/H] as
well as of 'weak-line' and 'strong-line' stars.

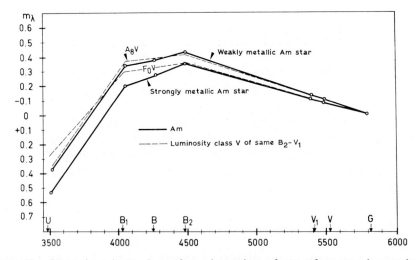

Fig. 22.   Comparison, in 7 colors, of pseudo-continua of stars of more or less marked
metallic characteristics.

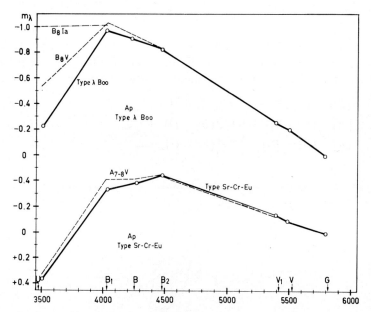

Fig. 23. Comparison, in 7 colors, of pseudo-continua of normal and peculiar stars.

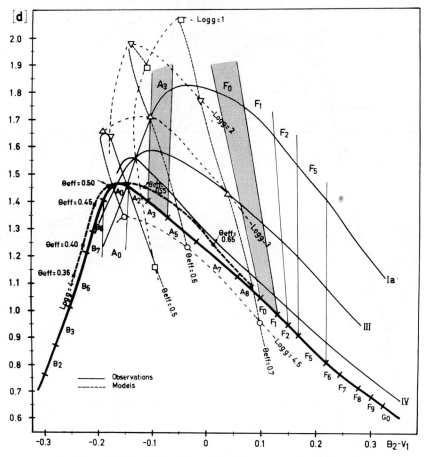

Fig. 24. Relation [d] vs. $(B_2 - V_1)$ of the seven-color photometry.

in the form of monochromatic colors which allow the construction of a pseudo-continuum in 7 points, or in the form of linear combinations. At the present, three combinations are indicated in the last study reported. They are:

$$[d] = (U - B_1) - 1.430(B_1 - B_2) + \varepsilon_d,$$
$$[\varDelta] = (U - B_2) - 0.832(B_2 - G) + \varepsilon_\varDelta,$$
$$[g] = (B_1 - B_2) - 1.357(V_1 - G) + \varepsilon_g. \tag{4.1}$$

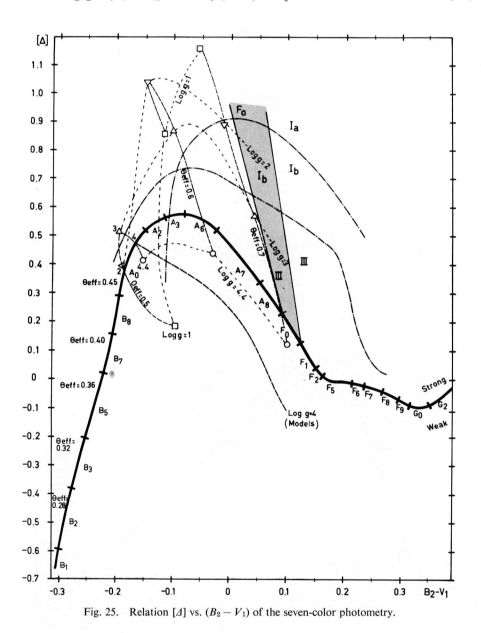

Fig. 25.   Relation $[\varDelta]$ vs. $(B_2 - V_1)$ of the seven-color photometry.

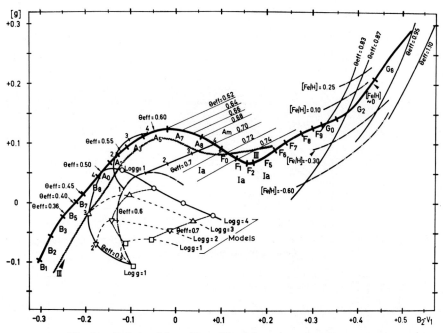

Fig. 26. Relation $[g]$ vs. $(B_2 - V_1)$ of the seven-color photometry.

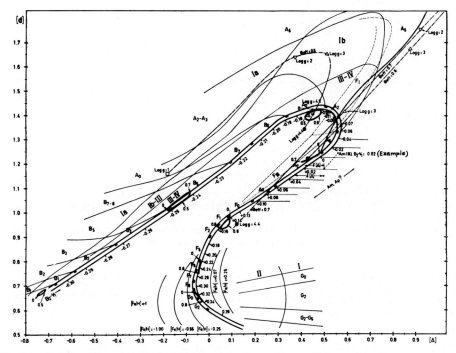

Fig. 27. Relation $[d]$ vs. $[\Delta]$. The hatched lines give the positions of binary stars, the components of which have mass ratios between 0 and 1. The dotted lines have been calculated for theoretical models by Mihalas with blocking by hydrogen lines.

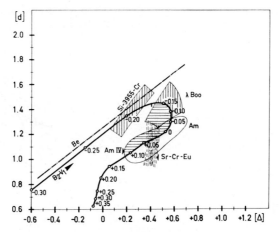

Fig. 28. Zone of the [d] vs. [Δ] diagram where the peculiar stars of the seven-color catalogue of the Geneva photometry are found.

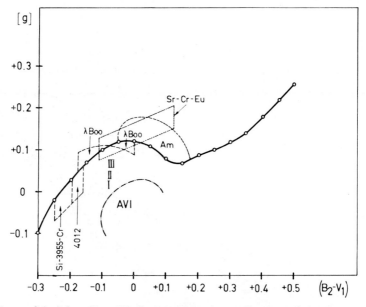

Fig. 29. Zones of the [g] vs. $(B_2 - V_1)$ diagram where the peculiar stars of the Geneva seven-color photometric catalogue are to be found.

The corrective terms ε_d, ε_Δ, ε_g depend on the position of the star in a $U-B$, $B-V$ or $U-B_2$, B_2-V_1 diagram and on the adopted law of interstellar extinction. These corrective terms have been calculated for a great number of spectral types on the basis of energy distributions measured by scanning and by introducing the Nandy laws of interstellar extinction. The basic color index of this photometry is B_2-V_1. The parameter [g] measures the line-blocking in the spectral region $\lambda > 3700$ Å. The parameters [d] and [Δ] measure the Balmer discontinuity. The difference

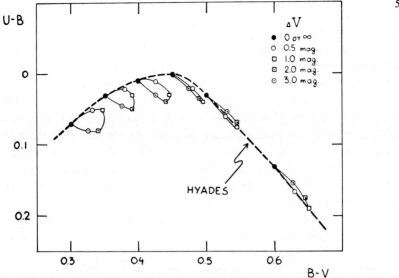

Fig. 30. Effect of binarity according to Smak (1967). In the upper right are indicated the differences in magnitude between the components.

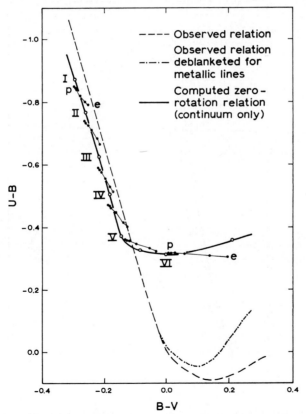

Fig. 31. Effect of rotation for stars rotating at almost the velocity of rupture and seen under different orientations. The cases for stars seen along the polar axis are the points at the far right. The cases for the stars seen along the equator are the points at the left.
(Figure taken from Hardorp and Strittmatter, 1968).

between $[d]$ and $[\varDelta]$ comes from the fact that these two parameters do not react in the same manner to the differences in line-blocking and to the various laws of interstellar extinction. Figures 24–29 illustrate the general properties of this photometry. We have superimposed (see Golay, 1969 (a), for the details of the operation) the lines of equal gravity and the lines of equal effective temperature obtained with theoretical models by Mihalas with blocking by hydrogen lines. In Figure 27, which has as coordinates the two parameters $[d]$ and $[\varDelta]$, we have plotted the loci occupied by binary stars with various mass ratios. These loci illustrate the dispersion that can result from unknown binary stars in photometric diagrams. Figure 30 shows, according to Smak (1967), the effect of binarity in part of the $(U-B)$ vs. $(B-V)$ diagram. In the rectilinear part, the couple is displaced along the sequence. In the curved part, the greatest divergence is obtained with the mass ratio which leads to a difference in magnitudes between the two components of 2. (Thus, binarity is very difficult to detect spectroscopically). Figures 31 and 32 show stellar rotation as another important cause of dispersion in photometric diagrams. These two figures, the one relating to the U, B, V photometry, the other to Strömgren's photometry, are taken from a study by Hardorp and Strittmatter (1968).

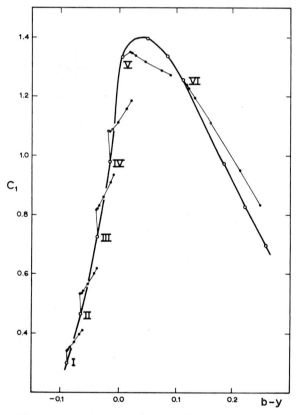

Fig. 32.   Effect of rotation on the relation C_1 vs. $(b-y)$ of Strömgren's photometry. Same remarks as for Figure 31.

TABLE I

Positions of pass-bands of selected photometric systems

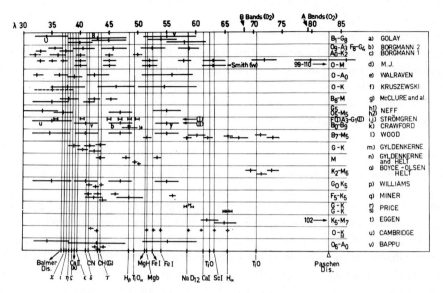

(a) Golay, M.: 1969, in *Vistas in Astronomy*, Vol. 14 (ed. by A. Beer), Pergamon Press, London and New York, in press.
(b) Borgman, J.: 1960, *Bull. Astron. Inst. Neth.* **15**, 255, 504.
(c) Borgman, J.: 1959, *Astrophys. J.* **129**, 362.
(d) Mitchell, R. I. and Johnson, H. L.: 1969, *Commun. Lunar Planetary Lab.*, No. 132.
(e) Walraven, Th. and Walraven, J. H.: 1960, *Bull. Astron. Inst. Neth.* **15**, 496.
(f) Kruszewski, A.: 1966, *Acta Astron.* **16**, 4.
(g) McClure, R. D., van den Bergh, S.: 1968, *Astron. J.* **73**, 313.
(h1) Neff, J. S.: 1968, *Univ. of Iowa*, No. 65-44.
(h2) Neff, J. S.: 1968, *Astron. J.* **73**, 75.
(i) Strömgren, B.: 1956, in *Vistas in Astronomy*, Vol. 2 (ed. by A. Beer), Pergamon Press, London and New York, p. 1336.
(j) Strömgren, B.: 1963, *Quart. J. Roy. Astron. Soc.* **4**, 8.
(k) Crawford, D. L.: 1958, *Astrophys. J.* **128**, 185.
(l) Wood, D. B.: 1969, *Astron. J.* **74**, 177.
(m) Gyldenkerne, K.: 1958, *Ann. Astrophys.* **21**, 77.
(n) Gyldenkerne, K. and Helt, B.: 1966, in *Spectral Classification and Multicolour Photometry*, IAU Symposium No. 24 (ed. by K. Lodén, L. O. Lodén, and U. Sinnerstad), Academic Press, London and New York, p. 162.
(o) Boyce, P. B., Olson, E. H., and Helt, B. E.: 1967, *Publ. Astron. Soc. Pacific* **79**, 469.
(p) Williams, J. A.: 1966, *Astron. J.* **71**, 615.
(q) Miner, E. D.: 1966, *Astrophys. J.* **144**, 1101.
(r) Price, M. J.: 1966, *Monthly Notices Roy. Astron. Soc.* **134**, 461.
(s) Price, M. J.: 1966, *Monthly Notices Roy. Astron. Soc.* **134**, 497.
(t) Eggen, O. J.: 1967, *Astrophys. J. Suppl. Ser.* **14**, 131.
(u) Redman, R. O.: 1966, in *Spectral Classification and Multicolour Photometry*, IAU Symposium No. 24 (ed. by K. Lodén, L. O. Lodén, and U. Sinnerstad), Academic Press, London and New York, p. 155.
(v) Bappu, M. K. V., Chandra, S., Sanwall, N. B., and Sinvhal, S. D.: 1962, *Monthly Notices Roy. Astron. Soc.* **123**, 521.

5. Narrow-Band Photometry

Expressions (2.10), (2.11), (2.12) of Section 2 become very simple when the term $(\mu/\lambda)^2$ becomes very small. Yet, when the pass-band decreases, the signal received becomes more and more dependent on the equivalent width of the lines found in the spectral interval covered by the pass-band. This signal can depend on one line only for given pass-bands. It is even possible to conceive pass-bands which are narrower than the equivalent width of an important line. The interpretation is evidently quite delicate in this case. Narrow-band photometry offers the great advantage of allowing the measurement of the energy lost in the line relatively to the nearby continuum, without the need of correcting for atmospheric and interstellar extinction (or very small corrections). This is of course possible only if we compare the intensity measured in a pass-band centered on the line to the intensity measured in a pass-band placed as near as possible, and in a spectral interval free of lines. The main disadvantage of narrow-band photometries is that the range of magnitudes covered by a given instrument is much smaller than that with wide and intermediate-band photometries. This can be a handicap for studies of galactic structure. In narrow-band photometry, the pass-band can be isolated either with an interferential filter, or by the aid of a slit placed on the image of a spectrum at the output of a spectrograph, or else from the recording of a spectrum obtained by direct scanning. Thus, there will be as many narrow-band photometric systems as there are interesting lines in stellar spectra. We have tried to summarize this in Table I. We do not intend to present there a complete survey of the existing narrow and intermediate-band photometries (one

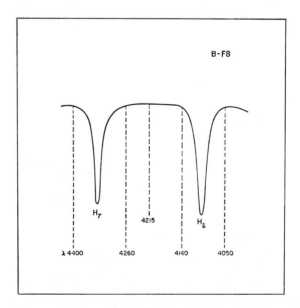

Fig. 33.   Position of the pass-bands used in Lindblad's criteria for spectral types B-F8. (Figure from
Strömgren, 1963.)

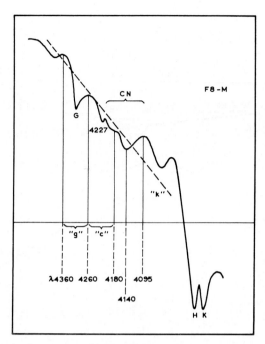

Fig. 34. Position of the pass-bands used in Lindblad's criteria for spectral types F8-M. (Figure from Strömgren, 1963.)

should for instance add the *U, P, X, Y, Z, V, S* system of Kakaras *et al.* (1968). This table gives the positions of the pass-bands of various photometric systems. The small letters placed before the names of the authors refer to the bibliography. We have given the spectral interval covered by each system, and sometimes underlined the spectral type which has more specially interested the author. The lines or bands which are of astrophysical interest and which have been used in these various photometries are given in the lower horizontal scale of the wavelengths. The main atmospheric absorption bands are given in the upper horizontal scale of the wavelengths. This table shows us that several narrow-band photometries rely on Lindblad's criteria. The lines and bands used to establish Lindblad's criteria are summarized in Figures 33 and 34, taken from the article by Strömgren (1963). Figure 33 gives the position of the bands containing Hγ and Hδ and of the comparison bands which do not contain any hydrogen lines. These bands have been used for the classification of spectral types B-F8. Figure 9 gives an example of classification carried out with the aid of the Hδ line (Hδ measured on a photographic spectrum). Figure 35 gives an example of the two-dimensional classification of Bappu *et al.* (1962 (v)) obtained with three narrow bands, one of which is a measurement of Hγ. Figure 34 gives the position of the pass-bands used in Lindblad's criteria for stars of spectral types F8-M. The bands at 4360 Å and 4260 Å allow the measurement of the importance of the CH G-band at 4300 Å. Those at 4260 Å and 4180 Å measure the intensity of the CN-bands in the region of wavelengths shorter than 4216 Å. These criteria are used in the narrow-band

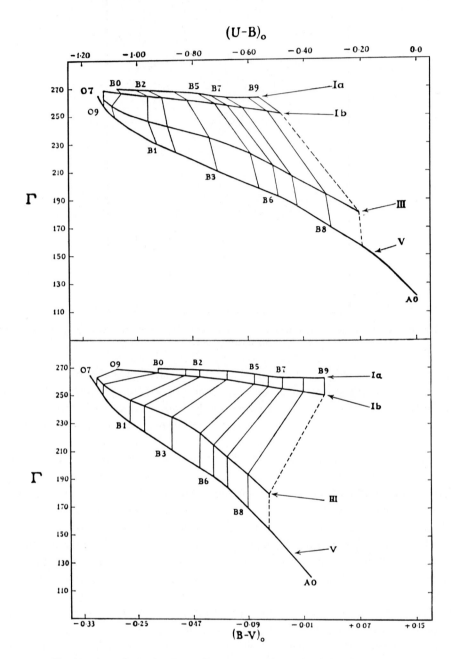

Fig. 35. Two-dimensional classification according to Bappu *et al.* (1962 (v)).

photometry developed at Cambridge (work of Griffin and Redman, reported by Redman, 1966 (u)). The Cambridge group has also studied narrow-band photometries using the Mgb- and Na D-lines, the Fe I triplet at 5250 Å, the Ca I triplet at 6100, the Sc I lines at 6306 Å and the Fe I lines. Figures 36 and 37, taken from an article referred to by Redman (1966 (u)) show the application of Lindblad's criteria by

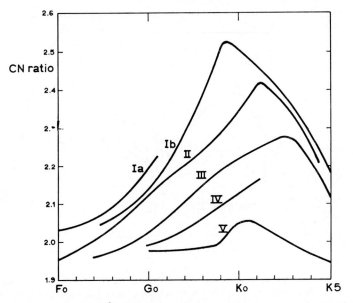

Fig. 36.   Application of the 4200 Å CN-band as criterion of luminosity in the Cambridge photometric work (reported by Redman 1966 (u)).

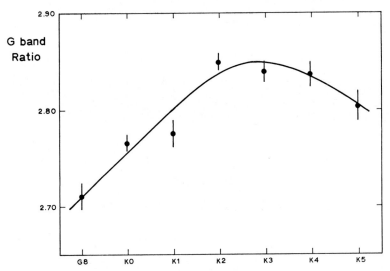

Fig. 37.   Application of the 4300 Å G-band as criterion of spectral class (Cambridge photometric work reported by Redman, 1966 (u)).

Griffin and Redman to narrow-band photoelectric photometry. The CN-band at 4200 Å is used as an indicator of luminosity, while the G-band is used as an indicator of spectral type. These two bands are also made use of in the intermediate-band photometry of McClure and van den Bergh (1968 (g)). In the latter photometry, the authors add pass-bands which allow the measurement of the discontinuity at

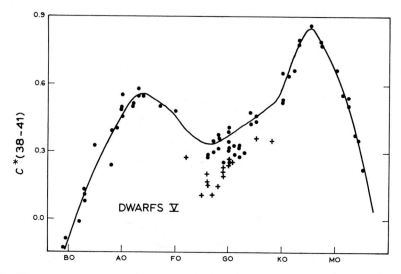

Fig. 38. The $C*(38-41)$ parameter measures the discontinuity close to 4000 Å due to 'blanketing'. It separates the normal dwarfs (dots) from the high velocity dwarfs (crosses). Figure from McClure and van den Bergh (1968 (g)).

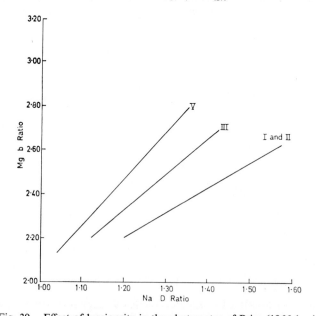

Fig. 39. Effect of luminosity in the photometry of Price (1966 (r, s)).

4000 Å in the energy distribution due to a strong difference in blanketing on either side of 4000 Å for stars of spectral type F, G, K. The index C^* (38-41) (38 and 41 mean 3800 Å and 4165 Å, wavelengths at maximum transmission for the two bands used) which measures this discontinuity is a parameter related to the abundance of metallic lines in the spectrum. Figure 38, taken from the article referred to above, shows a separation between normal dwarf stars and high-velocity dwarf stars obtained with the C^* (38-41) parameter corrected for interstellar reddening.

Let us go back to very narrow-band photometries and point out the photometry of Price (1966 (r, s)) for stars of spectral type G and K. This photometry uses in particular the D_1- and D_2-lines of Na which lie in a pass-band limited by 5885.0 Å and 5901.2 Å. The pass-band is adjusted so as to take account of the apparent radial velocity of the star. The same photometry also makes use of the Mgb triplet of Mg, the wavelengths of which are 5167.3 Å, 5172.7 Å and 5183.6 Å. The effects produced by these lines are measured with the three-channel spectrophotometer of the Cambridge Observatory.

Figure 39, taken from the article by Price (1966 (r, s)) shows the luminosity effect

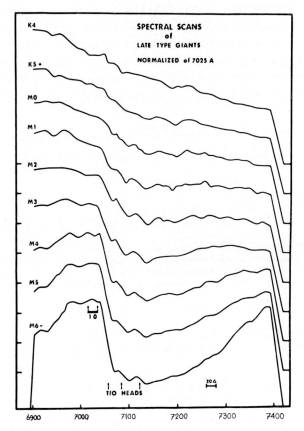

Fig. 40.   Energy distribution in the proximity of the TiO band, according to Boyce *et al.* (1967 (o)).

which appears in a diagram based on ratios measuring the Mgb-triplet and the Na D-lines. The development of photomultipliers having photocathodes sensitive to the red and the infrared has led several authors to the use of the TiO bands at 6200 Å and at 7100 Å.

Figure 40 shows the variation of the energy distribution in the neighborhood of the 7100 Å TiO band for spectral types K5 to M6. Eggen (1967 (t)) has made use of an index one of the pass-bands of which is sensitive to the TiO band (6250 Å). He plots this as a function of another index, the bands of which are centered on 6500 Å and 10200 Å so as to obtain a diagram permitting the separation of dwarfs from giants among stars more advanced than K5. Let us also point out here the photometric system of Wood (1969 (l)) which unites 12 pass-bands having widths between 375 Å and 56 Å covering the spectral interval 3400–7300 Å. This system contains the u, v, b, y-bands as well as narrower bands destined to measure all the lines and bands measured in all the formerly described photometric systems.

Thanks to the use of satellites, rockets or balloons, it is possible to extend photometric measurements towards the ultraviolet as well. Yet, the energy distribution as well as the laws of interstellar extinction in this region are still very little known. At the moment, and until the results of the OAO satellite are published, the number of measured stars is still limited. The few known results, which are very well summarized in an article by Wilson and Boksenberg (1969), show that there is a vast field of action for all types of photometry. Moreover, the complex electronic techniques developed for the photometers put on board spacecraft can be applied on the ground and contribute to increase the yield and the quality of the measurements made.

References

Baum, W. A.: 1962, in *Astronomical Techniques* (ed. by W. A. Hiltner), Univ. of Chicago Press, Chicago, p. 1.
Berger, J. and Fringant, A. M.: 1955, *J. Observateurs, Marseille* **28**, 12.
Chalonge, D.: 1958, in *Stellar Populations, Ric. Astron. Specola Astron. Vatic.* **5** (ed. by J. K. O'Connell), North-Holland Publishing Co., Amsterdam, p. 345.
Code, A. D.: 1960, in *Stellar Atmospheres* (ed. by J. L. Greenstein), Univ. of Chicago Press, Chicago, p. 50.
Divan, L.: 1954, *Ann. Astrophys.* **17**, 456.
Golay, M.: 1971, *Introduction à la photométrie astronomique*, in press.
Golay, M. and Goy, G.: 1965, *Publ. Observ. Genève*, No. 71.
Hack, M. and Struve, O.: 1969, *Stellar Spectroscopy, Normal Stars*, Osservatorio Astronomico, Trieste.
Hardie, R. H.: 1962, in *Astronomical Techniques* (ed. by W. A. Hiltner), Univ. of Chicago Press, Chicago, p. 178.
Hardop, J. and Strittmatter, P. A.: 1968, *Astrophys. J.* **151**, 1057.
Johnson, H. L.: 1962, in *Astronomical Techniques* (ed. by W. A. Hiltner), Univ. of Chicago Press, Chicago, p. 157.
Johnson, H. L.: 1965, *Astrophys. J*, **141**, 923.
Johnson, H. L. and Morgan, W. W.: 1953, *Astrophys. J.* **117**, 313.
Kakaras, G., Straizys, V., Sudzius, J., and Zdanavicius, K.: 1968, *Vilnius Astron. Observ. Bull.*, No. 22, p. 3.
Lallemand, A.: 1962, in *Astronomical Techniques* (ed. by W. A. Hiltner), Univ. of Chicago Press, Chicago, p. 126.

Lamla, E.: 1965, in *Landolt-Börnstein Zahlenwerte und Funktionen aus Naturwissenschaften und Technik, Gruppe VI: Astronomie – Astrophysik und Weltraumforschung, Band I: Astronomie und Astrophysik* (ed. by H. H. Voigt), Springer-Verlag, Berlin, Heidelberg, New York, p. 315.

Nandy, K.: 1964, *Publ. Roy. Observ. Edinburgh* **3**, 142.

Nandy, K.: 1965, *Publ. Roy. Observ. Edinburgh* **5**, 13.

Nandy, K.: 1966, *Publ. Roy. Observ. Edinburgh* **5**, 233.

Nandy, K.: 1967, *Publ. Roy. Observ. Edinburgh* **6**, 25.

Nandy, K.: 1968, *Publ. Roy. Observ. Edinburgh* **6**, 169.

Sinnerstad, U.: 1961, *Stockholm Observ. Ann.* **21**, 6.

Smak, J.: 1967, *Acta Astron.* **17**, 2.

Stebbins, J. and Whitford, A. E.: 1945, *Astrophys. J.* **107**, 102.

Stock, J. and Williams, A. D.: 1962, in *Astronomical Techniques* (ed. by W. A. Hiltner), Univ. of Chicago, Press, Chicago, p. 374.

Strömgren, B.: 1963, in *Basic Astronomical Data* (ed. by K. A. Strand), Univ. of Chicago Press, Chicago, p. 123.

Strömgren, B.: 1966, *Ann. Rev. Astron. Astrophys.* **4**, 433.

Wallerstein, G.: 1962, *Astrophys. J. Suppl. Ser.* **6**, 407.

Wolff, S. C.: 1967, in *The Magnetic and Related Stars* (ed. by R. C. Cameron), Mono Book Co, Baltimore, p. 421.

Wilson, R. and Boksenberg, A.: 1969, *Ann. Rev. Astron. Astrophys.* **7**, 421.

INTERSTELLAR DUST

H. ELSÄSSER

Max-Planck-Institut für Astronomie und Landessternwarte Heidelberg, Germany

Interstellar matter is an important component of our Galaxy although its mass may be less than 10% of the whole system. The structure of the Galaxy is strongly affected by the distribution of interstellar matter and its evolution is mainly determined by the amount of interstellar gas and dust which is available for star formation.

The investigation of interstellar matter is today a widespread field of astronomical activity and it is impossible to discuss within a short time all important aspects. In the following I shall mainly deal with problems of interstellar dust, which is in my opinion a much less well understood phenomenon than the gas, a phenomenon with many problems for the future.

1. Interstellar Reddening and Extinction

1.1. REDDENING CURVE $A(\lambda)$

This curve describes the reddening of starlight by interstellar extinction. In order to derive $A(\lambda)$ we may consider two stars of identical spectral type and luminosity located at the same distance, one of which is obscured. By measuring their magnitude difference at various wavelengths, $A(\lambda)$ can be obtained. In this way Whitford, for instance, determined the reddening curve in the visual part of the spectrum and in the near infrared (Figure 1).

1.2. RATIO R OF TOTAL TO SELECTIVE EXTINCTION

If stellar magnitudes are measured in two colors B, V, the color excess E is defined as the difference

$$E = (B - V) - (B - V)_0$$

of the color indices of reddened and unreddened (index 0) stars of the same type and corresponds to

$$E = A_B - A_V,$$

where A_B and A_V is the total extinction (absorption) in B and V. We can write

$$V_0 = V - A_V = V - R \cdot E,$$

where

$$R = A_V / E$$

is the ratio of total to selective extinction. R is an important figure for the determination of photometric distances; therefore, it is of interest to know the value of R and whether it is constant for different regions of the Milky Way.

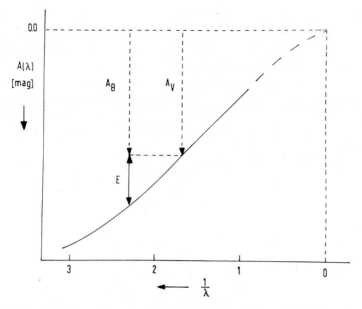

Fig.1.   Reddening curve with extrapolation in the infrared. A extinction, E color excess, λ wavelength (μ).

R can be derived from the reddening curve ('color difference method'), if the zero-point $A=0$ is known (see below). It can also be determined by the 'variable extinction method': If stars of identical spectral type and luminosity in clusters or associations, i.e. at the same distance, with variable extinction, are studied, R follows from the relation $V = V_0 + R \cdot E$ since V_0 is constant. This was done within the region of h $+\chi$ Per by Hiltner and Johnson. They obtained

$$R = 3.0 \pm 0.3,$$

whereas from Whitford's reddening law with a plausible extrapolation in the infrared Blanco also found $R=3.0$. In other regions values near 3.0 have been derived too.

1.3. $R>3$?

During the last years new data have been published, particularly by Johnson (1968), which indicate for many directions of the Milky Way larger values of R, between 3 and about 7. These results are based on the variable extinction method and the color difference method. This would mean that there is no universal law of interstellar reddening. Photometric distances calculated with $R=3$ would be too large and the interstellar extinction too low in many cases. The distance of the galactic center, as it follows from photometric observations, would be smaller than 8 kpc, if $R>4$. The distances of clusters and the position of the spiral arms had to be corrected as well.

The large R-values which Johnson has derived by the variable extinction method have been criticized by Becker (1966). He demonstrated that the intrinsic scatter in the color-magnitude diagram can simulate a large R; in addition, non-physical members of the investigated clusters increase R. In the case of III Cep and I Ara Johnson's R is considerably affected by non-physical members according to Becker.

What about the large R-values derived from the reddening curves? Johnson has made measurements in the infrared beyond $\lambda = 3\,\mu$, in order to reduce the uncertainties connected with the extrapolation to $1/\lambda = 0$ (see Figure 1). In several cases he found a

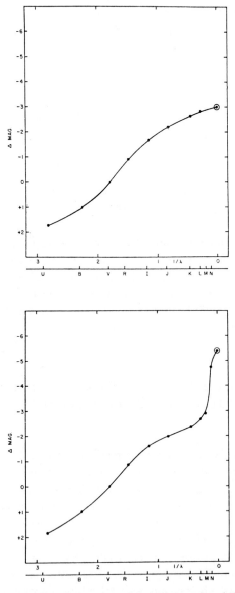

Fig. 2.   Extinction curve for the Perseus region (upper part) and the Cepheus region according to Johnson (1968).

bump in the reddening curve between 5 and 11 μ which gives rise to an increased R compared to a smoothly extrapolated reddening law. In the case of the Cepheus curve (Figure 2) this maximum in the infrared is mainly due to the star μ Cep, a M2 Ia star, which is bright enough for photometry at 11.5 μ. These results seemed to, confirm the large R-values determined by the variable extinction method for clusters, but we shall see (Section 4) that here, too, another explanation may be the correct one.

It cannot be excluded that deviations from $R=3$ occur. We cannot expect a priori that the composition and size distribution of the interstellar grains are everywhere the same, but at present it is not clear whether the deviations claimed are real and how large they are.

1.4. Reddening Law and Nature of Interstellar Grains

The reddening law contains information on the nature of interstellar dust. The extinction by small spherical particles can be calculated according to the theory of Mie and the similarity between the reddening curve $A(\lambda)$ and the extinction curve as a function of a/λ (a radius of particle, λ wavelength) in the range of small a/λ values is obvious (Figure 3). Considering such theoretical curves one should expect a smooth course near $1/\lambda=0$. At smaller wavelengths the cross-section Q_{ex} decreases again beyond the first maximum.

As was shown by van de Hulst twenty years ago, the extinction curve in the visual and near infrared can be explained by ice-grains with sizes below 1 μ. In the ultraviolet the dielectric grain extinction curve has a maximum of absorption. During the last years a number of UV-measurements from rockets have been performed at

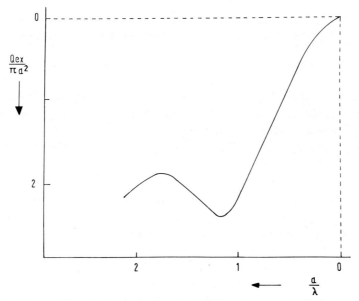

Fig. 3. Theoretical extinction curve (schematically). Q_{ex} cross-section for extinction, a radius of grain.

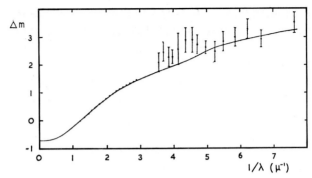

Fig. 4. Comparison of observations in the visual and ultraviolet and theoretical extinction curve for graphite sphere ($a = 0.054$ μ) with dirty ice mantle (outer radius $a = 0.16$ μ). Note different orientation of this figure (Lynds and Wickramasinghe, 1968).

wavelengths down to 1600 Å (Figure 4), but the slope of the empirical curve in the ultraviolet is not very different from that in the visual. The ice grain extinction model, therefore, must be abandoned.

Quite recently many papers have been published which try to explain the observed extinction curve. It is not possible to go into details here (compare for instance the review article of Lynds and Wickramasinghe, 1968). The best representation of the observations at present is accomplished on the basis of core-mantle grains (for example, graphite or metallic core with ice mantle), but on the other hand a unique solution is not obtainable. The nature of the interstellar grains cannot be derived from the extinction curve alone.

2. Diffuse Galactic Light and Albedo of Interstellar Grains

In a layer containing stars and interstellar grains diffuse light should be present, if the albedo of the particles is high and the extinction is mainly due to scattering (as for instance in the Earth's atmosphere). This diffuse galactic light contains independent additional information about the grains' nature, therefore, it is desirable to determine the albedo γ which is defined as the ratio

$$\gamma = Q_{sca}/Q_{ex},$$

where Q_{sca} is the grains' cross section for pure scattering and Q_{ex} that for extinction. For pure scatterers (i.e. no absorption) as ice grains $Q_{sca} = Q_{ex}$ and $\gamma = 1$, for pure absorbers $\gamma = 0$. A graphite grain of 0.05 μ radius has an albedo $\gamma = 0.25$ for $\lambda = 0.5$ μ, for a graphite core-ice mantle grain $\gamma \geqslant 0.5$ in the visual, and can be close to unity depending on the extension of the mantle.

Since the 1964 Lagonissi Summer School, where I also referred to this problem, several investigations have been performed in order to determine γ. The problem comprises two parts. At first the brightness of the diffuse light, if present, has to be measured. Then by considering the radiative transfer in the interstellar medium which depends on the scattering function and the albedo of the grains, the observed intensities are to be interpreted.

In an investigation similar to the former one of Henyey and Greenstein Witt (1968) derived the distribution of the diffuse light across the galactic equator in Cygnus and Taurus-Auriga. On the equator he finds for $\lambda = 6100$ Å intensities near $100 \, S_{10}/\Box^{\circ}$ ($= 100$ A0 V stars of $V = 10\overset{m}{.}0$ per square degree). They follow from observations which

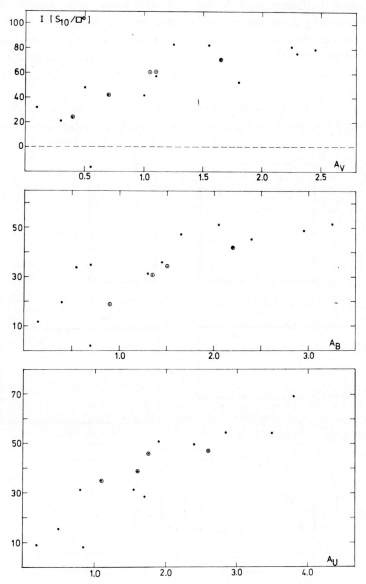

Fig. 5. Intensity of diffuse light in the Coalsack for V, B and U vs the extinction A according to measurements of Mattila (1970). Dots with ring have higher weight.

extend to galactic latitudes $b = 30°$ on both sides of the equator. The main difficulty is to eliminate the contributions of star light, zodiacal light and night airglow which vary with changing galactic latitude.

Mattila, a student of the Heidelberg Observatory, determined diffuse galactic light in a different way (1970). The idea was to use a near dust cloud which is illuminated by more distant stars and to obtain the diffuse light differentially by comparing obscured regions with unobscured fields besides the cloud. Mattila studied regions within and near the Coalsack of the Southern Milky Way and found a correlation between the varying extinction in the Coalsack and the brightness of diffuse light. For $A_V = 4^m$ he got about 80 $S_{10}/\square°$ in the visual region (Figure 5). The color of the diffuse light is more blue than the color of the illuminating stars. A model computation gives values of the albedo γ near 0.6. Similar results have been derived from Witt's observations.

Both investigations have demonstrated the existence of diffuse galactic light by using modern photoelectric equipment. The corresponding γ-values are consistent with

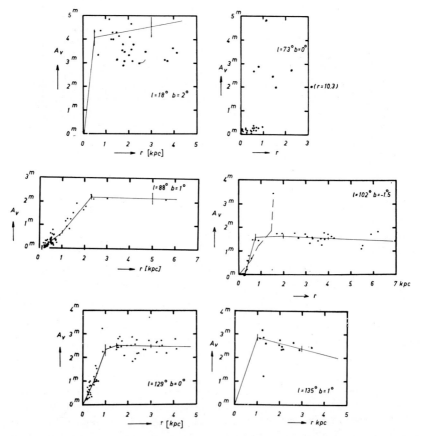

Fig. 6. Interstellar extinction A_V vs distance r from the Sun for different Milky Way fields according to Neckel (1967).

grains of the core-mantle or dirty ice type. Pure absorbers (metallic particles) are ruled out.

3. Space Distribution of Interstellar Dust

Our knowledge about the distribution of interstellar dust in the Galaxy is very incomplete; only for the immediate solar-neighborhood something reliable is known. I am referring to a thorough investigation of Neckel (1967) in which all available color-excess data have been critically studied in order to derive the extinction $A(r)$ as a function of distance r. Neckel used about 4700 measurements of stars and clusters, the ratio of total to selective extinction was assumed as $R=3.1$. For the determination of $A_V(r)$ it is essential to consider relatively small fields in galactic longitude and latitude, otherwise the large scattering which results from the cloudiness of the interstellar matter conceals any correlation of A_V and r. Therefore, the sphere was divided into 207 fields in such a way that the scatter within the fields is low.

Figure 6 shows $A_V(r)$ for some of these fields. In these diagrams quite often a strong absorption at small distances from the Sun is observed whereas at larger distances the extinction is rather small. This is no selection effect. Figure 7 shows the interstellar extinction along the Milky Way band within the distance $r=1$ kpc and demonstrates its patchiness. Between $l=140°$ and $260°$ $A(r \leqslant 1$ kpc$)$ is smaller than for the rest of the Milky Way. This may be an indication that the absorbing matter is concentrated at the inner edge of the local arm, but the picture is not very clear.

From this material a mean extinction for $r \leqslant 0.5$ kpc of $A_V = 2\overset{m}{.}46$/kpc in the galactic plane follows. The decrease with increasing distance z from the plane is fast $(e^{-1}$ at $z=40$ pc$)$.

In order to obtain information about the properties of the interstellar dust clouds Scheffler (1967) made a statistical analysis of Neckel's extinction catalogue. The observational data can be represented by a two-cloud model with the following parameters:

mean cloud extinction $\quad \bar{a}_1 = 0\overset{m}{.}26 \qquad \bar{a}_2 = 1\overset{m}{.}6$
mean cloud frequency $\quad \bar{\nu}_1 = 5$ kpc$^{-1} \qquad \bar{\nu}_2 = 0.5$ kpc^{-1}
mean cloud diameter $\quad \bar{\varLambda}_1 = 3$ pc $\qquad \bar{\varLambda}_2 = 70$ pc .

4. Infrared Stars and Interstellar Matter

The detection of infrared stars and their investigation during recent years added new aspects to our ideas of the interstellar medium. Infrared stars may be defined as objects with the maximum radiation-energy F_λ at wavelengths $\lambda > 1\ \mu$. Another definition in use is that for infrared stars $I-K>3^m$, when I and K are the star's brightness in Johnson's I $(\lambda \approx 0.9\ \mu)$ and K $(\lambda \approx 2.2\ \mu)$ spectral regions.

The first systematic survey for infrared stars by Neugebauer et al. (1965) was a photoelectric scan of the sky in the I- and K-region with a 60''-mirror and PbS-detectors for K. They detected very red objects, as for instance NML Cygnus with $I-K=6\overset{m}{.}6$, $V=16\overset{m}{.}6$, $K=0\overset{m}{.}4$ and the maximum of the spectral energy curve near 5 μ.

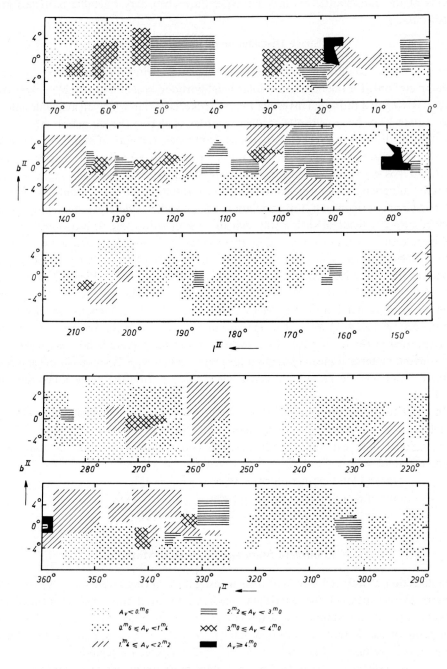

Fig. 7.   Interstellar extinction A_V within $r = 1$ kpc distance along the Milky Way
according to Neckel (1967).

But these extreme stars are relatively seldom; up to now about two dozen objects with $I-K>5^m$ are known.

A survey of several fields in the Milky Way has been performed at our observatory, making use of a combined photographic and photoelectric method. From objective prism spectra of low dispersion (3200 Å/mm) on IN-film and composite photographs obtained by the superposition of IN- and 103aE- plates suspicious stars were selected for photoelectric photometry in the 1 μ and 2.2 μ region. It is indeed possible to recognize infrared stars by photographic means as the photoelectric observations have demonstrated (Ackermann *et al.*, 1968). The limiting magnitude in K of this survey is about

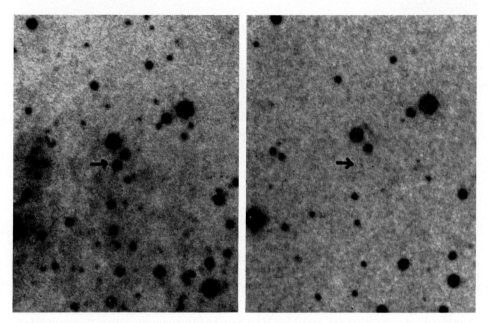

Fig. 8.   Infrared star on red (left) and blue (right) prints of the PSS (Ackermann *et al.*, 1968).

2^m more favourable than that of the CIT-scan despite of our moderate telescopes. Figure 8 shows one very red star found in this way. Within 160 $\square^\circ$ in Cygnus and Perseus, Ackermann (1969) identified by a photographic search 400 probable infrared stars (only a small part of which is observed with photoelectric equipment at present); the objects of these regions which were already in the CIT-catalogue were all found again. Figure 9 shows one of the surveyed fields in Cygnus. The new objects are marked by dots, the already known ones by crosses. Their distribution seems to be correlated with the distribution of interstellar matter and from these results one gets the impression that in highly obscured fields the number of infrared stars is rather high.

The interpretation of the observations is still rather difficult since several effects seem to be mixed up. One part of the known infrared stars may be highly reddened by interstellar dust, the required values of extinction are $A_V \gtrsim 10^m$. If this is true and

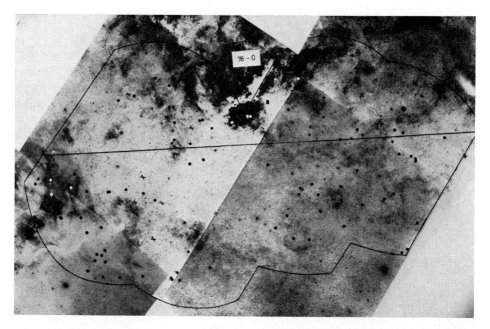

Fig. 9.   Extreme red stars in Cygnus ($l \approx 78°$, $b \approx 0°$) according to Ackermann (1969). Crosses: stars of the CIT-scan.

the results of Ackermann have to be explained in this way, very dense interstellar dust clouds would be much more numerous than it was believed up to now.

A number of infrared stars have special spectral energy distribution curves as demonstrated by Figure 10. R Mon for instance has a first maximum in the visual region (and the line spectrum of a Ge-star) and a second conspicuous one in the infrared. Curves of this type are explained now as due to a circumstellar dust cloud which is heated by a central star of a few thousand degrees photospheric temperature and reradiates this stellar energy at about 700°. The maximum energy of a Planck curve for $T = 700°$ appears at 4 μ wavelength. The mass of such a dust shell can be of the order of the planetary system (Low and Smith, 1966). These clouds around stars might represent important sources of interstellar dust. The grains could be blown by radiation pressure or carried away with stellar winds into interstellar space. There is little doubt today that condensation of solid materials in cool stellar atmospheres is to be expected; circumstellar shells might have their origin in this process.

Whether all infrared stars have dust shells is not clear at the moment. Also our knowledge about age and stage of evolution of these objects is still poor. There are certainly very young infrared stars, but it appears unlikely that the majority are stars in the contraction phase, especially since a number of infrared stars are Mira variables.

The detection of stellar dust shells deserves our attention with regard to the large R values discussed in Section 1. Woolf and Ney (1969) have found in the infrared spectra of several late-type giants and supergiants emission features which are most probably of circumstellar origin due to the radiation of mineral grains (Figure 11). Among their stars is μ Cep, the spectrum of which below 7 μ falls close to a black body curve of 3000°, whereas between 8 and 13 μ an emission almost five times as bright as the underlying continuum occurs. The infrared bump of the Cepheus curve

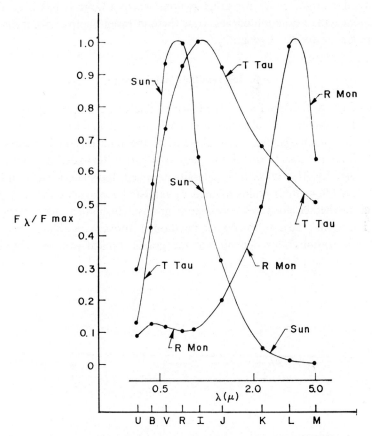

Fig. 10.   Spectral energy curves for the Sun, T Tau and R Mon.

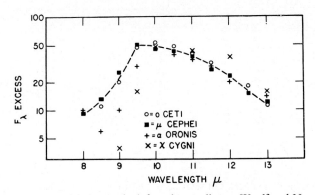

Fig. 11.   Excess radiation in the infrared according to Woolf and Ney (1969).

in Figure 2 is certainly caused by this emission which has here the consequence to increase R, if derived by the color-difference method. The extinction curve for Cepheus then consists of two parts, one which is due to interstellar extinction and which may be identical with the Perseus-curve of Figure 2 ($R=3.0$) and another one which is of circumstellar origin. Since for other regions where a large R has been found by color differences the extinction curves show these infrared bumps, too, it may be still best at present to use $R=3$ generally.

5. Interstellar Polarization

The light of many reddened stars is polarized due to an anisotropy of interstellar extinction. The amount of polarization is small. For the majority of stars measured up to now it remains below a few percent. If the polarization is expressed as a magnitude difference Δm_p the maximum ratio of polarization to extinction is according to observations $\Delta m_p/A_V=0.06$. It is generally assumed that the polarization is caused by anisotropic (elongated) grains aligned by interstellar magnetic fields. Today it is not quite clear whether elongated core-mantle grains, which are a satisfactory solution to other interstellar dust problems as mentioned above, are able to produce the observed polarization. This depends on the grains' optical properties and on the

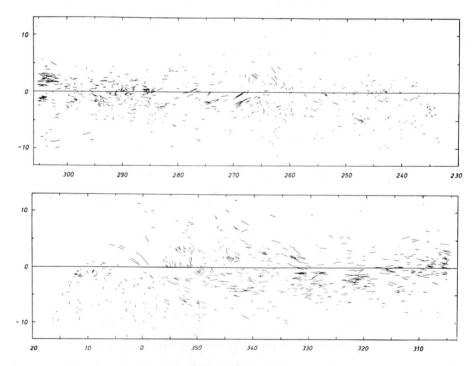

Fig. 12.   Polarization of southern OB-stars according to Klare and Neckel (1970). Longest lines correspond to 7% polarization.

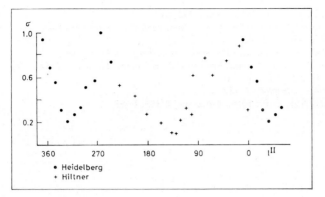

Fig. 13.   Deviation from the mean direction of polarization σ vs. galactic longitude l.

degree of alignment. Whether the necessary alignment can be achieved by magnetic fields with field strengths below 10^{-5} G, as indicated by independent information about the galactic field, is an open question, too.

If the particles are aligned according to the Davis-Greenstein theory, their long axes should be preferentially vertical to the magnetic field lines. Then we have to expect small amounts of polarization and no preferred orientation, if the line of sight is parallel to the field lines, and larger ones with approximately parallel directions, if we look perpendicular to the field. Bearing this in mind, some regularities in the distribution of polarization along the Milky Way are to be recognized. Figure 12 contains the results of a new investigation by Klare and Neckel (1970) who measured the polarization of about 1500 southern OB stars. Similar diagrams for the northern Milky Way have been published by Hiltner (1956). Apart from magnetic field effects, Figure 12 shows the influence of interstellar extinction. Between $l=230°$ and $260°$, for instance, the degree of polarization is small because the extinction is low.

The plane of vibration deviates from the mean direction of a certain galactic longitude in a systematic manner as is demonstrated in Figure 13. Low scatter of the polarization angles is observed at $l=140°$ and $320°$. At these two directions, which are $180°$ apart, the line of sight should be perpendicular to the magnetic field lines. That means that the local field lines should run in the direction $l=50°$. But a satisfactory model of the galactic magnetic field is not yet available and we don't know whether this distinguished direction is a local feature or indicates a large-scale structure.

References

Ackermann, G.: 1969, Thesis Heidelberg.
Ackermann, G., Fugman, G., Hermann, W., and Voelcker, K.: 1968, *Z. Astrophys.* **69**, 130.
Becker, W.: 1966, *Z. Astrophys.* **64**, 77.
Hiltner, W. A.: 1956, *Astrophys. J. Suppl. Ser.* **2**, 389.
Johnson, H. L.: 1968, in *Nebulae and Interstellar Matter* (ed. by. B. M. Middlehurst and L. H. Aller), Univ. of Chicago Press, Chicago, p. 167.
Klare, G. and Neckel, Th.: 1970, in *The Spiral Structure of Our Galaxy*, IAU Symposium No. 38 (ed. by W. Becker and G. Contopoulos), D. Reidel, Dordrecht-Holland, p. 449.

Low, F. J. and Smith, B. J.: 1966, *Nature* **212**, 675.

Lynds, B. T. and Wickramasinghe, N. C.: 1968, *Ann. Rev. Astron. Astrophys.* **6**, 215.

Mattila, K.: 1970, Thesis Heidelberg.

Neckel, Th.: 1967, *Veröffentl. Landessternw. Heidelberg-Königstuhl* **19**.

Neugebauer, G., Martz, D. E., and Leighton, R. B.: 1965, *Astrophys. J.* **142**, 399.

Scheffler, H.: 1967, *Z. Astrophys.* **65**, 60.

Witt, A. N.: 1968, *Astrophys. J.* **152**, 59.

Woolf, N. J. and Ney, E. P.: 1969, *Astrophys. J.* **155**, L181.

SPACE DISTRIBUTION AND STATE OF MOTION OF THE EARLY-TYPE STARS, OPEN CLUSTERS, ASSOCIATIONS AND STELLAR RINGS

TH. SCHMIDT-KALER

Astronomisches Institut der Ruhr-Universität Bochum, Germany

To find the space distribution and the kinematics of spiral-arm tracers in our Galaxy we have first to determine the distances, especially those of high-luminosity objects. Usually, fundamental far-reaching distance determinations are based on photometric and spectroscopic methods. The luminosity calibration cannot be separated, in general, from the more general problem of calibrating the spectral classifications in terms of luminosities, intrinsic colors, and population characteristics. Suppose a classification giving the parameters $S, L, P, Q \ldots$ depending on spectral type, luminosity, population type and so on, then these parameters should be related to the empirical parameters necessary for investigations of galactic structure and dynamics, namely, absolute magnitude, intrinsic color, age, abundance of metals [Fe/H], abundance of the CNO-group, and so on. These, in turn, should be related to the theoretical parameters mass, age, chemical composition, rotation... From these relationships we should be able to reach conclusions as to how mass, luminosity, and chemical composition vary with stellar age for different sub-groups of stars, that is, conclusions concerning galactic evolution.

In principle, distance determination is a simple problem. Difficulties enter into the problem mainly in two ways:

(1) direct, i.e. fundamental geometric methods of distance determination do not reach very far in space, so successive tie-in of other methods is necessary;

(2) the problems of interstellar extinction and reddening, especially the uncertainties in the ratio $R = A_V/E_{B-V}$, and the problem of neutral interstellar extinction.

1. Distance Determination in the Galaxy, Especially the Calibration of the Intrinsic Colors and the Absolute Magnitudes of High-Luminosity Objects

Previous work on this subject has been summarized earlier (Schmidt-Kaler, 1965a, b).

The methods of stellar distance determination may be divided into three classes.

1.1. FUNDAMENTAL GEOMETRIC METHODS

The *trigonometric parallaxes* have an average mean error of $\pm 0\overset{''}{.}011$ and systematic errors of the order of $\pm 0\overset{''}{.}005$. As a basis for absolute magnitudes accurate to $0\overset{m}{.}2$ this means a restriction to stars within 20 pc from the Sun, i.e. essentially the main-sequence stars later than about A5.

Accumulation of very long parallax series (van de Kamp and Lippincott, 1960), use of modern techniques (Vasilevskis, 1969) and the introduction of the astrometric reflector, especially for faint star work (Riddle *et al.*, 1969) appear to lead to an increased accuracy of about $\pm 0''.003$ as well as, hopefully, insignificant systematic errors.

Secular parallaxes (see Bok, 1965) may be based on:

a comparison of the mean solar motion derived from the radial velocities of a group of stars with that derived from the tangential velocities;

a comparison of the distribution of the velocity components in the direction of the solar motion with that of the v-components;

a comparison of the distribution of the velocity components perpendicular to the solar motion with that of the τ-components.

Substreamings and selection effects are carefully to be avoided.

Secular parallaxes provide only statistical distances, but no individual values. The distances are sensitive to systematic errors of the proper motion system. Fricke's (1965) comparison of the FK4 and N30 proper motion systems shows that the systematic error of the absolute magnitudes does not exceed $0''.2$, if the method is restricted to groups of stars within 200 pc. If particular care is exercised, and if the group considered extends over all or a major part of the sky, part of the systematic errors cancels out, and the method may be used to greater distance limits. An example of this kind is Geyer's (1970) recent analysis. From 94 population I cepheids with accurate proper motions on the FK4 system and good radial velocities he determined the solar motion and, adopting the slope of Kraft's (1961) period-luminosity relation, the zero-point in

$$M_V = - 1^m.88 - 2^m.54 \log P,$$
$$\pm 0^m.45$$

corresponding to a correction of $-0''.21$ to Kraft's zero-point from 5 cepheids in open clusters. The same data yielded a correction of $-0''.06$ to Fernie's (1967a, b) period-luminosity relation. Use of the N30 proper motion system instead of the FK4 would lead to an additional $-0''.34$. It was especially important to assign realistic weights to the proper motions which took into account the varying mean errors of the proper motion catalogues.

The τ-components of the proper motion perpendicular to the apex direction are much less sensitive to systematic errors in the proper motions. The implied assumption of random peculiar motions, however, does not seem to be justified in many instances. Another important point is made by Jung (1968, 1970) by the introduction of the principle of maximum likelihood to answer the more general question: for which mean absolute magnitude (or distribution of absolute magnitudes) may the corresponding tangential velocities on the one hand, and the radial velocities on the other hand, be considered as to result from the same general velocity pattern?

Observations of visual binary stars give the semi-major axis in angular measure, while from the radial velocity curve this same quantity may be obtained in kilometers. So a geometric distance determination is possible. However, reliable *dynamic parallaxes* are available for a few binaries only. New interferometric methods (Brown *et al.*, 1967) applied to close pairs with well-known radial velocity curves appear a great promise to the future.

Stream parallaxes still form the basis of the calibration of the absolute magnitudes (cf. Blaauw, 1963; Schmidt-Kaler, 1962 and Table I). Most important are the Hyades. They form the basis of all luminosities based on open clusters. This comprises most of the calibration of the MK-system in terms of absolute magnitudes, and the cepheids. Hodge and Wallerstein (1966) have suggested that the distance modulus of the Hyades, obtained by the convergent-point method should be increased by $0.''4$. A careful re-investigation (Wayman, 1967) shows, however, that errors in the observed motions and possible internal systematic motions can introduce an error of at most $\pm 0.''2$. Also, recent photometry of late-type dwarfs with large trigonometric parallaxes belonging to the young disk population results in a main sequence which coincides within less than $0.''1$ with that obtained from similar stars in the Hyades, using stream parallaxes (Eggen, 1969).

The Scorpio-Centaurus group necessitates a generalized treatment including internal expansion or differential galactic rotation. The most recent discussion of absolute magnitudes in the Scorpio-Centaurus association is given by Gutierrez-Moreno and Moreno (1968). An extension to later spectral types has been given by Garrison (1967b), while the southern parts of the association are being studied at Steward Observatory.

The Ursa Majoris stream, the third group with well-determined convergent point, has only about a dozen certain members. A detailed astrometric study of the Pleiades, Praesepe, and the α Per moving cluster, applying modern techniques, appears promising and would be most important.

1.2. OTHER GEOMETRIC METHODS

Distance estimates depending partly or totally on *galactic rotation* cannot be used for a basic luminosity calibration, otherwise one would proceed by a circular argument, since a distance scale has to be assumed in the calibration of that method. This remark applies also to the calcium parallaxes which so far depend mainly on galactic rotation.

The linear *diameters of open clusters* are correlated to the cluster type (Lynds, 1967), however, the scatter is much too large to derive accurate distances. An exception may be the stellar rings whose minor diameters seem to scatter less than 5% around the mean (Isserstedt, 1968).

1.3. PHOTOMETRIC METHODS

All photometric methods require eventually calibration by one of the fundamental methods. Because of the great distances involved, e.g. in the case of the supergiants, the linkage to the fundamental methods (which are severely restricted in distance) has to be done in several steps. Groups of stars of different luminosity at the same distance, in binaries, clusters, associations or other groupings, provide the tie-in ('Anschluss'). Because of the various amounts of interstellar reddening a tie-in method is also to be used for the determination of the intrinsic colors. The intrinsic colors are necessary to take the interstellar extinction into account; they should preferably be found in a determination independent of that to find the absolute magnitudes. A scheme used for a calibration of the MK-system in terms of intrinsic colors and absolute magnitudes (Blaauw, 1963; Schmidt-Kaler, 1965b, intrinsic colors: see also Serkowski, 1966 and FitzGerald, 1970) is described in Schmidt-Kaler (1962, 1963).

The *Q-method* of distance determination of open clusters based on U,B,V photometry leans upon the concept of a more or less universal reddening path E_{U-B}/E_{B-V}* and a general zero-age main sequence. Five steps lead, by fitting the unevolved parts of the main sequences in color-magnitude diagrams (Zero-Age Main Sequence), from the Hyades by way of the Pleiades, α Per moving cluster, NGC 2362, h+χ Per to the O-type clusters NGC 6231 and 6611. The tie-in errors accumulate up to about $\pm 0\overset{m}{.}3$.

Recent photometry of all h+χ Per stars in a $60' \times 36'$ field down to $V = 18\overset{m}{.}0$ (Vogt, 1969) made it possible to tie the two clusters directly to the Pleiades, confirming Blaauw's ZAMS within $0\overset{m}{.}15$ and reducing the total tie-in error considerably. An independent check (Borgman and Blaauw, 1964) is provided by fitting h+χ Per to the Sco-Cen association, using narrow-band photometry to measure absolute magnitudes of early-type stars. This alternative is, however, subject to the interpretation ambiguity of the motions of the Sco-Cen group mentioned above.

The main difficulties are:

(a) the accumulation of errors by the consecutive tie-ins (inaccurate intrinsic colors, questionable group membership, random errors etc.);

(b) the uncertainty and possible spatial variations of the ratio of reddening to visual extinction ($R = A_V/E_{B-V}$), and of the reddening path;

(c) the variations of initial chemical composition, and of effective rotational velocity and the corresponding changes in the ZAMS.

The effect of different chemical composition has been studied by means of intermediate- and narrow-band photometry (Golay, 1964; Strömgren, 1966). Fortunately, this effect rarely exceeds $0\overset{m}{.}1$. For instance, Crawford and Barnes (1969, 1970) found the reddening and the distance modulus of the Coma, Praesepe, α Per clusters, Pleiades, *h* and χ Per, NGC 752 and III Cep from u, v, b, y and Hβ photoelectric photometry in almost exact agreement with the earlier results from broad-band photoelectric U, B, V photometry.

Finally several pitfalls and statistical traps should be mentioned (see Schmidt-Kaler, 1965a):

(a) band-width effects in interstellar reddening and extinction (Schmidt-Kaler, 1961; Fernie, 1963): with increasing interstellar absorption the effective wavelengths of broad-band photometry are shifted to the red which causes the reddening, the reddening path and the ratio R to vary as a function of the amount of interstellar dust and of the spectral type (intrinsic color);

(b) bias due to cosmic dispersion and accidental errors in the measured luminosity criteria (Blaauw, 1963): since the luminosity criteria are affected by errors as well as the absolute magnitudes themselves, an impartial analysis is necessary which leads to

---

* The region of the Cygnus Rift probably presents a different reddening path.

improved values of the mean value and the dispersion of the absolute magnitudes;

(c) bias due to random errors in geometric distance determinations: a symmetric distribution of errors in distance results in an asymmetric distribution of the errors of the absolute magnitudes (Eddington's correction);

(d) difference of the mean absolute magnitudes of stars selected according to apparent magnitude respectively from a given volume of space (Malmquist's correction).

2. Examples of Luminosity Calibration

2.1. THE Ca II K-LINE REVERSAL WIDTH

The width of the bright reversals of the H and K absorption lines of late-type stars G0–M5 Ia–V is a linear function of their absolute magnitude over a range of more than 15 magnitudes. The function $M_V = -14\overset{m}{.}94 \log W + 27\overset{m}{.}59$ (Wilson and Bappu, 1957) is based on the values for the Sun and for the four Hyades giants; the mean error is $\pm 0\overset{m}{.}3$ for giants, $\pm 0\overset{m}{.}5$ for dwarfs.

This is the most simple luminosity effect regarding calibration problems since no dependence on spectral type and rotational velocity is observed; there seems to be a difference of up to 1^m for metal-deficient stars (Pagel and Tomkin, 1969). A luminosity determination of this kind requires, however, powerful optical means. Also, the theory of the effect is still very obscure. Dyck and Jennings (1969) working at a dispersion of 45 Å/mm, claim an accuracy of $\pm 0\overset{m}{.}7$.

2.2. THE BALMER LINE INTENSITIES

The equivalent width of the Balmer line Hγ or a correlated Balmer line quantity: the central depth of a Balmer line (Hack, 1953); the last visible Balmer line (Kopylov, 1961); the line intensity as measured by Öhman's method (Martin, 1964); the narrow-band index Hβ (Crawford and Barnes, 1969), is a well-known luminosity criterion for early-type stars. The most extensive series of equivalent widths is that of Petrie and his collaborators. Petrie's revised calibration (1965) starts from 7 good trigonometric parallaxes, 9 eclipsing and 8 visual binaries, yielding the zero-point. It is in fair agreement with a calibration based on open cluster members (Johnson and Iriarte, 1958; Schmidt-Kaler, 1965a), except for the luminosities of the O-stars which are still given too low (see also Walker and Hodge, 1968). The hydrogen line intensity depends on two parameters so that a single-valued spectral type correction introduces an additional error. Petrie partially takes account of this by separate calibration for stars of luminosity classes II-V and for supergiants. The mean error of an individual absolute magnitude is on the average $\pm 0\overset{m}{.}4$.

Binary character, existence of shells as shown by Hα emission, and stellar age, affect the equivalent widths of the Balmer lines as luminosity indicators. Contrary to many authors (Guthrie, 1963; Gutierrez-Moreno and Moreno, 1968; Andrews, 1968) Petrie found no dependence on rotation. A comprehensive study is needed, which takes into consideration all these stellar characteristics as well as the relation between the many different systems and their instrumental side-effects.

TABLE I

Scheme for the luminosity calibration

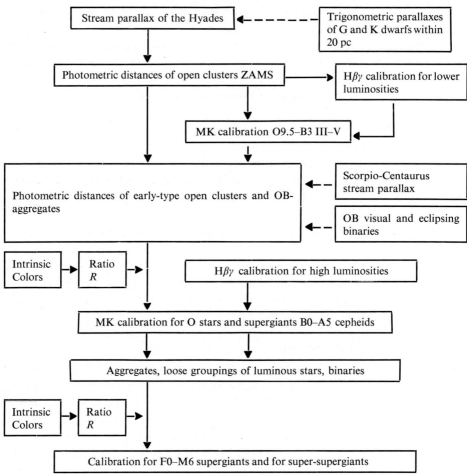

2.3. THE CALIBRATION OF THE MK-SYSTEM

About 15000 stars have been classified in the MK-system (Jaschek *et al.*, 1964). Since it is realizing a two-dimensional classification of more than 95% of the stars – it is the most comprehensive two-dimensional scheme anyway – it may be regarded as a *general reference system* for all spectroscopic work.

Main-sequence fitting of especially suitable open clusters has been used to find the absolute magnitudes of the most luminous and rare stars such as the supergiants. (See Table I.) The calibration (Blaauw, 1963; Schmidt-Kaler, 1965a) may be partially checked by the absolute magnitudes of all MK-classified stars in all open clusters with known photometric distance modulus (Weaver and Steinmetz, 1962; FitzGerald, 1969). With the exception of luminosity class III stars later than K3, and possibly the stars of type B8–A3 III–V, there is no evidence of systematic deviations exceeding 0."3.

An independent check is provided by Jung's (1968, 1970) derivation of absolute magnitudes from proper motions and radial velocities. Again, excellent agreement with Blaauw's calibration is noted, except for the K and M giants that appear to be brighter by about $0^m.6$ than assumed by Blaauw. In such comparisons the difference of the samples of stars (e.g. whether selected from a given volume of space or from apparent magnitude) is important.

The mean error of an MK-type absolute magnitude varies, of course, considerably with the type; it is on the average $\pm 0^m.6$. In certain parts of the Hertzsprung-Russell diagram, for instance B8–A5 III, IV and G5–7 III, the MK-system seems to suffer some lack of precise definition, and correspondingly larger scatter of the absolute magnitudes will result.

2.4. The Hamburg-Cleveland OB stars

Low-dispersion objective prism spectra provide, first of all, finding lists of very distant OB stars and supergiants for work at higher dispersion. One may ask, however, if additional photoelectric photometry together with the rough OB classifications may yield absolute magnitudes sufficient for studies of galactic structure. Schmidt-Kaler (1966b), Klare and Neckel (1967) and Herr (1969) investigated the absolute magnitudes and intrinsic colors of OB-stars. The result is summarized in Table II.

TABLE II

Group	M_V	$(B - V)_0$
Non-emission stars		
OB^+	$-5^m.9$	$-0^m.24$
OB	$-4^m.9$	$-0^m.26$
OB^-	$-3^m.8$	$-0^m.26$
Continuous emission only		
OB^+_{ce}	$-5^m.5$	$-0^m.27$
OB_{ce}	$-5^m.2$	$-0^m.28$
OB^-_{ce}	$-3^m.9$	$-0^m.26$
Hα-emission stars with or without continuous emission (ce) except $OB^+ h$	$-4^m.2$	$-0^m.24$

The dispersion of the absolute magnitudes is about $\pm 1^m$. Existing surveys contain 8703 OB stars, another catalogue for the Southern Milky Way is in preparation.

Data of this kind have been used to obtain rough information on the distribution of OB stars (Klare and Neckel, 1967).

If, in addition, photoelectric U,B,V- and possibly Hβ-photometry is obtained, absolute magnitudes with mean errors of $\pm 0^m.3$ to $\pm 0^m.7$ should result.

Especially interesting is the group of Hα-emission stars (excepting OB^+h which includes some supergiants), and the group $OB^+/OBce$ without Hα emission. These are the *natural groups* dBe respectively OcB1 with $M_V = -4^m.1$ respectively $-5^m.3$ and dispersions of $\pm 0.^{m}4$ respectively $\pm 0.^{m}7$ only. The stars of both groups belong to the rapid rotators with large shells (dBe) respectively without shells (OcB1). The last group includes the very youngest stars of the Galaxy.

3. The Ratio $R = A_V/E_{B-V}$

Since Johnson's disquieting papers (1965, 1968b; see also Borgman, 1966) on large *regional variations* of R this question is of pre-eminent importance for the study of galactic structure. Fortunately, in the course of the tie-in processes of the luminosity calibration individual and regional variations of R cancel out partially. This is so because the stars of different luminosities in the same group (e.g. a cluster) are in general subject to the same amount of interstellar extinction. In galactic structure, however, regional variations of R would stretch out or contract optical features up to about 50%.

Johnson's evidence was based on:

(1) the *cluster-diameter method* (comparing apparent photometric distance moduli and geometric distance estimates from apparent diameters of open clusters);

(2) the *variable-extinction method* (comparing the apparent magnitudes of stars of different reddening in one and the same cluster or aggregate so that $\Delta V/\Delta(B-V)$ may be determined);

(3) the *color-difference method* (based on the interstellar extinction curve from multi-color photometry up to the far infrared).

The first argument has been thoroughly discussed by Lynds (1967) with the result that a constant value $R = 3$ introduces no systematic differences between distances measured photometrically and geometrically. The second argument has been criticized by Becker (1966) who showed that inclusion of non-members in clusters and associations yields generally too high a value of R; also, evolutionary differences between subgroups of an association with different reddenings tend to make R too high (Schmidt-Kaler, 1961).

The color-difference method arouses suspicion in so far as late-type supergiants like μ Cep (M2Ia) and early-type shell stars like ϕ Per, e.g. stars with strong circumstellar envelopes lead to the highest R values. Since circumstellar shells will mostly affect the far infra-red, extrapolation to $\lambda \to \infty$ may be more appropriate than naive use of observations at $\lambda > 3\mu$. In this way it has been derived from Johnson's multicolor measurements that $R = 3.23$ is applicable to all galactic longitudes except perhaps very young H II regions like the Orion nebula (Schmidt-Kaler, 1967). Indeed, Johnson (1967, 1968a; see also Stein and Wolf, 1969) has later shown that the infrared excesses of M supergiants and various early-type stars are due to circumstellar shells.

Supporting evidence comes from detailed investigations of individual areas:

(a) Johnson found $R = 5.4$ for the association III Cep. New data show that $R = 3$ is the correct value, the high value resulting from inclusion of late B foreground stars (Garrison, 1967a). The adjoining OB-aggregates I Cep and IV Cep have $R = 3.0$ (Simonson, 1968) and $R = 3.2$ (McConnell, 1968).

(b) Walker (1968) claims that even in the vicinity of the Orion nebula the normal ratio prevails.

(c) The amount of reddening of stellar light can be compared to the absolute interstellar absorption obtained by comparing the theoretical and the observed ratio of the radio continuum to the Hα brightness of an HII region. In this way Ishida (1969) found for IC 1805 the general value $R = 3.8 \pm 0.7$, and, after subtraction of the effect of foreground absorption, for seven O stars $R = 5.0 \pm 0.5$. In a similar way Gebel (1968) found for the Rosette nebula (NGC 2244) $R = 3.3$ for which Johnson had given $R = 5.5$. From 15 nebulae $R = 3.26$ results without definite indication of a variation between longitudes $0°$ and $200°$. Again, Menon (1969) found from kinematical arguments that the ratio for NGC 2244 is more likely to be about 4.

Finally, investigations covering more or less all of the Milky Way indicate the same result:

If Johnson's regional variation of R is adopted the distribution of early-type clusters loses its arrangement in spiral features and becomes rather chaotic, and the kinematics of OB stars and cepheids leads to inconsistencies in galactic rotation (Schmidt-Kaler, 1967; for the cepheids see also Fernie and Hube, 1968). We conclude that, apart from a very few local deviations, the value $R = 3.25$ applies to all galactic longitudes with possible variations up to ± 0.5.

4. Space Distribution of Early-Type Stars, Open Clusters, Stellar Rings and Other Optical Spiral-Arm Tracers

The *spiral arms* are, first of all, the localities of the most luminous stars as ample evidence of other galaxies shows. Since these stars are very young the spiral arms must be the localities of massive star formation, and, therefore, most probably, also regions of concentration of gas and dust. Indeed, recent work suggests that the axis of the spiral arms are the places where the interstellar gas is particularly cool (Weaver, 1970; Schmidt-Kaler and Pospieszczyk, 1970).

In principle, the structure of the Galaxy may be found by *four different methods*:

(I) by determining positions and distances $(l, b; r)$ of the individual spiral tracers;

(II) by determining positions and intensities or number frequency distributions $(l, b; I)$;

(III) by determining positions and directions $(l, b; \theta)$, e.g. of elongated dark clouds and emission nebulae, and of magnetic fields;

(IV) by determining positions and velocities $(l, b; v_r \text{ or } v_t)$.

Here we will discuss in some detail optical investigations of the first type only. The second method has been applied to Milky Way surface photometry, but it is very sensitive against the assumptions on the distribution of the interstellar extinction (Isserstedt and Schmidt-Kaler, 1964; Behr, 1965; Pavlovskaya and Sharov, 1970). The magnetic fields as determined from optical and radio-astronomical observations seem to refer to rather local dust features. The fourth method implies reliable knowledge of the kinematics of the Galaxy.

The objects which may serve as *spiral tracers* should meet the following conditions:

(a) they should be very young since the random velocities of 10–15 km/sec will completely smear out the structure shown by objects originating in the spiral filaments within 5×10^7 years;

(b) they should be detectable and measurable to great distances in sufficient numbers;

(c) their distances should be determined with sufficient accuracy;

(d) they should be a representative, statistically homogeneous and complete sample of the space surveyed.

Condition (a) excludes, for instance, the cepheids with periods less than 10 days, condition (b) the T Tauri stars. Condition (c) excludes most single MK-classified early-type stars without additional information, if distances greater than 4 kpc are considered, since the usual dispersion of $\pm 0\overset{m}{.}6$ in M_V (MK) then corresponds to more than ± 1 kpc. Groups and associations of such stars, however, may still be useful. Condition (d) limits the usefulness of the WR stars at great distances.

4.1. LOCAL STRUCTURE

Earlier work has been reviewed by Schmidt-Kaler (1965b, 1966a) and Bok (1967, 1970).

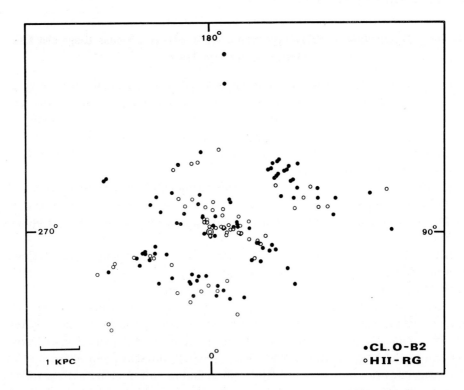

Fig. 1.   Distribution of young open clusters and H II regions projected on the galactic plane (Becker and Fenkart, 1970).

The most recent work will be described in the following, especially that reported at the 1969 Basel IAU Symposium No. 38 on 'The Spiral Structure of our Galaxy'.

Rohlfs (1970) and McCuskey (1970) reviewed the reliability of different spiral tracers. It appears that H$\textsc{ii}$ regions, early-type open clusters and associations, and early-type Be stars are the best individual spiral tracers. The most impressive picture of local spiral structure has recently been drawn by Becker and Fenkart (1970) with revised distance moduli of about 230 *open clusters* and 70 H$\textsc{ii}$ *regions*, some of these coinciding with young clusters. Instead of using the two-color diagram to determine the reddening, and the color-magnitude diagram to determine the distance modulus, they used both color-magnitude diagrams (in $B-V$ and in $U-B$) simultaneously. This procedure eliminates non-members more effectively, allows a more precise determination of the reddening, and – since the upper part of the main sequence is in $U-B$ much less steep than in the $B-V$ color-magnitude diagram – yields a more reliable distance modulus for early-type clusters. The distribution of the young clusters does not differ from that of the H$\textsc{ii}$ regions (Figure 1). The distribution of the older clusters seems quite random, although the number is too small for safe conclusions about the transition from spiral to random distribution (Figure 2).

Additional spiral tracers are Wolf-Rayet stars, R-associations (connected with reflexion nebulae), dark cloud complexes, OB$^+$ and OB^0 stars. The classical δ Cephei-

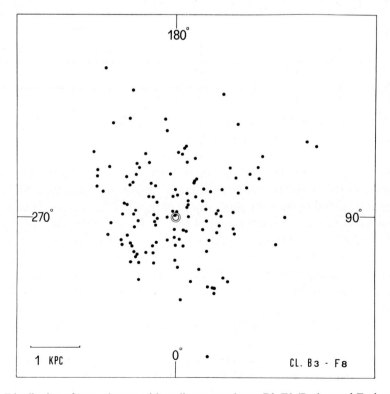

Fig. 2.   Distribution of open clusters with earliest spectral type B3–F8 (Becker and Fenkart, 1970).

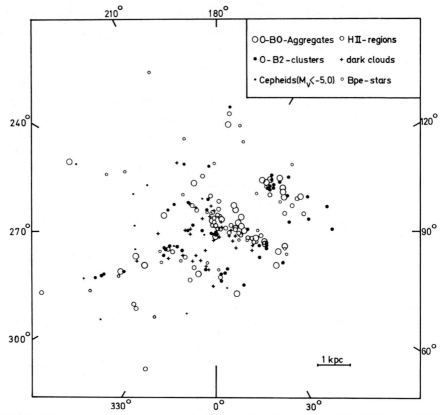

Fig. 3.   Distribution of selected types of objects in the galactic plane (Schmidt-Kaler, 1966a).

stars of highest luminosity, and some S and N (carbon) stars seem to be loosely associated to local spiral structure (Figure 3).

The classical *Wolf-Rayet stars* of the population I can be found, due to the strong emission bands, up to distances of more than 10 kpc. Their overall distribution (Smith, 1968) suggests spiral structure at large distances (Figure 4). More than 10% of these stars are at $|z| > 400$ pc from the plane, but nevertheless delineate spiral features.

H II-emission nebulae are excellent spiral tracers. Racine (1968) and van den Bergh (1968) have shown that stars embedded in reflection nebulae form often groups (called R-*associations*) which are arranged along the Local Feature in about the same way as the OB-associations (Figure 5). All these objects are within about 2 kpc from the Sun. Being more numerous than OB-associations, they delineate the Local Feature very well.

Dark cloud complexes are distributed preferentially along the inside edge of spiral features (Isserstedt and Schmidt-Kaler, 1964) just as the dust lanes in 17 Sc-type galaxies as reported by Lynds in Basel. The phenomenon is also suggested to some extent in the general distribution of the interstellar extinction (Neckel, 1966; Uranova, 1970).

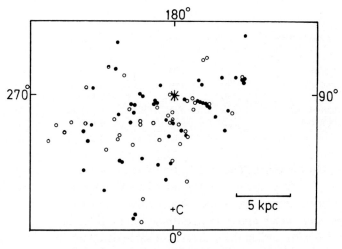

Fig. 4. Distribution of Wolf-Rayet stars after Smith (1968). Stars of class WN are denoted by dots, stars of class WC by circles.

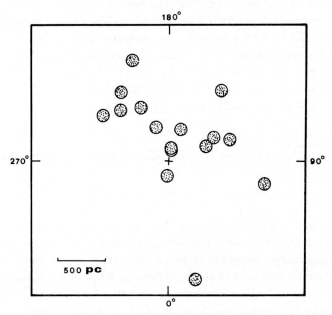

Fig. 5. Distribution of the R-associations in the galactic plane (van den Bergh, 1968).

From these investigations the following features (see Table IIIa) appear well established.

One of the most important questions relates to if and where the local feature is continuing. From all present optical evidence and from new work on the Carina region presented at Basel (Bok *et al.*; Graham; Seggewiss) it is evident that the local feature is not connected to the Carina section. Both are separated by a gap at least

TABLE IIIA

Spiral feature	Galactic longitude			Distance
Local	60°	...	210°	Sun at inside edge
Per	90	...	140	2–3 kpc
Sgr	330	...	30	2 kpc
Car(-Cen)	285	...	305 (...330)	1.5–5 kpc
Nor-Sct	325	...	30	4 kpc

1.4 kpc wide. On the other hand, the Carina section appears to be connected to the Sagittarius arm by the Centaurus feature at 2 kpc distance (which is, however, rather weak). The Carina section has a very sharp edge at $l=283°$ and is well separated from the Vela spur which appears to continue the local feature.

Dickel *et al.* (1970) studied the local arm in some detail investigating the shape and distribution of the gaseous nebulae in the Cygnus X region $l=70°...90°$. They find that the orientation of the nebulae (75% of which are highly elongated) suggests a helical magnetic field of the spiral arm. The distribution in space is approximated by a simple model: a straight cylinder with axes of about 400×400 pc in the plane respectively perpendicular to it; the cylinder axis makes an angle of 77° with the center direction (pitch angle of the spiral), and extends from about 0.5 kpc distance from the Sun to about 4 kpc. This cylinder is tilted by about 1° to the galactic plane. The width of the Cygnus feature compares rather well with the average width of all features of the local domain of about 550 pc (Schmidt-Kaler, 1966a).

4.2. LARGE-SCALE STRUCTURE

Now we enter an area which does not yet seem to belong to the realm of astronomical establishment.

All the spiral tracers discussed so far are – except for the WR stars – essentially restricted to the solar neighborhood within about 3 kpc distance. To extend our knowledge about the spiral structure to, say, 10 kpc distance, the accuracy of the distance determination must be considerably increased to see narrow features at such distance. Unexpectedly, geometric distance determinations of very high accuracy seem possible with a new kind of stellar aggregates different from clusters and associations. In the course of a joint investigation of dark clouds of regular form Isserstedt (1968) discovered the *stellar rings* which appear as regular elliptical aggregates of stars. Most important criteria for the definition of a stellar ring are the symmetry relative to the major axis, the small apparent thickness of the ring, and the sharp outer boundary. A systematic search of the Palomar Atlas (including the Whiteoak Extension) red prints yielded a total of 1067 objects. Three more very large objects were found on the Lick Observatory Sky Atlas.

A statistical test suggests that the majority of the stellar rings with diameter $D_m > 1'.5$ are not chance configurations. Also, the frequency of the rings and that of the faint stars as function of longitude respectively latitude are very different. The rings are

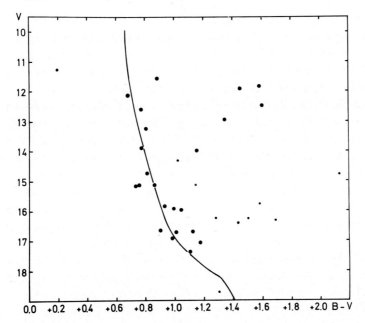

Fig. 6.   Color-magnitude diagram of the stellar ring No. 373 (Isserstedt and Schmidt-Kaler, 1970). Large dots are probable members, small dots are probable non-members according to the position in both color-magnitude diagrams.

much more concentrated to the plane, the absorption layer can well be recognized, and their surface density is *not* a monotonous function of the star density.

Many stars with known MK or Harvard spectral type were found in the three very large rings. It is found that early-type stars are much more frequent in the rings than in the surrounding fields. With these data and with additional U,B,V photometry photometric distance moduli have been determined for a number of objects (Schmidt-Kaler, 1968; Isserstedt, 1969; Isserstedt and Schmidt-Kaler, 1970). These observations seem to confirm the reality of the rings and indicate that the minor diameters of the rings scatter a little only around a mean value of $D_m = 7.1$ pc. Also, an ultraviolet anomaly is observed which seems to be characteristic of such very young open clusters like NGC 2264.

Ring No. 373 (Figure 6) is a typical case, especially with the dark cloud extending along the inner edge of the stellar ring. Another interesting case is ring No. 274 which contains 4 OB stars from the Hamburg-Cleveland survey and in the center the star P Cygni. The situation looks like an earlier stage of the Orion ring with ε Ori (whose *EUV* spectra show evidence of expansion and heavy mass loss) in the center; also, the Orion ring has the appearance of a disintegrating stellar ring. Yet another interesting object is the open cluster NGC 6683, studied and recognized in its ring form by Yilmaz (1966). There is much scatter in Yilmaz' color-magnitude diagram. If we subdivide the cluster area in three regions: the ring area 1.2 wide, the inside, and the outside of the ring, then we find the following proportion of cluster members (as such indicated by Yilmaz): outside the ring 22%, inside the ring 58%, in the ring area 71%.

Like the ellipsoidal gas shells which we will describe a moment later the rings are of course not 'rings' but stellar aggregates in the form of prolate ellipsoidal shells. The symmetry axis is the major diameter so that the minor diameter appears projected to the celestial sphere always without perspective foreshortening.

10 *ellipsoidal* H II *shells* with sharp filamentary boundaries are known which appear as ring nebulae (Schmidt-Kaler, 1968, 1970). The exciting stars are all of the same very peculiar type: broad-lined, definitely single Wolf-Rayet stars of class WN5, 6, 8-sequence B. This is remarkable since WR stars are a very rare class of stars, and about half of them are binaries. The minor diameter of the H II rings is on the average 6.8 pc. We have studied the ring nebula NGC 6888 in detail since a concentration of stars on the rim of the nebula is seen (the shape of the rim is completely determined by gas filaments, and not by the stars lined up along the edge). The color-magnitude diagram is that of a very young cluster. The minor diameter is 6.3 pc.

A wealth of dark cloud rings can be found in the Ross-Calvert and Palomar Atlas. For about 10 objects distance estimates were possible, for instance $D_m = 6.5$ pc for the dark cloud in Scutum north of M11, and $D_m = 6.9$ pc for the large dark ring (=Ceder-

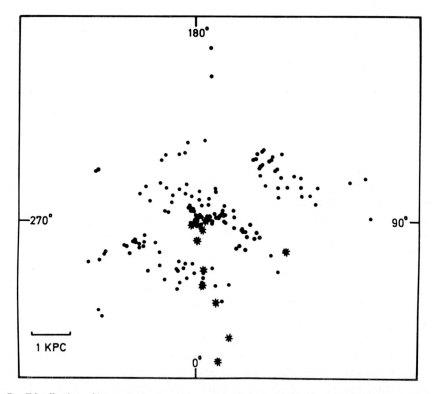

Fig. 7. Distribution of 'non-synchrotron' X-ray sources (asterisks) in the galactic plane together with that of young clusters and H II regions according to Becker and Fenkart (dots) (Haupt and Schmidt-Kaler, 1970).

blad 122) east of and probably connected to the Coal sack. On the outer rim of that ring-type dark cloud lies the star identified with the *X-ray source* Cen X-2. Again, the star identified with Sco X-1 lies on the outer rim of a ring-type dark cloud of 6.9 pc minor diameter (assuming the most probable distance of 300 pc). The star identified with G X3+1 lies on the outer rim of an H II ring nebula with 6.5 pc diameter. Cyg X-2 is the only X-ray star not associated to a conspicuous ring-type gas shell. For the remaining X-ray sources with well-established positions but no optical identifications (excluding one extragalactic source and three supernova remnants) in most cases correlated ring-type features are noted. When the distances of the 'thermal' X-ray sources are estimated from the photometric distance modulus of the associated ring-type feature or by assuming a unique diameter for it, the X-ray sources appear located on the main spiral arms of the Milky Way (Figure 7; Haupt and Schmidt-Kaler, 1970).

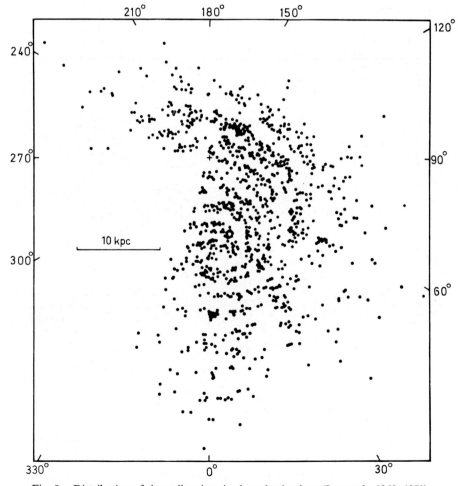

Fig. 8.   Distribution of the stellar rings in the galactic plane (Isserstedt, 1968, 1970).

In summary I would conclude that star formation seems to be going on in dense filamentary shells of gas of ellipsoidal shape whose characteristic diameter is 7 pc. The stellar rings may result from those gas shells (e.g. ring No. 271 contains the star HD 191765 WN6–B). To see whether the rings have always the same minor diameter the distances were determined from the apparent diameters with the assumption $D_m = 7.1$ pc. On this hypothesis a picture of the structure of our Galaxy up to 15 kpc from the Sun emerges (Figure 8; Isserstedt, 1970). The Galaxy appears of type Sb, perhaps Sb–c. The center lies in the direction $l = 0°$ at the distance $R_0 = 10.8$ kpc; the galactic plane is very well defined by the stellar rings.

The distance range seems unbelievably large. It is possible because most rings at great distances are far distant from the galactic plane. So the light passes through the galactic absorption layer for only 2 or 3 kpc. The problem is that stellar rings between $z = 500$ and 1000 pc contribute essentially to the picture. Oort (1970) has recently shown from 21 cm-line studies that matter far from the galactic plane is associated with spiral arms.

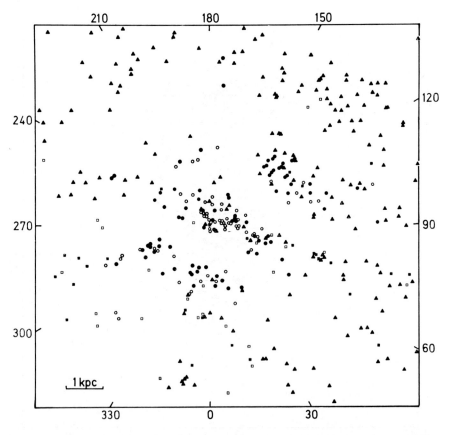

Fig. 9.   Distribution of stellar rings (triangles), young open clusters and H II regions according to Becker and Fenkart (circles) and WR stars according to Smith (squares) (Isserstedt, 1970).

In the solar neighborhood the distribution of the best spiral tracers is very similar to that of the rings (Figure 9). Taking the spiral features defined by Becker and Fenkart's (1970) paper on Hɪɪ-regions and young clusters as a basis we may ask how many objects of a given type fall inside that area around a spiral feature which covers just half of the galactic plane, (see Table IV).

TABLE IV

Type of objects	Total number	Percentage in spiral-arm area
Hɪɪ regions	64	100%
Young clusters	82	99
Stellar rings within 5 kpc	194	96
WR stars within 5 kpc	52	94
Bpe, dB0e stars within 2.5 kpc	75	83
X-ray sources	15	80
Cepheids with $P > 9$ days within 3 kpc	27	59

X-ray sources, WR stars and stellar rings seem to be the farthest-reaching spiral tracers of the Galaxy. Still, most of the photometric and spectroscopic work remains to be done.

5. Some Remarks on the State of Motion of Early-Type Stars, Open Clusters and Hɪɪ Regions from Optical Observations

We will briefly consider three questions:
(1) the Local Standard of Rest;
(2) the local characteristics of the galactic rotation curve;
(3) the coincidence of gas (Hɪ, Hɪɪ) and stellar spiral features.

Although it is current practice to use the 'standard' solar motion of 20 km/sec toward 18^h, $+30°$ (1900) there are several new investigations suggesting a revision. Systematic motions of the stars in Gould's Belt and other moving groups indicate that stars within about 500 pc of the Sun cannot be depended upon to define a reference system moving in a circular orbit. Objects distributed over a large region of space should, therefore, be used to define the *local solar motion*. This implies simultaneous solutions of solar motion and galactic rotation parameters. Recent studies of this kind are summarized in Table V.

The mean weighted with the number N of the objects is

$$U_0 = -8.9 \pm 0.7, \qquad V_0 = +13.4 \pm 0.8 \text{ km/sec}.$$

This agrees very well with the weighted mean solar motion of early-type stars (B0, dA0) and supergiants according to Delhaye's (1965) compilation:

$$U_0 = -8.7, \qquad V_0 = +13.8, \qquad W_0 = +6.6 \text{ km/sec},$$

TABLE V

Objects	N	U_0 (km/sec)	V_0 (km/sec)	A (km/sec per kpc)	References
dB0-2e stars $0.3 < r < 1.5$ kpc	104	$-\ 5.3 \pm 1.9$	$+\ 12.7 \pm 2.1$	11.8 ± 1.7	Crampton, 1968a
Cepheids (r.v. and p.m.)	146	$-\ 8.1$	$+\ 13.9$	14.3	Takase, 1969
Cepheids (r.v.)	145	$-\ 8.0 \pm 1.5$	$+\ 14.5 \pm 1.5$	12.5 ± 0.9	Crampton and Fernie, 1968
B stars $0.25 < r \leq 2$ kpc	376	-11.1 ± 0.9	$+\ 14.2 \pm 0.8$	14.3 ± 0.8	Feast and Shuttleworth, 1965
Young galactic clusters, $|R - R_0| < 2$ kpc	63	$-\ 7.6 \pm 2.9$	$+\ 15.3 \pm 4.1$	13.6 ± 2.0	Crampton, 1968b
H II regions $r < 5$ kpc	91 (174)	$-\ 7.5 \pm 1$	$+\ 13.1 \pm 1.5$	14.3 ± 0.5	Courtès et al., 1969; Georgelin and Georgelin, 1970
H II regions	30	$-\ 6.4 \pm 2.1$	$+\ 13.3 \pm 1.7$	15.3 ± 1.4	Miller, 1968
Interstellar Ca II $0.25 < r \leqq 1$ kpc	317	$-\ 8.4 \pm 0.6$	$+\ 12.1 \pm 0.5$	(14.3)	Feast and Shuttleworth, 1965

and is closer to the Standard Solar Motion $(-10.5; +15.4)$ than to the Basic Solar Motion $(-10.2; 10.1)$. It is gratifying that the average of Fricke's (1967) two determinations of the apex from McCormick-Cape and from fundamental proper motions $\alpha = 269°9, \delta = +29°7$ coincides with the Standard Apex.

The most recent determinations of the *galactic rotation constants* from proper motion data reduced to the FK4 system are due to Fricke, Dieckvoss and van Schewick. The unweighted average from FK4 proper motions and from distant stars is $A = 13.4 \pm 1.7, B = -10.2 \pm 1.7$ km/sec per kpc. It appears that Oort's constant A can now be derived from proper motions with essentially the same accuracy as from radial velocities (Fricke, 1967). Dieckvoss (1967) determined the solar motion and Oort's constants from the 163 811 stars of the revised AGK3, assuming Standard Solar Motion, and obtained $A = +14.8, B = -11.3$ km/sec per kpc with formal mean erors of ± 0.7 respectively ± 0.6 km/sec per kpc. Van Schewick (1970) found $B = -7.7$ km/sec per kpc from 56 open clusters with well-determined proper motions reduced to the FK4. The dispersion corresponds to about ± 10 km/sec. – Radial velocities have been extensively used to determine the galactic rotation curve. In addition to the investigations quoted in Table V Petrie and Petrie's (1968) paper using the Hγ calibration of 688 stars of spectral type O9–B8 should be noted, resulting in $A = 15.1 \pm 0.4$ km/sec per kpc.

A weighted average of all values is $A = 13.9 \pm 0.5, B = -10.1 \pm 1.2$ km/sec per kpc.

A correction to these values would be necessary because the value $R = 3.25$ seems more probable than the conventional value $R = 3$. The resulting correction is (Schmidt-Kaler, 1967)

$$dA/A = 0.46\, \bar{E}_{B-V} \cdot dR = 0.36\, dR.$$

The average reddening of the stars in question is approximately constant from 1 to 5 kpc distance and equal to $\bar{E}_{B-V} = 0.^m75$. So, we have finally

$$A = 15.1 \pm 0.7, \qquad B = -11.0 \pm 1.3 \text{ km/sec per kpc}.$$

The revised galactic rotation curve based on B stars, H$_{II}$-regions and cepheids is in good agreement with the most recent 21 cm-line rotation curve (Schmidt-Kaler, 1967; Feast, 1967; Courtès *et al.*, 1969; Georgelin and Georgelin, 1970).

Does spiral structure from young stars and gas coincide or not? The answer was – notwithstanding the apparent anticorrelation of H$_I$ and optical data at that time – in the affirmative (Schmidt-Kaler, 1965a, 1966a). It was based on model-independent comparisons of the two distributions:

(a) by direct comparison of the frequency distribution of the radial velocities of the stars and the corresponding 21 cm-profiles (Rubin *et al.*, 1962), or by locating the stars according to their radial velocity in exactly the same way as the H$_I$ pattern plotted on the Leiden-Sydney map (Fletcher, 1963);

(b) by comparison of the results of the surface intensity distribution of the Milky Way with the total number of H$_I$ atoms as a function of galactic longitude (Schmidt-Kaler, 1967).

There is now additional evidence that the velocity difference between the motion of gas and stars in the galactic plane is less than about 3 km/sec:

(a) recent solar motion solutions from neutral hydrogen and from young population I stars agree within about 2 km/sec (Kerr, 1969);

(b) the recombination line velocities of the ionized H II regions lie approximately on the ridge-lines of the neutral H I longitude-velocity distribution (Kerr et al., 1968). The thermal sources of small diameter coincide in many cases with optically visible H II regions which in turn are known as excellent spiral-arm tracers;

(c) in several features OB stars and neutral hydrogen gas deviate from the circular motion by very nearly the same amount, e.g. in the Perseus arm (Rickard, 1968), as has been noted before (Schmidt-Kaler, 1965a);

(d) a revised overall analysis of the H I distribution in the galactic plane agrees pretty well with the optical spiral structure (as given in Figure 3) (Weaver, 1970). Similar conclusions result from work in progress (Schmidt-Kaler and Pospieszczyk, 1970).

Finally, it may be noted that the stellar rings would open new avenues to galactic dynamics up to very great distances.

Note Added in Proof

To bring this review up to date (1970, November) the following supplement is added to § 4.1:

Quite recently Courtès et al. (1969, 1970) traced the spiral structure based on some 174 optical H II-regions whose radial velocities were measured (Georgelin and Georgelin, 1970). Their results are condensed in Table IIIb.

TABLE IIIB

Spiral feature	Galactic longitude		Distance
Local (O)	59°	... 254°	Quite poorly defined, 0.5 kpc at 180°, Sun at inner edge
Per (+ I)	103	... 190	very conspicuous between 140° and 168°, 3 kpc at 180°
Sgr (− I)	274	... 32	1.5 kpc at 330°
Nor (− II)	305	... 333	3.5 kpc at 330°, clearly separated from − I

These features are continuous regarding the positions as well as the velocities of the H II-regions. The Centaurus link (305° ⋯ 330°) between the strong Carina and Sagittarius H II-regions consists, however, mostly of small weak Hα-nebulae.

Another promising spiral tracer are the *supergiants*, especially the M-Supergiants since their luminosities and radial velocities may be measured in the near infrared without strong interference of interstellar extinction (Sharpless, 1965; Humphreys, 1970).

The potential reach of the M-supergiants might eventually be 5–10 kpc. About 70% of the supergiants belong to associations and open clusters, small wonder, therefore, that their space distribution suggests the same characteristics as these major spiral tracers.

Humphreys (1970) determined $A = 13.8 \pm 0.5$ (formal mean error) km/sec kpc from 401 09–M4 supergiants based on our luminosity calibration.

References

Andrews, P. J.: 1968, *Mem. Roy. Astron. Soc.* **72**, 35.

Becker, W.: 1966, *Z. Astrophys.* **64**, 77.

Becker, W. and Fenkart, R.: 1970, in *The Spiral Structure of Our Galaxy*, IAU Symposium No. 38 (ed. by W. Becker and G. Contopoulos), D. Reidel, Dordrecht-Holland, p. 205.

Behr, A.: 1965, *Z. Astrophys.* **61**, 182.

Bergh, van den, S.: 1968, *Astrophys. Lett.* **2**, 71.

Blaauw, A.: 1963, in *Basic Astronomical Data* (ed. by K. Aa. Strand), University of Chicago Press, Chicago, p. 383.

Bok, B. J.: 1965, in *Observational Aspects of Galactic Structure* (ed. by A. Blaauw and L. N. Mavridis), Athens, p. XI, 1.

Bok, B. J.: 1967, *Am. Scientist* **55**, 375. See also Invited Discourse at the General Assembly of the IAU, Brighton 1970.

Borgman, J.: 1966, in *Proceedings of the Twelfth General Assembly*, Hamburg, 1964, IAU Transactions **XIIB**, (ed. by J.-C. Pecker), Academic Press, London and New York, p. 397.

Borgman, J. and Blaauw, A.: 1964, *Bull. Astron. Inst. Neth.* **17**, 358.

Brown, H. R., Davis, J., Allen, L. R., and Rome, J. M.: 1967, *Monthly Notices Roy. Astron. Soc.* **137**, 375, 393.

Courtès, G., Georgelin, Y. P., Georgelin, Y. M., and Monnet, G.: 1969, *Astrophys. Lett.* **4**, 129.

Courtès, G., Georgelin, Y. P., Georgelin, Y. M., and Monnet, G.: 1970, in *The Spiral Structure of Our Galaxy*, IAU Symposium No. 38, p. 209, and *Vistas in Astronomy* (in press).

Crampton, D.: 1968a, *Astron. J.* **73**, 338.

Crampton, D.: 1968b, *Publ. Astron. Soc. Pacific* **80** 443.

Crampton, D. and Fernie, J. D.: 1968, *Astron. J.* **74**, 53.

Crawford, D. L. and Barnes, J. V.: 1969, *Astron. J.* **74**, 407; 1970, *Astron. J.* **75**, 822, 946, 952.

Delhaye, J.: 1965, in *Galactic Structure* (ed. by A. Blaauw and M. Schmidt), University of Chicago Press, Chicago, p. 71.

Dickel, H. R., Wendker, H. J., and Bieritz, J. H.: 1970, in *The Spiral Structure of Our Galaxy*, IAU Symposium No. 38 (ed. by W. Becker and G. Contopoulos), D. Reidel, Dordrecht-Holland, p. 213.

Dieckvoss, W.: 1967, *Astron. Nachr.* **290**, 141.

Dyck, A. M. and Jennings, M. C.: 1969, *Publ. Astron. Soc. Pacific* **81**, 536.

Eggen, O. J.: 1969, *Astrophys. J.* **158**, 1109.

Feast, M.: 1967, *Monthly Notices Roy. Astron. Soc.* **136**, 141.

Feast, M. and Shuttleworth, M.: 1965, *Monthly Notices Roy. Astron. Soc.* **130**, 245.

Fernie, J. D.: 1963, *Astron. J.* **68**, 780.

Fernie, J. D.: 1967a, *Astron. J.* **72**, 422.

Fernie, J. D.: 1967b, *Astron. J.* **72**, 1327.

Fernie, J. D. and Hube, J. O.: 1968, *Astrophys. J.* **153**, L111.

FitzGerald, M. P.: 1969, *Publ. Astron. Soc. Pacific* **81**, 71.

FitzGerald, M. P.: 1970, *Astron. Astrophys.* **4**, 234.

Fletcher, E. S.: 1963, *Astron. J.* **68**, 407.

Fricke, W.: 1965, *Z. Astrophys.* **61**, 20.

Fricke, W.: 1967, *Astron. J.* **72**, 642, 1368.

Garrison, R. F.: 1967a, *Astron. J.* **72**, 797.

Garrison, R. F.: 1967b, *Astrophys. J.* **147**, 1003.

Gebel, W.: 1968, *Astrophys. J.* **153**, 743.

Georgelin, Y. P. and Georgelin, Y. M.: 1970, *Astron. Astrophys.* **6**, 349; **8**, 117.

Geyer, U. F.: 1970, *Astron. Astrophys.* **5**, 116. See also Jung, J.: 1970, *Astron. Astrophys.* **6**, 130.

Golay, M.: 1964, *Publ. Observ. Genève*, No. 66.

Guthrie, B. N. G.: 1963, *Publ. Roy. Observ. Edinburgh* **3**, 83, 261.

Gutierrez-Moreno, A. and Moreno, H.: 1968, *Astrophys. J. Suppl. Ser.* **15**, 459; Santiago Publ. No. 5.

Hack, M.: 1953, *Ann. Astrophys.* **16**, 417.

Haupt, W. and Schmidt-Kaler, Th.: 1970, *Astron. Astrophys.* **7**, 481.

Herr, R. B.: 1969, *Astron. J.* **74**, 200.

Hodge, P. W. and Wallerstein, G.: 1966, *Publ. Astron. Soc. Pacific* **78**, 411.

Humphreys, R.: 1970, *Astron. J.* **75**, 602; *Astrophys. J.* **160**, 1149.

Ishida, K.: 1969, *Monthly Notices Roy. Astron. Soc.* **144**, 55.

Isserstedt, J.: 1968, *Veröff. Astron. Inst. Univ. Bochum*, No. 1.

Isserstedt, J.: 1969, *Astron. Astrophys.* **3**, 210.

Isserstedt, J.: 1970, *Astron. Astrophys.* **9**, 70.

Isserstedt, J. and Schmidt-Kaler, Th.: 1964, *Z. Astrophys.* **59**, 182.

Isserstedt, J. and Schmidt-Kaler, Th.: 1970, *Astron. Astrophys.*, **7**, 446.

Jaschek, C., Conde, H., and de Sierra, A. C.: 1964, *Publ. Astron. Univ. Nac. La Plata* **28** (2).

Johnson, H. L.: 1965, *Astrophys. J.* **141**, 923.

Johnson, H. L.: 1967, *Astrophys. J.* **149**, 345.

Johnson, H. L.: 1968a, *Astrophys. J.* **150**, L39.

Johnson, H. L.: 1968b, in *Nebulae and Interstellar Matter* (ed. by B. M. Middlehurst and L. H. Aller), University of Chicago Press, Chicago, p. 167.

Johnson, H. L. and Iriarte, B.: 1958, *Lowell Obs. Bull.* **4**, 47.

Jung, J.: 1968, *Bull. Astron. Paris, Ser. 3*, **3**, 461.

Jung, J.: 1970, *Astron. Astrophys.* **4**, 53.

Kerr, F. J.: 1969, *Ann. Rev. Astron. Astrophys.* **7**, 39.

Kerr, F. J., Burke, B. F., Reifenstein III, E. C., Wilson, T. L., and Mezger, P. G.: 1968, *Nature* **220**, 1210.

Klare, G. and Neckel, Th.: 1967, *Z. Astrophys.* **66**, 47.

Kraft, R. P.: 1961, *Astrophys. J.* **134**, 616.

Kopylov, J. M.: 1961, *Commun. Crimean Observ.* **26**, 232.

Lynds, B. T.: 1967, *Publ. Astron. Soc. Pacific* **79**, 448.

Martin, N.: 1964, *J. Observateurs, Marseille* **47**, 125.

McConnell, D. J.: 1968, *Astrophys. J. Suppl. Ser.* **16**, 275.

McCuskey, S. W.: 1970, in *The Spiral Structure of Our Galaxy*, IAU Symposium No. 38 (ed. by W. Becker and G. Contopoulos), D. Reidel, Dordrecht-Holland, p. 189.

Menon, T. K.: 1969, *Bull. Am. Astron. Soc.* **1**, 253.

Miller, S.: 1968, *Astrophys. J.* **151**, 473.

Neckel, Th.: 1966, *Z. Astrophys.* **63**, 221.

Oort, J. H.: 1970, in *The Spiral Structure of Our Galaxy*, IAU Symposium No. 38 (ed. by W. Becker and G. Contopoulos), D. Reidel, Dordrecht-Holland, p. 142. See also Kepner, M.: 1970, *Astron. Astrophys.* **5**, 444.

Pagel, B. E. J. and Tomkin, J.: 1969, *Quart. J. Roy. Astron. Soc.* **10**, 194.

Pavlovskaja, E. D. and Sharov, A. S.: 1970, in *The Spiral Structure of Our Galaxy*, IAU Symposium No. 38 (ed. by W. Becker and G. Contopoulos), D. Reidel, Dordrecht-Holland, p. 222.

Petrie, R. M.: 1965, *Publ. Dominion Astrophys. Obs., Victoria, B.C.* **12**, 317.

Petrie, R. M. and Petrie, J. K.: 1968, *Publ. Dominion Astrophys. Obs., Victoria, B.C.* **13**, 253.

Racine, R.: 1968, *Astron. J.* **73**, 233. See also *Astron. J.* **74**, 816, 1969 and *The Spiral Structure of Our Galaxy*, IAU Symposium No. 38, p. 219.

Rickard, J.: 1968, *Astrophys. J.* **152**, 1019.

Riddle, R. K., Priser, J., Worley, C. E., and Strand, K. A.: 1969, *Bull. Am. Astron. Soc.* **1**, 293.

Rohlfs, K.: 1967, *Z. Astrophys.* **66**, 225.

Rubin, V., Burley, J., Kiasatpoor, A., Klock, B., Pease, G., Rutscheidt, E., and Smith, C.: 1962, *Astron. J.* **62**, 281.

Schewick, H. van: 1970, private communication.

Schmidt-Kaler, Th.: 1961, *Astron. Nachr.* **286**, 113.

Schmidt-Kaler, Th.: 1962, *Sterne* **38**, 220.

Schmidt-Kaler, Th.: 1963, *Mitt. Astron. Ges.* 1962, 67.

Schmidt-Kaler, Th.: 1965a, in *Observational Aspects of Galactic Structure* (ed. by A. Blaauw and L. N. Mavridis), Athens, p. X, 1.

Schmidt-Kaler, Th.: 1965b, in *Landolt-Börnstein Zahlenwerte und Funktionen aus Naturwissenschaften und Technik, Gruppe VI: Astronomie-Astrophysik und Weltraumforschung, Band I: Astronomie und Astrophysik* (ed. by H. H. Voigt), Springer-Verlag, Berlin, Heidelberg, New York, p. 297.

Schmidt-Kaler, Th.: 1966a, in *Proceedings of the Twelfth General Assembly, Hamburg, 1964,* IAU Transactions **XIIB** (ed. by J.-C. Pecker), Academic Press, London and New York, p. 416.

Schmidt-Kaler, Th.: 1966b, unpubl., abstract in *ESO Bull.* **1**, 50 (see also Haug, U., *Astron. Astrophys. Suppl.* **1**, 40).

Schmidt-Kaler, Th.: 1967, in *Radio Astronomy and the Galactic System,* IAU Symposium No. 31 (ed. by H. van Woerden), Academic Press, London, p. 161, 171.

Schmidt-Kaler, Th.: 1968, *Publ. Astron. Inst. Univ. Bochum,* No. 1.

Schmidt-Kaler, Th.: 1970, in *The Spiral Structure of Our Galaxy,* IAU Symposium No. 38 (ed. by W. Becker and C. Contopoulos), D. Reidel, Dordrecht-Holland, p. 284.

Schmidt-Kaler, Th. and Pospieszczyk, A.: 1970, in preparation.

Serkowski, K.: 1966, *Astrophys. J.* **144**, 857.

Sharpless, S.: 1965, in *Galactic Structure* (ed. by A. Blaauw and M. Schmidt), University of Chicago Press, Chicago, p. 131.

Simonson III, S. C.: 1968, *Astrophys. J.* **154**, 923.

Smith, L. F.: 1968, *Monthly Notices Roy. Astron. Soc.* **141**, 317.

Stein, W. A. and Woolf, N. J.: 1969, *Astrophys. J.* **155**, L3.

Strömgren, B.: 1966, *Ann. Rev. Astron. Astrophys.* **4**, 433 (see also *Mitt. Astron. Ges.* **27**, 15, 1969).

Takase, B.: 1969, *Publ. Astron. Soc. Japan* **21**.

Uranova, T. A.: 1970, in *Spiral Structure of Our Galaxy,* IAU Symposium No. 38 (ed. by W. Becker and G. Contopoulos), D. Reidel, Dordrecht-Holland, p. 228.

Vasilevskis, S.: 1969, *Bull. Am. Astron. Soc.* **1**, 209.

Vogt, N.: 1969, Ph.D. Thesis, Bochum Univ., *Astron. Astrophys.,* in press.

Walker, M. F.: 1968, *Astrophys. J.* **155**, 447.

Walker, G. A. H. and Hodge, S. M.: 1968, *Publ. Astron. Soc. Pacific* **80**, 290.

Wayman, P. A.: 1967, *Publ. Astron. Soc. Pacific* **79**, 156.

Weaver, H. F.: 1970, in *The Spiral Structure of Our Galaxy,* IAU Symposium No. 38 (ed. by W. Becker and G. Contopoulos), D. Reidel, Dordrecht-Holland, p. 126.

Weaver, H. F. and Steinmetz, D. L.: 1962, *Publ. Astron. Soc. Pacific* **74**, 125.

Wilson, O. C. and Bappu, M. K. V.: 1957, *Astrophys. J.* **125**, 661.

Yilmaz, F.: 1966, *Z. Astrophys.* **64**, 54.

SPACE DISTRIBUTION OF THE LATE-TYPE STARS

L. N. MAVRIDIS

Dept. of Geodetic Astronomy, University of Thessaloniki,
Thessaloniki, Greece

1. Introduction

In the present paper a discussion is made of the space distribution of the stars of spectral types gM, C and S. The discussion is limited to those stars of these spectral types that may be discovered by spectroscopic observations, regardless of whether they are variable or not. Because of the intrinsic redness of these stars their study is more effectively done with infrared techniques.

The study of the space distribution of the stars of the spectral types gM, C and S presents considerable interest from many points of view:

(a) The number of gM stars per square degree is generally large at low galactic latitudes. Frequently 20 to 40 M2–M4 stars are found brighter than $I=12^m$, thus presenting rich material for statistical analysis.

(b) The interstellar absorption in the infrared is one half the visual absorption. On the other hand all the C- and S-type stars and practically all the M stars in low galactic latitudes are giants. Therefore, they can be identified with ease to great distances comparable with the distance to the galactic center. This advantage makes the M-, C- and S-type stars extremely useful for the study of galactic structure in great heliocentric distances. In this connection it should be mentioned that only a few very luminous OB stars and some rare supergiants have been identified, so far, at distances exceeding 4 kpc.

(c) Interesting results concerning stellar evolution may be found through the study of the space distribution of those very late-type stars which lie to the extreme right of the Hertzsprung-Russell diagram, where the more important and less known changes in the structure of the stars occur. Furthermore, the infrared surveys present a very effective method for the discovery of new infrared objects (infrared stars etc.).

An infrared survey usually includes the following steps:

(a) Discovery and spectral and luminosity classification of the stars considered. Because of the great numbers of stars involved this work can be more effectively done with the help of objective prism spectra supplemented, when necessary, by slit spectra.

(b) Determination of the magnitudes and colors. Photoelectric observations of the individual stars are certainly the most accurate method for the determination of these quantities. In most cases, however, photographic interpolation between properly distributed photoelectric standards can also give satisfactory results.

(c) Study of the surface distribution. For this purpose at least an approximate knowledge of the variation of the interstellar absorption over the region of the sky studied should be available.

L. N. Mavridis (ed.), Structure and Evolution of the Galaxy, 110–134. All Rights Reserved
Copyright © 1971 by D. Reidel Publishing Company, Dordrecht-Holland

(d) Study of the space distribution. In order to do this one should first determine the heliocentric distances of the stars studied. Therefore, a knowledge of the absolute magnitudes and the intrinsic colors of these stars is necessary.

So far a number of extensive surveys of the red stars have been carried out in different observatories with the help of objective prism low-dispersion infrared spectra. In the following we shall discuss briefly these surveys starting with a description of the techniques used for the spectral classification of the M-, C- and S-type stars with the help of these low-dispersion spectra.

2. The Spectral Classification

Most of the infrared surveys discussed in this paper have been carried out with the 24/36-inch Schmidt telescope of the Warner and Swasey Observatory, Case Institute of Technology in Cleveland. The photographic plates used were circular in shape with a diameter equal to $5°16$ and each covers an area of the sky equal to about 20 square degrees. The telescope can be equipped with one $2°$ or one $4°$ prism.

In all the infrared surveys at Cleveland the Eastman Kodak IN plates in combination with a Wratten No. 89 filter were used. The spectral area cut off by this plate-filter combination is contained between 6800 Å and 8800 Å and is approximately centered at the atmospheric A-band (7590 Å). The spectra taken with the $2°$ and $4°$ prism have a dispersion at the A-band respectively equal to 3400 Å/mm and 1700 Å/mm. Furthermore, all the spectra used in the infrared surveys were unwidened in order to distinguish better the fine spectral features used for the spectral classification, especially in the case of the fainter stars, and at the same time reduce the overlapping in the crowded areas.

The spectral classification of the M-, C- and S-type stars is based on a number of molecular bands that fall in the spectral area under consideration. The most interesting among these bands are the telluric bands and the TiO, CN, LaO and the VO bands.

The criteria for the spectral classification of the M-, C- and S-type stars with the help of the objective prism infrared spectra considered have been developed by Nassau and his collaborators and have been extensively described elsewhere (Mavridis, 1967).

With the help of these criteria one can classify the stars of the spectral classes M2–M10. As regards the stars of the spectral classes M0 and M1, their classification on the basis of these criteria should be avoided, because of its uncertainty.

In many statistical studies a system of classification of the M stars in three natural groups of spectral classes, i.e. the groups of the M2–M4, the M5–M6 (or M5–M6. 5) and the M7–M10 stars, has proved very useful.

The low-dispersion infrared spectra considered here do not in general allow a luminosity classification of the M stars. For statistical studies, however, this is not very serious, because of the fact that the great majority of the M stars found in low galactic latitudes are giants. The only difference between giants and dwarfs that has been established, so far, with the help of the low-dispersion spectra considered is that the late-type dwarfs do not seem to show the VO bands at 7900 Å observed in the giants,

Carbon stars are customarily classified as C0 to C9 following the criteria established by Keenan and Morgan (1941), or as R and N according to the Henry Draper system. Approximately, the R stars (R0–R9) correspond to the classes C0–C4 and the N stars (N0–N9) to the classes C5–C9. A classification of the carbon stars by individual sub-classes has not proven possible with the help of the objective prism infrared spectra considered here. Also the stars of the classes C0–C2 are usually not detected with these spectra.

The S-type stars sometimes show well developed the LaO bands at 7403 Å and 7910 Å by which they can be identified. These stars, however, are missed if these bands are weak. Moreover, a confusion of these stars with the reddened early M-type stars is possible. Nassau and Stephenson (1960) have shown, however, that the S-type stars can be better sought in the red region of the spectrum where ZrO at 6474 Å can be recognized with low dispersions. Merrill (1952) found in the spectra of the S-type stars spectral lines of the unstable element TcI, a fact that suggests that this group of stars may belong to a short-lived phase of stellar evolution.

It is important to notice here, that the M-, C- and S-type stars cannot be classified to the same magnitude on a given plate. Thus, the C-type stars can be recognized almost to the limiting magnitude on the plate. This is, however, not valid for the S-type stars. Also the stars M5 and later can be classified to much fainter limiting magnitude than the early M-type stars (M2–M4) on the same plate (Blanco and Münch, 1955; Blanco, 1963).

The system of spectral classification of the M-, C- and S-type stars described above is known as the Case system for the classification of these stars. The corresponding standards have been published by Nassau and Velghe (1964). Blanco (1964) has studied the relation of the Case system for the spectral classification of the M stars with the Mt. Wilson and the MK systems described respectively by Adams *et al.* (1926) and Keenan (1963). His results are summarized in Table I. From this table we see that the stars with the Mt. Wilson classes M0–M2 cannot be readily recognized with the help of the low-dispersion infrared spectra used in Cleveland. For the spectral classes M8–M10 on the other hand no standards exist in the Mt. Wilson and the MK systems and their classification is, therefore, possible only in the Case system.

TABLE I

Relation between the Case and the Mt. Wilson systems for the spectral classification of the M stars

Case system	M0	M1	M2	M3	M4	M5	M6	M7
Mt. Wilson system	M1.4	M2.1	M3.3	M3.9	M4.6	M5.6	M6.1	M7.0
Mt. Wilson system	M0	M1	M2	M3	M4	M5	M6	M7
Case system	–	–	M0.8	M1.9	M3.1	M4.6	M5.9	M7.0

3. The Galactic Belt Surveys

Starting with the pioneer work of Nassau in Cleveland a number of infrared surveys have been carried out in different observatories along the old (Lund) and the new (1958) galactic equator. A description of these surveys has been given elsewhere (Mavridis, 1967). Table II, which is self-explanatory, summarizes the main information concerning these galactic belt surveys. Following remarks should be added:

(a) In all the surveys contained in Table II with the only exception of the survey No. 6 the M-stars were classified only in the three natural groups M2–M4, M5–M6, M7–M10 or even M5–M10 and no classification in the individual subclassesM2, M3 etc. was made. Also in most of the cases no information about the individual stars but only the total number of the stars belonging to each of these natural groups was given.

(b) The detailed numbers of the M2–M4, M5–M6 and M7–M10 stars found in the survey No. 9 have not yet been published. Only the results concerning some parts of the belt surveyed have been given (Westerlund, 1964). Also a catalogue containing the positions and estimated visual and infrared magnitudes of the 1124 carbon stars found in the longitude interval $l=235°$ through 0° to 7° has been prepared so far (Westerlund, 1970).*

In all the galactic belt surveys contained in Table II with the only exception of the survey No. 6** no apparent magnitudes of the stars studied were measured but all the apparent infrared magnitudes given were estimated on the spectral plates used for the spectral classification. Therefore, great differences in the limiting infrared magnitudes of these surveys may exist from one region of the sky to another even in the same survey. Still greater differences may exist between the magnitude scales of different surveys. Westerlund (1970) for example found that the m_{i_r} magnitudes given by Blanco and Münch (1955) are on the average 1^m4 fainter than the infrared magnitudes I given by himself for the stars in common.

But even if the limiting magnitudes were strictly the same for all the regions of the sky studied, the surface densities found in these surveys for the different types of stars and the different directions in the sky would hardly be comparable to each other because of the following reasons:

(a) The limiting magnitudes to which the stars of the different spectral types considered can be classified on the same plate are, as mentioned already, different from type to type.

(b) The interstellar absorption varies considerably from one region of the sky to another.

(c) The absolute magnitudes of the stars of the different spectral types considered are not the same.

For these reasons the volume of space in which the stars studied in these surveys are contained varies considerably both with the spectral type and the region of the sky examined.

From the above discussion it is evident that in order to determine the real distribution of the M-, C- and S-type stars in the Galaxy one should study the variation

---

* The author is very much indebted to Dr. Westerlund for a prepublication copy of this catalogue.
** The visual magnitudes given by Neckel (1958) are the BD magnitudes corrected with the help of the tables given in *Harvard Annals* 5, 72.

TABLE II

Galactic belt surveys of the M-, C- and S-type stars

No.	Region	Number of stars M	classified C	S	Limiting magnitude	Observatory	References
1	$l^I = 333°-0°-201°, b^I = -2°$ to $+2°$	M5–M10 = 3010	271	31	$I = 10^m.2$	Cleveland	Nassau and Blanco (1954a, b, c); Nassau et al. (1954)
2	Ten 4°-wide belt-shaped regions at right angles to the Lund galactic equator at intervals of 21° in galactic longitude l^I (centered at $l^I = 342°, 3°, ..., 171°$) and extending between $b^I = -18°$ to $-6°$ and $b^I = +6°$ to $+18°$	M5–M10 = 2179	56	?	$I = 10^m.2$	Cleveland	Nassau and Blanco (1954c); Blanco (1958)
3	$l^I = 333°-0°-201°, b^I = +2°$ to $+6°$	M5–M6 = 1873, M7–M10 = 446	230	12	$I = 10^m.2$	Cleveland	Nassau and Blanco (1957); Nassau et al. (1956); Blanco and Nassau (1957)
4	$l^I = 333°-0°-201°, b^I = -6°$ to $-2°$	M5–M6 = 2139, M7–M10 = 495	192	25	$I = 10^m.2$	Cleveland	Nassau and Blanco (1957); Nassau et al. (1956); Blanco and Nassau (1957)
5	Twenty-five one-square degree areas relatively free from interstellar absorption lying between $b^I = -6°$ to $+6°$ and spread as evenly as possible in galactic longitude between $l^I = 350°-0°-200°$.	M2–M4 = 653, M5–M6 = 242, M7–M10 = 374	–	–	–	Cleveland	Sanduleak (1957)
6	All BD stars of spectral types M2–M10, contained in the region $l^I = 333°-0°-199°, b^I = -6°$ to $+6°$.	M2–M4 = 1124, M5–M6 = 161, M7–M10 = 5	–	–	$V = 10^m.5$	Cleveland	Neckel (1958)
7	$l^I = 200°-270°, b^I = -2°$ to $+2°$	M2–M4 = 1313, M5–M6 = 938, M7–M10 = 234	172	13	$I = 10^m.4$	Tonantzintla	Blanco and Münch (1955)
8	A more or less continuous belt at least 3°-wide centered at the Lund galactic equator and extending between $l^I = 180°-360°$	M5.5–M10 = 2291	186	8	$I = 10^m.3$	Bloemfontein	Smith and Smith (1956)
9	$l = 230°-0°-10°, b = -5°$ to $+5°$	M2–M4, M5–M6, M7–M10 = 1326	87		$I = 12^m.5$	Mt. Stromlo	Westerlund (1964, 1970)

of the space densities rather than of the surface densities of these stars throughout the Galaxy. To this purpose a knowledge of the absolute magnitudes and the intrinsic colors of these stars is necessary.

4. The Absolute Magnitudes and the Intrinsic Colors of the M-, C- and S-Type Stars

Blanco (1964, 1965) has given a calibration of the absolute visual magnitudes (M_V) and the intrinsic $(B-V)_0$ and $(V-I)_0$ colors of the normal M giants. His values are given in Table III together with the corresponding values of the absolute infrared magnitudes M_I computed with the help of the values of M_V and $(V-I)_0$ given in the same table. The spectral types given in Table III refer to the Mt. Wilson system. For the classes M8 and M9 for which no standards exist in the Mt. Wilson system the data of Table III refer to stars classified in the Case system. According to Blanco (1965) the values of M_V (and therefore also the values of M_I) given in Table III should be made fainter by about $0^m.1$ when used for space density analysis.

TABLE III

Calibration of the absolute magnitudes and the intrinsic colors of the normal M giants classified according to the Mt. Wilson system (parenthesis indicate uncertain values)

	M0	M1	M2	M3	M4	M5	M6	M7	M8	M9
M_V	$-0^m.3$	$-0^m.5$	$-0^m.8$	$-1^m.1$	$-1^m.0$	$(-0^m.9)$	$(-0^m.9)$	$(-0^m.9)$	–	–
M_I	-1.6	-2.0	-2.6	-3.3	-3.6	(-3.9)	(-4.4)	(-4.9)	–	–
$(B-V)_0$	1.60	1.60	1.61	1.61	1.62	1.61	1.60	1.73	$(2^m.0)$	–
$(V-I)_0$	1.35	1.45	1.75	2.16	2.58	3.02	3.46	4.03	(4.7)	$(5^m.4)$

The relation of the Mt. Wilson and the Case systems of spectral classification of the M stars has been given in Table I. With the help of this table and Table III the absolute magnitudes and the intrinsic colors of the normal M giants classified according to the Case system have been computed and are given in Table IV.

TABLE IV

Calibration of the absolute magnitudes and the intrinsic colors of the normal M giants classified according to the Case system

	M0	M1	M2	M3	M4	M5	M6	M7
M_V	$-0^m.6$	$-0^m.8$	$-1^m.1$	$-1^m.0$	$(-0^m.9)$	$(-0^m.9)$	$(-0^m.9)$	$(-0^m.9)$
M_I	-2.2	-2.6	-3.4	-3.5	(-3.7)	(-4.2)	(-4.4)	(-4.9)
$(B-V)_0$	1.60	1.61	1.61	1.62	1.61	1.60	1.61	1.73
$(V-I)_0$	1.57	1.79	2.29	2.54	2.84	3.28	3.52	4.03

TABLE V

Intrinsic colors, bolometric corrections (BC) and effective temperatures (T_e) of the normal M giants

	M0	M1	M2	M3	M4	M5	M6	M7	M8	M9
$(B-V)_0$	$1^m.57$	$1^m.60$	$1^m.60$	$1^m.60$	$1^m.63$	$1^m.71$	$(1^m.70)$	$(1^m.80)$	$1^m.93$	$2^m.53$
$(V-I)_0$	2.23	2.32	2.42	2.65	3.35	4.05	5.50	7.00	8.51	10.02
BC	−1.29	−1.37	−1.43	−1.74	−2.51	−3.33	−4.8	−7.7	−8.6	−10.8
T_e(K)	3680	3600	3600	3370	3060	2800	2550	2150	1900	1650

Again the values of M_V and M_I given in Table IV should be made fainter by about $0\overset{m}{.}1$ when used for space density analysis.

The dispersion σ of the absolute magnitudes is also required in any star count analysis of giant M stars. Among the M giants the effect of variability and of spectral classification errors must be taken into account when estimating σ. Blanco (1965) found the following results:

(a) for classes M2 and earlier $\sigma = \pm 0\overset{m}{.}3$; (b) for classes M3–M5 more $\sigma = \pm 0\overset{m}{.}35$; (c) for classes M6 and later the effect of variability is predominant and values of σ over $\pm 0\overset{m}{.}4$ must be used. Other investigators, however, use greater values for σ. McCuskey (1969), for example, uses the value $\sigma = \pm 0\overset{m}{.}7$. The same author uses also greater value than the one $(+0\overset{m}{.}1)$ proposed by Blanco for the correction to be applied to the absolute visual magnitudes given in Tables III and IV, in order to reduce them to values per unit volume of space. For the mean absolute visual magnitude per unit volume of space of the M2–M4 stars (Mt. Wilson system), for example, McCuskey uses the value $M_V = -0\overset{m}{.}5$ while the corresponding value according to Blanco would be $M_V = -0\overset{m}{.}9$.

With the rapid development of infrared photometry considerable work has been done in the field of narrow- and wide-band photometry of the gM stars during the recent years. Johnson (1964) and Mendoza and Johnson (1965), for example, have measured a great number of normal M giants in the $U, B, V, R, I, J, K, L, M, N$ system. With the help of this material the values of the intrinsic colors, the bolometric corrections and the effective temperatures of the normal M giants given in Table V have been found. It should be noted that the I magnitudes used throughout this paper with the exception of Table V are the magnitudes in the infrared system of Kron, Gascoigne, and White (1957) and, therefore, they do not coincide with the I magnitudes given in Table V.

A considerable percentage of the gM-, C- and S-type stars are variables (mainly long-period, semiregular, irregular etc.). Therefore, it would be interesting to compare the absolute magnitudes given above for the normal M giants with the corresponding values found for the variable stars of the same spectral type.

The absolute magnitudes of the semiregular and some irregular variables have been investigated by Joy (1942) from spectroscopic criteria and by Wilson (1942) from proper motions and radial velocities. The results of both investigations agree very well. Table VI gives the results found by Joy (the stars have been arranged according to the types of variability used in the General Catalogue of Variable Stars).

TABLE VI

Absolute visual magnitudes of the semiregular and irregular variables

Type	M_V	σ	n	Remarks
SRa	$-0^m.9$	$\pm 0^m.4$	24	Supergiants not included
SRb	-0.9	± 0.3	59	Supergiants not included
I	-1.0	± 0.4	6	Supergiants not included

The absolute magnitudes of the long-period variables of the spectral types M, C and S have been discussed by Osvalds and Risley (1961) with the help of McCormick proper motions and Mt. Wilson radial velocities. Their results are given in Table VII.

TABLE VII

Absolute visual magnitudes of the long-period
variables at mean maximum

Group	$\langle P \rangle$ days	n	$\langle Sp \rangle$	$\langle M_V \rangle$
M	131	14	M1.9	$-1^m.67$
M	176	29	M2.7	-2.74
M	223	55	M3.7	-2.10
M	273	65	M4.2	-2.03
M	324	73	M5.3	-0.93
M	376	42	M6.2	-1.05
M	419	23	M6.5	-0.31
M	508	18	M6.0	-1.17
C	404	26	C	-1.44
Se	364	22	Se	-1.57

Clayton and Feast (1969) made recently a new determination of the absolute visual magnitudes of the long-period variables of spectral type M as a function of period. Their results which refer both to the mean maximum (M_m) and to the mean light intensity (M_l) are given in Table VIII.

TABLE VIII

Absolute visual magnitudes of the long-period
variables of spectral type M at mean maximum
(M_m) and at mean light intensity (M_l)

$\langle P \rangle$ days	$\langle M_m \rangle$		$\langle M_l \rangle$	
132	$-1^m.6$	$\pm 0^m.6$	$-0^m.1$	$\pm 0^m.6$
177	-3.0	± 0.4	-1.5	± 0.4
225	-1.8	± 0.3	-0.5	± 0.3
274	-1.6	± 0.3	$+0.2$	± 0.3
324	-1.3	± 0.3	$+0.5$	± 0.2
376	-0.8	± 0.3	$+0.9$	± 0.3
419	-1.0	± 0.5	$+0.4$	± 0.6
503	-1.0	± 1.3	$+2.1$	± 0.8

From Tables VII and VIII we see that considerable differences exist between the absolute visual magnitudes of the normal M giants and those of the long-period variables of the same spectral type. It is, however, an open question at which phase of the light variation one should compare the absolute magnitudes of the variables with those of the constant stars.

The absolute visual magnitudes of the R stars have been determined by Vandervort (1958) by the secular parallax method, as well as from radial velocities and τ-components of proper motion. His results are: $M_V = +0\overset{m}{.}44 \pm 0\overset{m}{.}29$ for the R0 to R2 stars and $M_V = -1\overset{m}{.}10 \pm 0\overset{m}{.}49$ for the R5 to R8 stars. Gordon (1968) has discussed recently the absolute magnitudes of the carbon stars using double stars with one member a carbon star, carbon stars in clusters, the Wilson-Bappu effect, carbon stars in the Large Magellanic Cloud, interstellar lines in carbon stars and radial velocities and proper motions. His results are given in Table IX.

TABLE IX

Absolute visual magnitudes of the carbon stars

Spectral type	R $-$ K0 to R $-$ K3	N $-$ M3.5	N $-$ M4	N $-$ M5	N $-$ M6
M_V	$+0^m.4$	$-1^m.5$	$-2^m.0$	$-3^m.0$	$-3^m.5$

The spectral types are on the system developed by Gordon (1967). The first letter R or N corresponds to the type of carbon star. The second part gives the equivalent blue spectral type. For example, the classification N–M5 refers to an N star closest in type to an M5 star (except for the bands due to carbon).

The $B - V$ colors of R stars have been determined by Vandervort (1958). The values found range from $1\overset{m}{.}22$ at class R0 to $2\overset{m}{.}10$ for class R8. Mendoza and Johnson (1965) made photoelectric observations in the $U, B, V, R, I, J, K, L, M, N$ system of 39 carbon stars. The $B - V$ colors found range between $1\overset{m}{.}04$ and $5\overset{m}{.}52$. On the basis of this material the authors computed also the bolometric corrections (range $0^m - 5^m$) and the effective temperatures (range 5500–2270 K) of the carbon stars. Another discussion of the colors of the carbon stars has been published recently by Smak (1968).

Dahn (1964) has determined the velocity ellipsoid for the N and R stars separately. He found that the velocity distributions of these two carbon star subgroups are different in the sense that the N-star ellipsoid is considerably flatter than the R-star ellipsoid and that the dispersion in the direction of galactic rotation is smaller for the R stars.

From the above results it is clear that the carbon stars belong to two distinct groups of stars (R and N stars) with different luminosities and very probably different ages and different space distribution. According to Blanco (1965) the N stars are, in their majority, apparently concentrated in spiral arms. The R stars, on the contrary, apparently are not concentrated in the spiral arms, but so far no evidence exists of their presence in the galactic nuclear bulge.

The absolute magnitudes of the S-type stars have been studied by Keenan (1954) and Keenan and Teske (1956). Keenan (1954) divided the known S stars into two groups: (a) the S-type stars which are long-period variables and (b) the S-type stars which are constant in brightness or are variables of low amplitude. Using secular parallaxes as well as radial velocities and τ proper motion components for 17 stars of the first group Keenan found for the mean absolute visual magnitude of the stars of this group a value $M_V = -1\overset{m}{.}0$. The galactic concentration of the stars of the second group on the other hand is higher than that of the first group. Furthermore, this concentration is probably slightly less than that of the similar N-type stars but appreciably greater than that of the R-type stars. Based on these facts Keenan accepted tentatively the value $M_V = -1\overset{m}{.}5$ for the mean absolute visual magnitude of the S stars of the second group, which are considered by him to be associated with the spiral arms. Analogous results have been obtained also by Takayanagi (1960). Takayanagi, however, found the values $M_V = -0\overset{m}{.}1 \pm 0\overset{m}{.}6$ for the less variable S stars and $M_V = -3\overset{m}{.}0 \pm 0\overset{m}{.}5$ for the long-period S-type variables at maximum brightness.

Therefore, the S stars, which are a low-velocity class of stars (none is known to have a radial velocity exceeding 100 km/sec), should also be separated into two distinct groups with different luminosities and probably different ages and different space distribution.

A discussion of the colors of the S-type stars has been published recently by Smak (1968).

The above results are of special interest for the discussion of the results of any infrared survey of the C- and S-type stars. As mentioned already the C-stars discovered in an infrared survey are a mixture of R and N stars. Also the S stars are a mixture of stars of the two subgroups of S-type stars discussed above. Therefore, special attention should be paid in order to avoid any erroneous conclusions when studying the space distribution of the C- and S-type stars discovered in the infrared surveys.

Another conclusion from the discussion given above is that our present knowledge of the absolute magnitudes and the intrinsic colors of the C- and S-type stars is very incomplete. Much more work is, therefore, urgently needed in this field in the future.

5. The Deep Surveys in Selected Regions of the Milky Way

In all the infrared surveys included in Table II with the only exception of the survey No. 9 (Westerlund, 1964), the limiting infrared magnitude was equal to about $I = 10\overset{m}{.}2$ and, therefore, relatively small distances from the Sun were reached. In order to avoid this disadvantage a number of deep infrared surveys have been carried out in recent years in strategically selected regions of the Milky Way. The limiting infrared magnitude in all these surveys was approximately equal to about $I = 13^m$ and, therefore, much greater distances could be reached. In addition, the following precautions were taken in order to avoid the bias affecting the results of the galactic belt surveys mentioned above:

(a) The magnitudes (I or V) of all the stars studied were measured on direct plates using appropriate photoelectric sequences. In this way the apparent magnitudes of the individual stars as well as the limiting magnitude of the survey could be accurately determined.

(b) The interstellar absorption was determined with the help of Wolf diagrams or color excesses of stars with known color indices and intrinsic colors.

TABLE X

Deep infrared surveys in selected regions of the Milky Way

No.	Region		l	b	Area	Number of stars found per s.d.			References
						M2 — M4	M5 — M6	M7 — M10	
1	Cygnus	1	64°.8	0°.0	2 s.d.	24.5	10.0	13.5	Westerlund (1959a)
		2	64.8	+1.0	3 s.d.	33.7	16.7	15.3	
		3–6	64.9	+3.6	9 s.d.	92.0	34.2	19.0	
		7	64.9	+7.6	4 s.d.	14.8	8.8	7.0	
2	Aquila	A	44.2	−6.6	2 s.d.	20.5	20.5	11.0	Westerlund (1959b)
		B	44.2	−5.2	2 s.d.	24.5	31.5	19.0	
		C	44.2	−3.6	2 s.d.	32.0	46.5	33.0	
3	Crux	A	301.9	−0.2	3 s.d.	17.0	23.7	44.0	Westerlund (1960a, b)
		B	301.9	−1.2	3 s.d.	13.7	12.3	18.0	
		S	302.8	+3.2	1 s.d.	34.0	38.0	29.0	
		R	303.8	+1.3	2 s.d.	12.0	7.0	10.5	
4	Scutum	A	26.3	−1.5	1 s.d.	215.0	170.0	146.0	Albers (1962)
		B	28.3	−2.6	1 s.d.	156.0	168.0	94.0	
		C	26.6	−2.5	0.5 s.d.	88.0	66.0	40.0	
		D	26.1	−3.7	1 s.d.	67.0	64.0	13.0	
5	NGC 6522		0.5	−2.9	390 s.m.	664.6	849.2	258.5	Nassau and Blanco (1958)
6	SA 158	1+8+7	3.9	−8.1	3 s.d.	–	90.7	41.0	McCuskey and Mehlhorn (1963)
		2+9+6	3.9	−9.1	3 s.d.	–	69.7	28.3	
		3+4+5	3.9	−10.1	3 s.d.	–	55.0	20.3	
7	SA 193		293.5	+0.8	2 s.d.	42.5	45.5	42.0	Westerlund (1965)
8	Groningen–Palomar	1	0	+29	5 s.d.	–	–	–	Hidajat and Blanco (1968)
	Variable Star Fields	2	4	+12	5 s.d.	–	–	–	
		4	82.0	+11.0	5 s.d.	–	–	–	
9	Galactic Anticenter		186.0	+1	8 s.d.	32.1	10.1 [a]	–	McCuskey (1969)

[a] M5 — M8 stars

(c) Finally, by adopting a reasonable set of values for the mean absolute magnitudes per unit volume of space of the stars of the different spectral classes considered and the dispersion of these magnitudes, the space densities of the stars of the corresponding spectral classes for different heliocentric distances could be determined.

Table X summarizes the main information concerning the deep infrared surveys carried out in some selected regions of the Milky Way. A discussion of the results of these surveys together with the results of the galactic belt surveys included in Table II will be given in the next section.

Besides the surveys included in the Tables II and X a number of additional surveys of gM-, C- and S-type stars have been carried out in different observatories with infrared or different other techniques (Albers, 1967; Barbier, 1965; Barbier and Maiocchi, 1966; Bertiau et al., 1964; Dolidze, 1964, 1965, 1968a, b; Lodén and Sundman, 1966; McCarthy, 1968; McConnell, 1967; Price, 1968; Slettebak et al., 1969; Stephenson, 1965; Stephenson and Terrill, 1967; Upgren and Grossenbacher, 1968; Wyckoff, 1968 etc.). These surveys are, however, of more or less limited interest and, therefore, will not be discussed here in more detail.

6. Discussion of the Results

6.1. DISTRIBUTION IN GALACTIC LONGITUDE AND LATITUDE

A first study of the distribution of the gM-, C- and S-type stars in galactic longitude has been made with the help of the results of the galactic belt surveys included in Table II. Blanco (1965) has used the results of the surveys Nos. 1, 7 and 8 of Table II in order to construct the diagrams given in Figures 1 and 2. In these diagrams the numbers of the M5–M10 stars* and the carbon stars observed per area of 16 square degrees in the 4°-wide belt centered at the Lund galactic equator $\left(-2° \leqslant b^{\mathrm{I}} \leqslant +2°\right)$ are plotted respectively as a function of the old galactic longitude l^{I}. Next to the values of the old galactic longitude l^{I}, to which the original results were referred, the new galactic coordinates l, b of the corresponding points of the Lund galactic equator have been added in parentheses. The striking difference in the longitude distribution of the M5–M10 and the carbon stars is clearly seen from Figures 1 and 2. Besides the local irregularities, which have been attributed by Blanco (1965) to the influence of the interstellar absorption, it is clearly seen from these figures that the number of the M5–M10 stars increases as we approach the direction of the galactic center, while the number of the carbon stars increases toward the galactic anticenter. The importance of this result becomes greater, if we take into account that both Figures 1 and 2 are based on the same observational material. In evaluating these results, however, one should always keep in mind the fact that the 4°-wide galactic belt considered in plotting Figures 1 and 2 is not any longer symmetrical with respect to the new (1958) galactic equator.

---

* The numbers of the M5.5 – M10 stars found in the survey No. 8 were multiplied by the factor 2.2 in order to become comparable with the numbers of the M5 – M10 stars found in surveys Nos. 1 and 7.

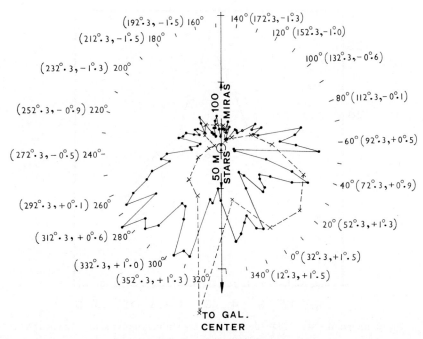

Fig. 1.   Distribution of the M5—M10 stars with $|b^I| \leqslant 2°$ (continuous line) and the long-period variables with photographic magnitude at average maximum $< 15^m$ and $|b^I| < 10°$ (dashed line) in old galactic longitude l^I (the values given in parentheses next to each value of l^I are the new galactic coordinates l, b of the corresponding points of the Lund galactic equator).

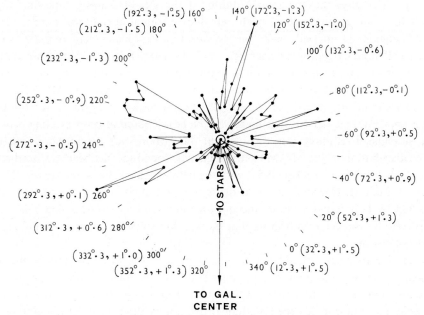

Fig. 2.   Distribution of the carbon stars with $|b^I| \leqslant 2°$ in old galactic longitude l^I (the values given in parentheses next to each value of l^I are the new galactic coordinates l, b of the corresponding points of the Lund galactic equator).

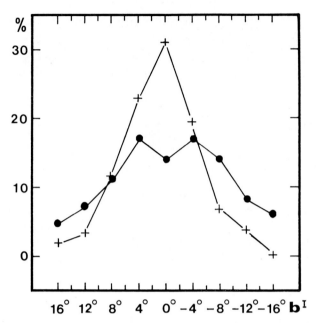

Fig. 3.   Distribution of the M5 — M10 (dots) and the carbon stars (crosses) in old galactic latutide b^{I}.
For explanations see text.

A first study of the distribution of the gM-, C- and S-type stars in galactic latitude was made with the help of the results of the surveys Nos. 1 and 2 of Table II. Figure 3 (Blanco, 1965) gives the latitude distribution of the M5–M10 and C-type stars found during these surveys in the ten 4°-wide belt-shaped regions at right angles to the Lund galactic equator at intervals of 21° in old galactic longitude l^{I} (centered at $l^{\mathrm{I}}=342°$, 3°, ..., 171°) and extending between $b^{\mathrm{I}}=-18°$ to $+18°$. From this figure we see that both types of stars considered show a marked galactic concentration. This concentration is, however, more pronounced for the carbon stars than for the M5–M10 stars. The decrease in the number of the M5–M10 stars shown in Figure 3 for $b^{\mathrm{I}}=0°$ is rather apparent and caused presumably by the heavy obscuration prevailing near the galactic plane. It is interesting to note that the carbon stars do not show this dip.

The distribution of the M2–M4 stars in galactic longitude has been first studied by Sanduleak (1957) and Neckel (1958) with the help of the results of the surveys Nos. 5 and 6 of Table II. Neckel found that the longitude distribution of the M2–M4 stars classified by him shows a pronounced maximum at about $l \simeq 62°$. This result was interpreted by Neckel as indicating that the M2–M4 stars are concentrated along the local spiral arm.

According to Neckel the longitude distribution of the M2–M4 stars in the area studied by him is quite similar to the distribution of the long-period variables in the same area that are brighter than $10^{m}.5$ (visual) at maximum light. In order to study this point more thoroughly the longitude distribution (Plaut, 1965) of the long-period variables with photographic magnitude at average maximum $<15^{m}.0$ and $|b^{\mathrm{I}}|<10°$ has been added to Figure 1 (dashed line). From this figure we see that the longitude distribution of the long-period variables considered presents a maximum in the direction $l \simeq 50°$ as well as another more pronounced maximum in the direction $l \simeq 350°$.

For comparison the spiral arms of the Galaxy as indicated by the space distribution of the young clusters and H II regions (Becker and Fenkart, 1970) are plotted in Figure 4. In this figure parts of three spiral arms can be seen, i.e. the *local arm* ($l = 60°-210°$, with the Sun on its inside edge), the *Perseus arm* ($l = 90°-140°$, $r = 2-3$ kpc) and the *Sagittarius arm* ($l = 330°-0°-30°$, $r = 2$ kpc). According to McCuskey (1970) two more spiral arms should be added i.e. the *Carina section* ($l = 280°-310°$, $r = 1.4-5.5$ kpc) and the *Centaurus link* ($l = 310°-330°$, $r = 2$ kpc).

Westerlund (1964) from a preliminary discussion of the results of the survey No. 9 in Table II reached the following conclusions:

(a) The M2–M4 type stars show clusterings with preference for spiral-arm regions.

(b) The M5–M10 type stars are evenly distributed in arm and interarm regions. Their density increases appreciably as the galactic center is approached.

(c) The carbon stars are spiral arm objects. Their more even distribution per unit volume of space as compared with the OB stars makes them extremely useful for tracing distant spiral arms.

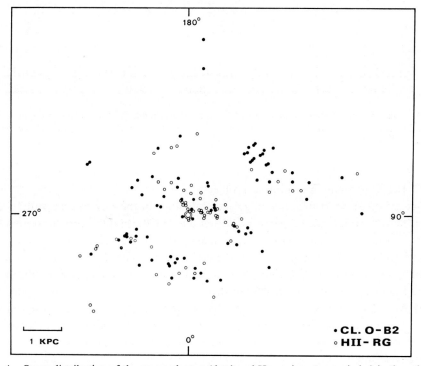

Fig. 4. Space distribution of the young clusters (dots) and H II regions (open circles) in the galactic plane according to Becker and Fenkart (1970). Parts of three spiral arms can be seen, the local arm, the Perseus arm and the Sagittarius arm. The position of the sun is also indicated [⊙].

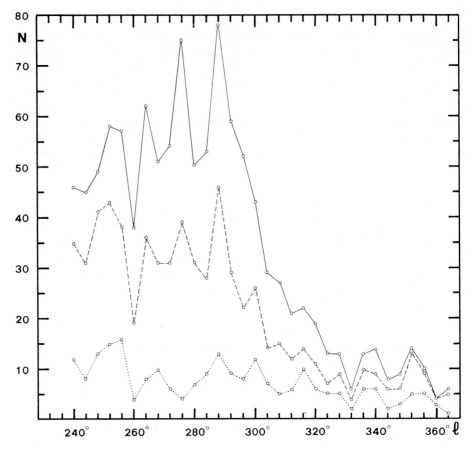

Fig. 5.   Distribution of the carbon stars with $|b| \leqslant 5°$ (Westerlund, 1970) in galactic longitude. The dotted, dashed and continuous lines refer respectively to stars with apparent infrared magnitude $I \leqslant 8^m.5$, $I \leqslant 10^m.5$ and $I \leqslant 12^m.5$.

(d) The S-type stars are spiral arm objects.

Figure 5 gives the distribution in galactic longitude of the carbon stars with $|b| \leqslant 5°$ included in the catalogue prepared by Westerlund (1970). In this figure the dotted line refers to the carbon stars with infrared magnitude $I \leqslant 8^m.5$ (bright stars), while the dashed and continuous lines refer to the carbon stars with $I \leqslant 10^m.5$ (bright + intermediate stars) and $I \leqslant 12^m.5$ (bright + intermediate + faint stars) respectively. From this figure following conclusion can be drawn:

(a) At $l = 248°$, where the line of sight passes first through the local spiral arm and then extends to an interarm region (Figure 4), the number of bright and especially of intermediate carbon stars assumes high values, while the number of faint carbon stars is relatively low.

(b) At $l = 264°$, where the line of sight extends along the local spiral arm, the numbers of both the intermediate and faint carbon stars are high.

(c) At $l=276°$, the numbers of the intermediate and faint carbon stars assume peak values, while the number of bright carbon stars is very low.

(d) In the interval $l=288°-296°$, where the line of sight extends along the Carina section, the number of faint carbon stars remains high, while the numbers of the bright and especially of the intermediate carbon stars drop rapidly after $l=288°$.

6.2. SURFACE DISTRIBUTION

The surface distribution of the S stars has been investigated by Blanco and Nassau (1957) who found that these stars are found in and out of the regions of heavy obscuration. These results have been verified later by Blanco (1965), who presented a number of cases where the S stars are found in the periphery or inside of young clusters, nebulosities or obscured regions, while in many other cases S stars are found in relatively clear areas. Westerlund (1964) on the contrary found that in the Southern Milky Way the surface distribution of the S stars shows an overall agreement with that of the OB-associations. If we take into account that, as mentioned already, the S stars belong to two groups with distinct galactic distribution, these results seem to indicate that very probably *only* the one of these two groups is associated with the spiral arms. More accurate distances of the S stars are, however, needed before any definitive conclusion can be drawn.

The surface distribution of the carbon stars also presents many interesting problems. Nassau and Blanco (1954c) have noted that carbon stars seem to avoid the clearest regions of the Milky Way and often show concentration in areas of evident interstellar obscuration. Their affinity for somewhat obscured regions may explain why the latitude distribution of carbon stars shown in Figure 3 does not show at $b^l=0°$ a dip similar to that shown by the M stars. The surface distribution of the carbon stars, however, is much less irregular than that of the OB stars. Furthermore, a number of pairs of carbon stars with separations of less than $0°.2$ have been found. The clustering tendency of the carbon stars has been verified by Blanco and Münch (1955) and by Smith and Smith (1956). Westerlund (1964) presented a discussion of the surface distribution of the carbon stars discovered in the survey No. 9 of Table II. He found that between $l=260°-300°$ most of the carbon stars and the Wolf-Rayet stars are found south of the galactic equator. Also the carbon stars seem to avoid OB-associ-

TABLE XI

Distribution of the pairs of carbon stars with a distance of less than 0°.2 between the components

l	b	Number of pairs	
		Observed	Expected
230°–260°	−6° to +4°	26	9
260°–300°	−5° to +5°	101	20
30°–10°	−4° to +6°	10	3
Total number		137	32

ations. Westerlund found also, like the earlier investigators, a great number of pairs of carbon stars with separation less than $0°.2$ (137 pairs). For a more thorough study of this phaenomenon he divided his material into zones of $10 \times 1°$ and computed for each zone the number of pairs to be expected from a random distribution of stars of the average density observed in the zone. His results are given in Table XI.

From this table it is evident that the number of pairs of carbon stars observed is much greater than the one predicted from a random distribution. It appears extremely important to investigate further this phaenomenon with regard to the magnitude differences between the components and with regard to spiral structure.

6.3. SPACE DENSITIES

A number of determinations of the space densities of the gM-, C- and S-type stars in the solar neighborhood as well as in different directions and distances from the Sun have been carried out so far.

Neckel (1958) computed the space densities of the M2–M4 and M5–M10 stars (Case system) using the value $-0^m.4$ for the absolute visual magnitudes of all these stars. He obtained a value $D(r)=6.2$ stars per 10^6 pc^3 for the M2–M4 stars within 430 pc from the Sun. For the M5–M10 stars Neckel found a practically constant space density equal to about 0.5 stars per 10^6 pc^3 in the galactic plane out to about 1000 pc. If, instead of the value $M_V = -0^m.4$, one uses the absolute visual magnitudes of Table IV, these values become $D(r)=2.7$ stars per 10^6 pc^3 for the M2–M4 stars and about 0.3 stars per 10^6 pc^3 for the M5–M10 stars (Blanco, 1965).

Blanco (1965), using the values of the absolute visual magnitudes and their dispersion given by him (section 4), found the following space densities for the gM stars classified according to the Mt. Wilson system within 100 pc from the Sun (see Table XII).

TABLE XII

Space densities (stars per 10^6 pc^3) found by Blanco (1965) for the gM stars within 100 pc from the Sun

Type	M0	M1	M2	M3	M4	M5
$D(r)$	4.1	1.6	1.4	1.1	0.5	0.2

McCuskey and Mehlhorn (1963) made a space density analysis using the data found in the survey No. 6 of Table X and the values $M_I = -4^m.3 \pm 0^m.7$, $M_I = -5^m.3 \pm 0^m.7$ and $M_I = -4^m.5 \pm 0^m.7$ for the absolute infrared magnitudes per unit volume of space of the M5–M6.5, M7–M10 and M5–M10 stars (Case system) respectively. In this way they came to the conclusion that the late M stars (M5–M10) are concentrated for the most part in a layer extending to about 1000 pc from the galactic plane. In this layer the density averages 0.2 stars per 10^6 pc^3. Superposed on this layer and beginning at approximately 6 kpc from the Sun toward the galactic center there appears to be an increasing population due to a concentration of the late M stars in the galactic nucleus. Here the density may be 5 to 6 times that in the basic stratum.

A comparison of the space densities of the M5–M6.5 and M7–M10 stars separately showed on the other hand that the M7–M10 stars show a greater concentration toward the galactic plane than the M5–M6.5 stars. The M5–M6.5 stars predominate by factors of about 10 or 15 at larger values of z.

Westerlund (1965) has carried out an analysis of the data obtained in the surveys Nos. 1, 2, 3, 4 and 7 of Table X using the following values for the absolute infrared magnitudes of the gM- and C-type stars per unit volume of space: $M_I = -3\overset{m}{.}0 \pm 0\overset{m}{.}6$ for the M2–M4 stars, $M_I = -4\overset{m}{.}3 \pm 0\overset{m}{.}75$ for the M5–M6.5 stars, $M_I = -5\overset{m}{.}3 \pm 0\overset{m}{.}75$ for the M7–M10 stars and $M_I = -5\overset{m}{.}0 \pm 0\overset{m}{.}5$ for the C stars. The spectral types used are in the Case system. The space densities found for the M5–M6.5 and M7–M10 stars are given in Table XIII.

TABLE XIII

Space densities (stars per 10^6 pc³) found by Westerlund (1965) for the M5 — M6.5 and the M7 — M10 stars in some selected regions of the Milky Way

Region r (kpc)	2–4	5	6	7	8	10
			M5 — M6.5			
SA 193	0.36	0.33	0.27	0.20	0.13	0.09
Crux (T₅) [a]	0.30	0.13	0.10	0.08	0.06	–
Crux (S)	0.47	0.23	0.19	0.14	0.11	–
Cygnus (3, 4) [b]	0.34	0.14	0.12	0.10	0.09	–
Aquila	0.32	0.31	0.28	0.27	0.26	–
			M7 — M10			
SA 193	0.008	0.030	0.036	0.038	0.038	0.029
Crux (T₅) [a]	0.013	0.025	0.026	0.024	0.022	0.017
Crux (S)	0.051	0.054	0.044	0.036	0.030	0.023
Cygnus (3, 4) [b]	0.071	0.034	0.025	0.022	0.020	0.017
Aquila	–	0.012	0.016	0.019	0.020	0.019

[a] $l = 302°$, $b = -0°.7$; [b] $l = 64°.5$, $b = +3°$.

Westerlund (1965) has also computed the space densities of the M2–M4 and the C-type stars in SA 193 and found the following values: $D(r) = 4.6$ stars per 10^6 pc³ for the M2–M4 stars at $r = 2$–4 kpc and $D(r) = 0.015, 0.12, 0.007, 0.005$ and 0.004 stars per 10^6 pc³ for the carbon stars at $r = 4, 6, 8, 10$ and 12 kpc respectively. Finally, from a rediscussion of Albers data for the region A in Scutum, Westerlund found for this area the value $D(r) = 7$ stars per 10^6 pc³ for the M2–M4 stars and $r = 1.2$–4 kpc.

From the above discussion Westerlund reached following conclusions:

(a) The M2–M4 stars appear to form a thin layer the half-density points being 250 pc from the galactic plane. They show a tendency to concentrate in spiral arms: the density in the Carina-Cygnus arm is about 5 stars per 10^6 pc³ and in the Sagittarius arm about 7 stars per 10^6 pc³.

(b) The space densities of the M5–M6.5 stars increase toward the galactic center.

(c) The M7–M10 stars appear to form a fairly homogeneous thick layer with negligible variations in density in the surveyed fields.

Hidajat and Blanco (1968) have carried out a space density analysis of the data obtained during the survey No. 8 of Table X using the absolute magnitudes given in Table III properly corrected in order to give mean absolute magnitudes per unit volume of space. Their results are given in Table XIV. The spectral types used are in the Mt. Wilson system.

TABLE XIV

Space densities (stars per 10^6 pc^3) found by Hidajat and Blanco (1968) for the giant M stars in the Groningen–Palomar variable star fields

Distance	Field 1	Field 2		Field 4	
(kpc)	M2 – M4	M2 – M4	M5 – M10	M2 – M4	M5 – M10
1.0	(3.7)	(5.8)	–	(1.3)	–
1.5	1.9	6.1	–	3.1	0.9
2.0	1.3	4.4	0.5	3.0	0.6
2.5	0.9	3.6	0.4	2.7	0.4
3.0	0.7	2.7	0.4	2.3	0.3
4.0	0.4	1.7	0.3	1.5	0.2
5.0	0.3	1.0	0.2	0.9	0.2
6.0	0.2	0.7	0.2	0.5	0.1
8.0	–	(0.3)	(0.1)	–	–

On the basis of these data as well as of the data found by McCuskey and Mehlhorn (1963) Hidajat and Blanco reached following conclusions:

(a) The M2–M4 stars show a plane-parallel distribution extending approximately to a distance of 5 kpc from the Sun. The space density D (stars per 10^6 pc^3) of these stars can be represented as a function of the distance z from the galactic plane with the help of the following relations:

$$D(z) = 5.0 \exp(-1.7z^2), \quad \text{for} \quad |z| < 1 \text{ kpc}$$

and

$$D(z) = 2.6 \exp(-0.91|z|), \quad \text{for} \quad |z| > 1 \text{ kpc}.$$

(b) The M5–M10 stars are symmetrically distributed about the galactic plane. The results obtained at $r=8$ kpc by McCuskey and Mehlhorn $(l=3°.9)$ suggest that near the galactic nucleus, the space density D of the M5–M10 stars can be represented by the relation:

$$D(R, z) = [202/(R^2 + 102)] \exp(-5.1z^2), \quad \text{for} \quad |z| < 0.5 \text{ kpc},$$

where R is the distance to the galactic center $(R_0 = 10 \text{ kpc})$. On the other hand the results obtained for the M5–M10 stars in Field No. 4 $(l=8°.20)$ are satisfied by the relation

$$D(z) = 1.0 \exp(-3.8z^2), \quad \text{for} \quad |z| < 1.0 \text{ kpc}.$$

McCuskey (1969) carried out a space density analysis of the data obtained during the survey No. 9 of Table X, using the values $M_V = 0^m.0 \pm 0^m.7$, $M_V = -0^m.5 \pm 0^m.7$ and $M_V = -1^m.0 \pm 0^m.7$, for the absolute visual magnitudes per unit volume of space of the M0–M1, M2–M4 and M5–M8 stars (Mt. Wilson system) respectively. The results are given in Table XV.

TABLE XV

Space densities (stars per 10^6 pc³) found by McCuskey (1969) for the giant M stars toward the galactic anticenter

Distance (kpc)	M0 – M1	M2 – M4	M5 – M10
0.5	9.5	7.1	–
0.8	9.3	4.5	2.6
1.0	9.7	4.1	2.3
1.2	10.1	3.9	2.1
1.4	9.3	3.6	1.7
1.6	8.2	3.2	1.4
1.8	7.6	3.0	1.2
2.0	6.2	2.6	0.9
2.5	5.0	2.3	0.65
3.0	4.3	2.3	0.51
4.0	3.4	2.4	0.34
5.0	2.8	2.7	0 26
6.0	2.5	3.3	0.20

On the basis of these data McCuskey reached following conclusions:

(a) The M0–M1 stars decrease in number substantially with distance from the Sun within the first 3 kpc toward the galactic anticenter.

(b) The space density of the M2–M4 stars decreases rapidly from the solar vicinity to about 2 kpc toward the galactic anticenter and thereafter remains substantially constant at about 2.5 stars per 10^6 pc³.

(c) The space density of the M5–M10 stars decreases rapidly in the range $1 < r < 4$ kpc from the Sun toward the galactic anticenter. It should be pointed out, however, that the high density of these stars in $1 < r < 2$ kpc is quite uncertain because of the scarcity of such objects, and hence the likelihood of large statistical fluctuations in the data. At distances greater than 4 kpc the space density of the M5–M10 stars appears to be about the same as in the galactic regions along the local spiral arm (Table XIII).

McCuskey (1969) carried out also a space density analysis of the data obtained by Wehinger (1965) in an area of 2.25 square degrees centered upon the radio source Sagittarius A at $l = 0°$, using the value $M_V = -1^m.0$ for the absolute visual magnitudes of the stars of spectral classes M2–M10. He found the values $D(r) = 5.3$ and 3.0 stars per 10^6 pc³ for the space density of the M2–M4 stars in the distance intervals $r = 2$–3 kpc and $r = 3$–4 kpc respectively. Also the space density of the M5–M10 stars was found equal to $D(r) = 1.1$ stars per 10^6 pc³ over the interval $2 < r < 5$ kpc.

7. Conclusions

From the above discussion following conclusions could be drawn:

(a) The normal M2–M10 giants are not an homogeneous class of stars but rather of mixture of stars belonging to three subgroups of stars with different space distribution. These subgroups are currently identified with the subgroups of the M2–M4, M5–M6 and M7–M10 stars. The M2–M4 stars are distributed somewhat irregularly along the galactic equator with concentrations suggestive of membership in the local spiral arm population. The M5–M6 stars are uniformly distributed between spiral arm and interarm regions but also show an appreciable increase in numbers toward the galactic center. The M7–M10 stars are more or less uniformly distributed in a thin disk centered on the galactic plane. These stars do not seem to be as concentrated toward the galactic center as do the M5–M6 stars. They are, however, more concentrated toward the galactic plane than the M5–M6 stars.

(b) The carbon stars are a mixture of stars belonging to two subgroups i.e. the R- and N-type stars. The N stars exhibit some tendency toward concentration in spiral arm regions. The same is not valid for the R stars.

(c) The S stars are very rare in space. These stars also seem to be a mixture of two subgroups of stars with different space distribution. Only the one of these subgroups seems to be associated with spiral structure.

Also following remarks concerning future work in the field could be added:

(a) The deep infrared surveys are still a very promising method for the study of the space distribution of the gM-, C- and S-type stars. These surveys should, therefore, be continued in strategically selected regions of the Milky Way. Simultaneously, recent efforts for the development of new techniques for the discovery of red stars (infrared photometry etc.) should be continued.

(b) In order to make the infrared surveys more effective, efforts should be made to separate the carbon stars found in such a survey into the two groups of the R- and N-type stars as well as the S stars into the two subgroups of the S stars mentioned in section 4. Also the long period variables should be segregated from the rest of the gM-, C- and S-type stars.

(c) More information is urgently needed concerning the absolute magnitudes and the intrinsic colors of the stars under consideration and especially of the different subgroups of C- and S-type stars. The infrared techniques developed during recent years will be of great importance in this respect, too.

(d) The parallel use of the Mt. Wilson (or the MK) and the Case system for the spectral classification of the M stars can cause considerable confusion in the discussion of the results of the infrared surveys. For example, the group of the M2–M4 stars in the Mt. Wilson system corresponds to the stars M0.8–M3.1 in the Case system. Therefore, efforts should be made in order to publish the results of all the infrared surveys in a uniform system.

(e) The values of the space densities found depend very critically on the values of the absolute magnitudes per unit volume of space and their dispersions used in the space density analysis. Therefore, before making any intercomparison of the space densities found in different surveys one should reduce all these densities in a common system of absolute magnitudes and dispersions.

References

Adams, W. S., Joy, A. H., and Humason, M. L.: 1926, *Astrophys. J.* **64**, 225.
Albers, H.: 1962, *Astron. J.* **67**, 24.
Albers, H.: 1967, *Publ. Astron. Soc. Pacific* **79**, 259.
Barbier, M.: 1965, *J. Observateurs, Marseille* **48**, 149.
Barbier, M. and Maiocchi, R.: 1966, *J. Observateurs, Marseille* **49**, 290.
Becker, W. and Fenkart, R.: 1970, in *The Spiral Structure of Our Galaxy*, IAU Symposium No. 38 (ed. by W. Becker and G. Contopoulos), D. Reidel, Dordrecht-Holland, p. 205.
Bertiau, F. C., McCarthy, M. F., and Treanor, P. J.: 1964, *Ric. Astron. Specola Astron. Vatic.* **6**, 571.
Blanco, V. M.: 1958, *Astrophys. J.* **127**, 191.
Blanco, V. M.: 1963, *Astrophys. J.* **137**, 513.
Blanco, V. M.: 1964, *Astron. J.* **69**, 730.
Blanco, V. M.: 1965, in *Galactic Structure* (ed. by A. Blaauw and M. Schmidt), Univ. of Chicago Press, Chicago, p. 241.
Blanco, V. M. and Münch, L.: 1955, *Bol. Observ. Tonantznintla Tacubaya* No. 12, 17.
Blanco, V. M. and Nassau, J. J.: 1957, *Astrophys. J.* **125**, 408.
Clayton, M. L. and Feast, M. W.: 1969, *Monthly Notices Roy Astron. Soc.* **146**, 411.
Dahn, C. C.: 1964, *Publ. Astron. Soc. Pacific* **76**, 403.
Dolidze, M. V.: 1964, *Bull. Astrophys. Observ. Abastumani* **30**, 71, 81.
Dolidze, M. V.: 1965, *Bull. Astrophys. Observ. Abastumani* **32**, 53.
Dolidze, M. V.: 1968a, *Astron. Cirk.*, No. 457, 2.
Dolidze, M. V.: 1968b, *Astron. Cirk.*, No. 464, 5.
Gordon, C. P.: 1967, Dissertation, The University of Michigan.
Gordon, C. P.: 1968, *Publ. Astron. Soc. Pacific* **80**, 597.
Hidajat, B. and Blanco, V. M.: 1968, *Astron. J.* **73**, 712.
Johnson, H. L.: 1964, *Bull. Observ. Tonantzintla Tacubaya* **3**, 305.
Joy, A. H.: 1942, *Astrophys. J.* **96**, 344.
Keenan, P. C.: 1954, *Astrophys. J.* **120**, 484.
Keenan, P. C.: 1963, in *Basic Astronomical Data* (ed. by K. Aa Strand), Univ. of Chicago Press, Chicago, p. 78.
Keenan, P. C. and Morgan, W. W.: 1941, *Astrophys. J.* **94**, 501.
Keenan, P. C. and Teske, R. G.: 1956, *Astrophys. J.* **124**, 499.
Kron, G. E., Gascoigne, S. C. B., and White, H. S.: 1957, *Astron. J.* **62**, 205.
Lodén, L. O. and Sundman, A.: 1966, *Stockholm Observ. Ann.* **22**, 215.
McCarthy, M. F.: 1968, *Publ. Astron. Soc. Pacific* **80**, 100.
McConnell, D. J.: 1967, *Publ. Astron. Soc. Pacific* **79**, 266.
McCuskey, S. W.: 1969, *Astron. J.* **74**, 807.
McCuskey, S. W.: 1970, in *The Spiral Structure of Our Galaxy*, IAU Symposium No. 38 (ed. by W. Becker and G. Contopoulos), D. Reidel, Dordrecht-Holland, p. 189.
McCuskey, S. W. and Mehlhorn, R.: 1963, *Astron. J.* **68**, 319.
Mavridis, L. N.: 1967, in *Colloquium on Late-Type Stars* (ed. by M. Hack), Trieste, Oss. Astron. Trieste, p. 420.
Mendoza, E. F. and Johnson H. L.: 1965, *Astrophys. J.* **141**, 161.
Merrill, P. W.: 1952, *Astrophys. J.* **116**, 21.
Nassau, J. J. and Blanco, V. M.: 1954a, *Astrophys. J.* **120**, 118.
Nassau, J. J. and Blanco, V. M.: 1954b, *Astrophys. J.* **120**, 129.
Nassau, J. J. and Blanco, V. M.: 1954c, *Astrophys. J.* **120**, 464.
Nassau, J. J. and Blanco, V. M.: 1957, *Astrophys. J.* **125**, 195.
Nassau, J. J. and Blanco, V. M.: 1958, *Astrophys. J.* **128**, 46.
Nassau, J. J. and Stephenson, C. B.: 1960, *Astrophys. J.* **132**, 130.

Nassau, J. J. and Velghe, A. G.: 1964, *Astrophys. J.* **139**, 190.
Nassau, J. J., Blanco, V. M., and Morgan, W. W.: 1954, *Astrophys. J.* **120**, 478.
Nassau, J. J., Blanco, V. M., and Cameron, D. M.: 1956, *Astrophys. J.* **124**, 522.
Neckel, H.: 1958, *Astrophys. J.* **128**, 510.
Osvalds, V. and Risley, A. M.: 1961, *Publ. Leander McCormick Observ.* **11**, Part 21.
Plaut, L.: 1965, in *Galactic Structure* (ed. by A. Blaauw and M. Schmidt), Univ. of Chicago Press, Chicago, p. 267.
Price, S. D.: 1968, *Astron. J.* **73**, 431.
Sanduleak, N.: 1957, *Astron. J.* **62**, 150.
Slettebak, A., Keenan, P. C., and Brundage, R. K.: 1969, *Astron. J.* **74**, 373.
Smak, J.: 1968, *Astron. Acta* **18**, 317.
Smith, E. v. P. and Smith, H. J.: 1956, *Astron. J.* **61**, 273.
Stephenson, C. B.: 1965, *Astrophys. J.* **142**, 712.
Stephenson, C. B. and Terrill, C. L.: 1967, *Astrophys. J.* **147**, 148.
Takayanagi, W.: 1960, *Publ. Astron. Soc. Japan* **12**, 314.
Upgren, A. R. and Grossenbacher, R.: 1968, *Publ. Astron. Soc. Pacific* **80**, 342.
Vandervort, G. L.: 1958, *Astron. J.* **63**, 477.
Wehinger, P. A.: 1965, Ph.D. Dissertation, Case Institute of Technology (unpublished).
Westerlund, B. E.: 1959a, *Astrophys. J. Suppl. Ser.* **4**, 73.
Westerlund, B. E.: 1959b, *Astrophys. J.* **130**, 178.
Westerlund, B. E.: 1960a, *Ark. Astron.* **2**, 429.
Westerlund, B. E.: 1960b, *Ark. Astron.* **2**, 451.
Westerlund, B. E.: 1964, in *The Galaxy and the Magellanic Clouds*, IAU–URSI Symposium No. 20 (ed. by F. J. Kerr and A. W. Rodgers), Australian Academy of Science, Canberra, Australia, p. 160.
Westerlund, B. E.: 1965, *Monthly Notices Roy. Astron. Soc.* **130**, 45.
Westerlund, B. E.: 1970, *Astron. Astrophys.*, in press.
Wilson, R. E.: 1942, *Astrophys. J.* **96**, 371.
Wyckoff, S.: 1968, *Astron. J.* **73**, S41.

SURVEYS OF RADIO EMISSION FROM THE GALAXY

F. J. KERR

Astronomy Program, University of Maryland, College Park, Md., U.S.A.

In this lecture, I shall review the various types of radio-frequency radiation from the Galaxy, and their main observed characteristics. Then, in the second presentation, I shall interpret these astronomically under five main headings.

As this is a large field to cover, I wish to stress the overall perspectives rather than give a very detailed account. I shall cover the similarities and differences between the various types of radiation, and also sketch the astrophysical applications and possibilities. I will, however, concentrate attention on a few particular aspects.

1. Continuum

Galactic noise is readily observable over a wide wavelength range with any reasonably good receiver and antenna. The observed spectrum extends from a few millimeters to about 30 m (10^{11}–10^7 Hz), with the cutoff at each end produced by the terrestrial atmosphere; most of the observational work has been carried out in the range 3 cm–3 m.

At long wavelengths, there is a detectable background from the Galaxy over the whole sky, with many superposed 'sources', some of which are of small angular diameter, while others are extended. A recent example of a long-wavelength map is that of Landecker and Wielebinski (1969), which is mainly derived from observations at Parkes and Cambridge near 2 m, with an angular resolution of about 2° (Figure 1).

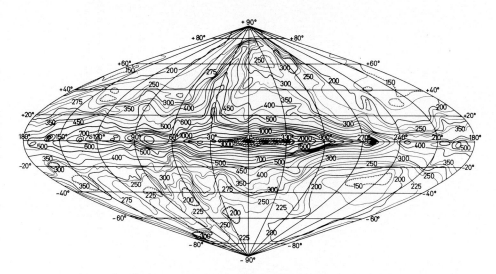

Fig. 1.   150-MHz map of the sky in galactic coordinates, from observations at Parkes and Cambridge (Landecker and Wielebinski, 1969).

L. N. Mavridis (ed.), Structure and Evolution of the Galaxy, 135–144. *All Rights Reserved*
Copyright © 1971 by D. Reidel Publishing Company, Dordrecht - Holland

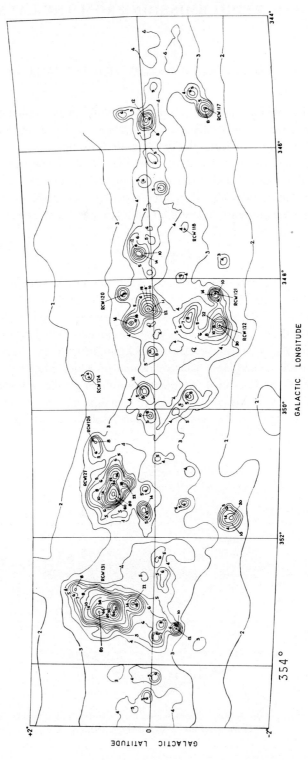

Fig. 2.   11-cm map of portion of the Southern Milky Way (Beard *et al.* 1969). The unit is brightness temperature in Kelvin. Visible HII regions are marked with RCW catalog numbers.

The radiation extends to the galactic poles, but there is a clear concentration towards the galactic plane, as well as a concentration towards the center, with the main emission extending over the longitude range 275° to 100°. We can see 'steps' in the longitude distribution near the plane, which correspond to end-on spiral arms, and also there are a number of prominent 'spurs', starting off from the galactic plane feature in nearly perpendicular directions.

By contrast, at wavelengths below 50 cm the background is detectable only in a narrow Milky Way strip about 4° wide in latitude (Figure 2). For example, recent 11-cm surveys at Green Bank and Parkes (Altenhoff, 1968; Beard *et al.*, 1969) show many sources and an apparent background in the thin disk layer, but outside this strip we only find discrete sources, most of which are extragalactic.

Because the continuum distribution over the sky is different at different wavelengths, or in other words, the spectral index varies from place to place, we can say that there must be more than one mechanism of continuum emission operating with different spectral characteristics. We recognize two important mechanisms, free-free emission and the synchrotron process. The former occurs in ionized hydrogen, and is detectable from discrete H$_{II}$ regions (some of which are visible, while more distant ones are optically obscured), and from an apparent distributed background concentrated to a thin layer. The synchrotron process is seen in the disk, the spurs, the radio halo (if one exists), and in supernova remnants, of which 40 or so are known as individual objects.

For both types of mechanism, we tend to speak in terms of a background and sources, but the relative roles of these are not clear. We do not really know whether there is an actual distributed background, or whether it is merely composed of unresolved sources. Present probabilities point to the existence of some non-thermal background, but this seems much less certain in the thermal case.

1.1. POLARIZATION

Radio polarization can be studied in three ways, the first being the weak linear polarization of much of the galactic background radiation, which was first detected at Dwingeloo at 75 cm (Westerhout *et al.*, 1962). There are three regions of comparatively high polarization, near $l = 140°$, $b = +8°$, near $l = 150°$, $b = -50°$, and the region of the north polar spur. Mathewson and Milne (1965) showed that most of the areas with observable polarization lie in a band about 60° wide, containing a great circle through the galactic poles, intersecting the plane at longitudes 340° and 160°, a direction which is presumably perpendicular to that of the main local field.

The polarization looks more ordered as we go to shorter wavelengths, such as 21 cm (Bingham, 1966), while in the other direction it drops away completely at wavelengths above a meter or so. These effects are due to Faraday rotation and depolarization of the background polarized emission, which is the second way in which we can study the galactic polarization.

Thirdly, the Faraday rotation of the radiation from extragalactic radio sources provides evidence on the mean longitudinal component of the magnetic field along the line of sight through the nearby galactic disk (Gardner and Whiteoak, 1966). The effect is greatest near the equator, and there is also a longitude effect. As the 'rotation measure' varies fairly smoothly over the sky, the effects must occur mainly in the Galaxy, rather than being due to intrinsic properties of the sources.

In all three cases, polarization studies can only be used to explore the local region.

2. 21-cm HI Line

This line, which is produced by hyperfine transitions in neutral hydrogen atoms, is detectable from galactic hydrogen all over the sky, with a major concentration along the Milky Way, but no increase of intensity towards the direction of the center. The profiles show considerable fine structure, some of which is related to young stars or optically observable interstellar features.

The peak brightness temperatures in the observed profiles are 100–150 K in the Milky Way, dropping to 20–30 K at higher latitudes. The line has a very small natural width at 1420.406 MHz, and the broadening of galactic profiles is essentially Doppler in origin, in part due to thermal motions in the gas clouds, but mainly due to galactic rotation and mass motions. The main peaks of the profiles are closely related to spiral structure.

Several extensive surveys have been carried out at low latitudes in recent years (Westerhout, 1969; Kerr, 1969; Weaver, 1970), and an enormous amount of data is now available for study. No physically anomalous effects have been observed, and

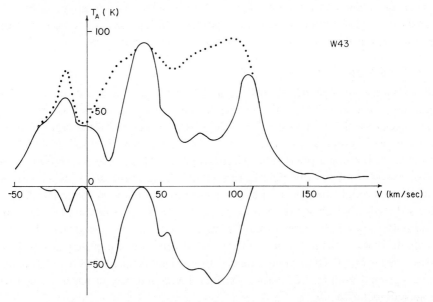

Fig. 3.   21-cm absorption for the source W43 (Kerr and Knapp, 1970). The three curves, reading from the top, are (i) the 'expected' emission profile, (ii) the observed profile, (iii) the derived true absorption profile.

no variability, and therefore the 21-cm line is very suitable as a tracer for studies of galactic structure and the physics of HI regions.

Absorption effects are also seen in the same frequency range in directions of discrete continuum sources (Figure 3). These observations can be used to estimate distances to the sources, which at low latitudes are mostly HII regions, and also to investigate the excitation temperature of the hydrogen, which cannot be determined from emission measures alone. Absorption observations have also been used for detection of Zeeman splitting of the line; these measurements are difficult, as the separation of the two components with opposite senses of circular polarization is only 2.8 Hz/μG.

3. OH Lines

This group of lines is produced by the Λ-doubling process, an interaction between the spin of the unpaired electron and the molecular rotation, with additional effects produced by hyperfine splitting. The frequencies of the lines are 1612, 1665, 1667 and 1720 MHz (near 18 cm), with intensity ratios of 1:5:9:1 under LTE conditions. As these are electric dipole transitions, the transition probability of the main lines is 10^4 times that for the 21-cm HI line, which is a magnetic dipole transition. Two recent reviews of the observations have been given by Robinson and McGee (1967) and Barrett (1967).

3.1. ABSORPTION

Absorption phenomena will be treated first, because they are simpler, and also they came first historically. The OH lines can be seen in absorption in front of many continuum sources (perhaps 40–50), most of which are HII regions, but also for Taurus A, Cassiopeia A, and two extragalactic sources (Goss, 1968). The line-intensity ratios depart somewhat from the LTE values, but not grossly so.

The absorption must be primarily produced by OH which is distributed along the line of sight because the observed, velocities are generally close to those found for HI, and also similar results are obtained for sources 1°–2° apart in the sky. The observations can, therefore, be used for studying the general OH which is apparently widely distributed throughout HI regions, and for obtaining additional information about galactic structure.

3.2. NONTHERMAL EMISSION

Non-thermal OH emission has been detected from 20–30 HII regions. Very complex phenomena have been observed, and it has been said that "no single aspect of the OH emission observations could have been predicted, or is yet understood" (Palmer and Zuckerman, 1967; McGee et al., 1967; Weaver et al., 1968).

Most of the line components are very narrow ($\sim$1–5 kHz), and the line-intensity ratios are highly anomalous. The sources are very small in angular size ($<0\rlap{.}''02$ in W3, corresponding to a physical size <35 AU), and they are located in positions in HII regions which appear to have no special significance optically. Some of the sources show marked time variations, over periods of months, and many components show strong circular or linear polarization, with complex relationships between the different narrow components and the four lines for a single source.

There is general agreement that a maser phenomenon is responsible for the non-thermal emission, but the detailed mechanism is not yet understood.

In addition to the complex emission from H II regions, non-thermal emission has also been observed in a number of infrared stars (Wilson and Barrett, 1968). For example, NML Cygni is a strong radiator, producing an antenna temperature of 200 K in the 1612-MHz line. In addition, the Swedish group (Ellder *et al.*, 1969) has found several new sources in a survey of an arbitrarily chosen grid of points in Cygnus.

3.3. THERMAL EMISSION

In the early period of OH observations, several unsuccessful attempts were made to detect thermal emission, leading to the conclusion that the LTE excitation temperature of the general OH must be low. Heiles (1968) has recently detected low-level thermal emission from nearby dust clouds, with brightness temperatures 0.18–0.83 K, half-widths of 3–11 km/sec, and central velocities of -4 to $+6$ km/sec.

4. Recombination Lines

Many of the high-level transitions in excited hydrogen atoms have been detected at radio wavelengths from H II regions. These transitions occur when an electron cascades down through a series of levels during the recombination of a hydrogen atom. For example, a change of the principal quantum number from 110 to 109, which is known as the H109α transition, occurs at 5008.9 MHz, near 6 cm.

The recombination lines normally show a simple profile shape (Figure 4), and, in the case of nearby H II regions, the velocities and other characteristics agree with those determined optically. However, the recombination-line observations enable H II regions to be seen all over the Galaxy, and their kinematics investigated. The observed line temperatures are found to be a little different from those expected on an LTE basis, but the departures from equilibrium conditions appear to be small. Line broadening has been shown to be entirely Doppler in origin, and no Stark-broadening effects have been observed, as had originally been expected from the large size of the atomic radius in these highly-excited states (e.g., 0.5μ for $n=100$). This result has been attributed to the fact that the perturbations produced by nearby atoms will be essentially the same in the adjacent levels involved in these transitions.

Helium recombination lines have also been observed from H II regions, at strengths that lead to reasonable He/H ratios of about 0.10 (Palmer *et al.*, 1969a), and also a possible carbon line. In addition, transitions in which the quantum number changes by 2, 3, or 4 (β, γ or δ transitions) have been observed.

The recombination lines are valuable for studying the astrophysical conditions in H II regions, and they also provide a new approach to the structure and kinematics of the Galaxy. The subject has been reviewed by Mezger and Palmer (1968).

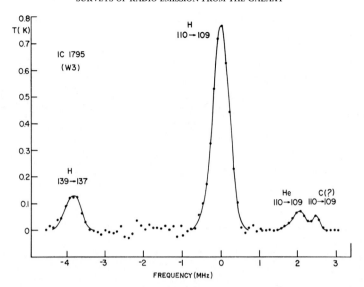

Fig. 4.   Portion of the recombination-line spectrum for W3 = IC1795 (Zuckerman and Palmer, private communication).

5. NH₃ Emission

The inversion transition of the $J=1$, $K=1$, rotation level in the vibrational ground state at 23 694.5 MHz (1.25 cm) was first detected astronomically in the galactic center direction by Cheung *et al.* (1968), following an earlier suggestion by Townes (1957) in a classical discussion of radio line possibilities. The line was seen in emission with an antenna temperature 0.6 K, and a line width of 40 km/sec, centered on +35 km/sec, which is inside the velocity range of the strongest OH absorption component in this direction. Some higher level transitions were also seen.

Subsequently, the NH_3 line has been observed in a larger region surrounding Sgr B2, with an irregular distribution (Cheung *et al.*, 1969b) and it has also been seen in several other sources.

6. H₂O Emission

Shortly after the NH_3 detection, the same group detected the $6_{16} \rightarrow 5_{23}$ rotational transition of water, which occurs at 22 235.22 MHz or 1.35 cm (Cheung *et al.*, 1969a). This transition, which occurs through a chance near-coincidence of two levels, has been detected from about 10 sources, all of which are close to OH sources, though not all OH sources are also H_2O sources. High antenna temperatures have been observed (up to 1700 K in W49), and the profiles are very complex, with some components varying enormously within weeks (Knowles *et al.*, 1969). The rapid time variations obviously imply very small sizes for the sources, but no VLB measurements have been reported so far. Some of the features are slightly polarized. It is clear that maser action must be involved, as with the OH-emission sources. A sample profile is shown in Figure 5.

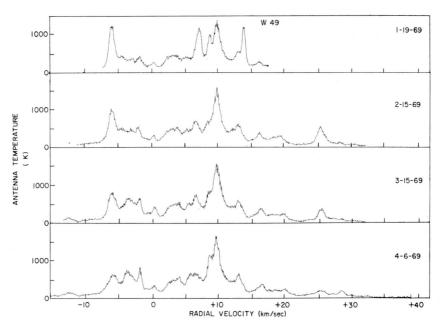

Fig. 5. Set of profiles for W49 in the H₂O line at 1.35 cm, illustrating the time variations
(Knowles *et al.*, 1969).

7. CH₂O Absorption

The $1_{11} - 1_{10}$ ground-state rotational transition from formaldehyde at 4829.659 MHz
(6.2 cm) was first detected in absorption in the direction of Sgr A by Snyder *et al.*
(1969). This was the first detection of an organic polyatomic molecule in interstellar
material.

The line has now been seen against numerous thermal and non-thermal sources,
often corresponding in velocity with OH-absorption features, indicating that a large
part of the Galaxy is filled with CH_2O clouds, with a concentration to the spiral arms.
The widespread distribution indicates that interstellar processes are more complicated
than has usually been assumed.

The line has also been unexpectedly observed in absorption in several dust clouds,
with absorption brightness temperatures of 0.2–0.8 K and velocities near zero (Palmer
et al., 1969b). As there are no significant discrete sources or distributed galactic
continuum in the background of these dust clouds, the phenomenon must be due
to absorption of the general 2.8 K microwave background.

8. Other Lines

Two isotope lines have been found, namely the O^{18} isotope of OH (Rogers and
Barrett, 1966; Robinson, 1967), and the C^{13} isotope of CH_2O (Zuckerman *et al.*,
1969), and also the Λ-doubling of a metastable state of OH $(^2\pi_{\frac{1}{2}}, J = \frac{1}{2})$ at 4765 MHz

(Zuckerman *et al.*, 1968). These lines were all observed at low intensity, and all in absorption towards Sgr A.

Many other lines have been looked for but not yet seen. These include transitions in deuterium, CH and other hydrides, and other simple molecules. In some cases the laboratory frequencies are not well known, in which case a negative result does not give a firm upper limit on the abundance of the relevant atomic or molecular species.

9. Pulsars

Pulsars can be considered as continuum emitters from one point of view, as their radiation covers a very broad spectral range, but also as line emitters, because the very precise pulse repetition rate corresponds to a sharp low-frequency spectral line. Their detailed characteristics will not be considered here (a recent review has been published by Maran and Cameron, 1969). Studies of pulsars are often concerned with the intrinsic properties of the emitting objects, but they can also be used for studying the mean electron density in the intervening medium through measurements of the pulse dispersion effects. Distances to pulsars can be estimated from 21-cm absorption observations (Gordon *et al.*, 1969), or by making some assumption about the mean electron density. As more and more pulsars are detected, they should contribute information on galactic structure, additional to that obtainable in other ways.

References

Altenhoff, W.: 1968, *Interstellar Ionized Hydrogen* (ed. by Y. Terzian), Benjamin, N.Y., p. 519.
Barrett, A. H.: 1967, *Science* **157**, 881.
Beard, M., Day, G. A., and Thomas, B. M.: 1969, *Australian J. Phys., Astrophys. Suppl.* **11**, 19.
Bingham, R. G.: 1966, *Monthly Notices Roy. Astron. Soc.* **134**, 327.
Cheung, A. C., Rank, D. M., Townes, C. H., Thornton, D. D., and Welch, W. J.: 1968, *Phys. Rev. Letters* **21**, 1701.
Cheung, A. C., Rank, D. M., Townes, C. H., Thornton, D. D., and Welch, W. J.: 1969a, *Nature* **221**, 626.
Cheung, A. C., Rank, D. M., Townes, C. H., and Welch, W. J.: 1969b, *Nature* **221**, 917.
Ellder, J., Rönnäng, B., and Winnberg, A.: 1969, *Nature* **222**, 67.
Gardner, F. F. and Whiteoak, J. B.: 1966, *Ann. Rev. Astron. Astrophys.* **4**, 245.
Gordon, C. P., Gordon, K. J., and Shalloway, A. M.: 1969, *Nature* **222**, 129.
Goss, W. M.: 1968, *Astrophys. J. Suppl. Ser.* **15**, 131.
Heiles, C. E.: 1968, *Astrophys. J.* **151**, 919.
Kerr, F. J.: 1969, *Australian J. Phys. Astrophys. Suppl.* No. 9, 1.
Kerr, F. J. and Knapp, G. R.: 1970, *Australian J. Phys. Astrophys. Suppl.*, in press.
Knowles, S. H., Mayer, C. H., Cheung, A. C., Rank, D. M., and Townes, C. H.: 1969, *Science* **163**, 1055.
Landecker, T. L. and Wielebinski, R.: 1969, in press.
Maran, S. P. and Cameron, A. G. W.: 1969, *Proc. Fourth Texas Symposium on Relativistic Astrophysics, Dallas, Texas, December 16–20, 1968.*
Mathewson, D. S. and Milne, D. K.: 1965, *Australian J. Phys.* **18**, 635.
Mezger, P. G. and Palmer, P.: 1968, *Science* **160**, 29.
McGee, R. X., Gardner, F. F., and Robinson, B. J.: 1967, *Australian J. Phys.* **20**, 407.
Palmer, P. and Zuckerman, B.: 1967, *Astrophys. J.* **148**, 727.
Palmer, P., Zuckerman, B., Penfield, H., Lilley, A. E., and Mezger, P. G.: 1969a, *Astrophys. J.* **156**, 887.

Palmer, P., Zuckerman, B., Buhl, D., and Snyder, L. E.: 1969b: *Astrophys. J.* **156**, L147.

Robinson, B. J. and McGee, R. X.: 1967, *Ann. Rev. Astron. Astrophys.* **5**, 183.

Robinson, B. J.: 1967, in *Radio Astronomy and the Galactic System*, IAU Symposium No. 31 (ed. by H. van Woerden), Academic Press, London, p. 49.

Rogers, A. E. E. and Barrett, A. H.: 1966, *Astron. J.* **71**, 868.

Snyder, L. E., Buhl, D., Zuckerman, B., and Palmer, P.: 1969, *Phys. Rev. Letters* **22**, 679.

Townes, C. H.: 1957, in *Radio Astronomy*, IAU Symposium No. 4 (ed. by H. C. van de Hulst), Cambridge University Press, Cambridge, p. 92.

Weaver, H., Dieter, N. H., and Williams, D. R. W.: 1968, *Astrophys. J. Suppl. Ser.* **16**, 219.

Weaver, H. F.: 1970, in *The Spiral Structure of Our Galaxy*, IAU Symposium No. 38 (ed. by W. Becker and G. Contopoulos), D. Reidel, Dordrecht-Holland, p. 126.

Westerhout, G., Seeger, C. L., Brouw, W. N., and Tinbergen, J.: 1962, *Bull. Astron. Inst. Neth.* **16**, 187.

Westerhout, G.: 1969, *Maryland-Green Bank Galactic 21-cm Line Survey*, 2nd ed.

Wilson, W. J. and Barrett, A. H.: 1968, *Science* **161**, 778.

Zuckerman, B., Palmer, P., Penfield, H., and Lilley, A. E.: 1968, *Astrophys. J.* **153**, L69.

Zuckerman, B., Palmer, P., Snyder, L. E., and Buhl, D.: 1969, *Astrophys. J.* **157**, L167.

SPACE DISTRIBUTION AND STATE OF MOTION OF THE INTERSTELLAR MATTER AS REVEALED FROM RADIO-ASTRONOMICAL OBSERVATIONS

F. J. KERR

Astronomy Program, University of Maryland, College Park, Md., U.S.A.

1. Large-Scale Structure

The large-scale *structure* of the Galaxy can be studied through the non-thermal and thermal continuum, the 21-cm H I line, the hydrogen recombination lines, and the OH and CH_2O absorption, while in addition any line observations can contribute to knowledge of the large-scale *kinematics*. The 21-cm line has still provided the most information, with the other components entering the picture mainly through confirmatory or comparative data.

Both continuum components are concentrated to the disk, and to the spiral arms, as shown by the 'steps' in the continuum distribution along the equator. An important difference is that the non-thermally-emitting layer has a greater width than the thermal (H II) layer or the H I layer.

Early in the history of galactic continuum studies, Baldwin and Mills introduced the concept of a large spheroidal halo, in addition to the disk component. More recent observations and discussions have thrown considerable doubt on the existence of such a radio halo, partly because it is difficult to separate a smoothly-distributed halo component from the spurs and other irregularities. If a halo exists, it must be more elliptical and lower in emissivity than was previously thought.

Our main picture of the overall spiral structure comes from 21-cm observations, but the interpretation has two main difficulties, arising from the restriction to kinematic distances, which are based on a limited understanding of the large-scale velocity field, and from the existence of a large amount of fine structure superposed on the large-scale pattern. Two detailed reviews of the subject have recently been published (Kerr, 1969; Kerr, 1970), and only a brief outline will be given here.

Conversion of the observed Doppler velocities to the distances of the hydrogen concentrations is generally considered to depend mainly on differential galactic rotation. However, the rotation curves derived from the terminal velocities of 21-cm profiles are different on the two sides of the galactic center, and in addition there are many signs of departures from circular motion. The overall 'bumpiness' of the rotation curves can be related to the density-wave theory of spiral structure, thereby indicating the directions in which major arms are seen end-on. The tangential directions can also be derived from the longitude distribution of the integrated hydrogen, and from that of the low-latitude continuum sources, which are mainly H II regions. The pitch angle of the main spirals can be inferred fairly directly from examination of the detailed

L. N. Mavridis (ed.), Structure and Evolution of the Galaxy, 145–151. All Rights Reserved
Copyright © 1971 by D. Reidel Publishing Company, Dordrecht - Holland

data, and appears to be about 8° for some sections of the Sagittarius and Norma-Scutum arms.

One interpretation of the spiral structure is given in Figure 1 which has been derived on a circular-orbit assumption. A large number of features can be identified inside the solar position. These lead to a complex pattern in the velocity-longitude plane, but it is difficult to draw a detailed map in the inner region, and only the major arms are represented in the diagram.

An alternative interpretation of 21-cm data has been given by Weaver (1970), mainly on the basis of a new survey at Hat Creek, California. He derives a much more

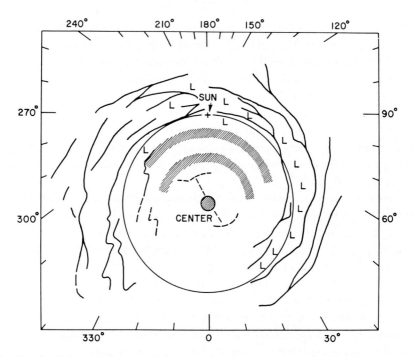

Fig. 1.   Sketch of the main features of the neutral hydrogen spiral structure, from observations of Kerr, Hindman and Henderson (Kerr, 1969). Structural details are not shown in the inner region, owing to the large uncertainty in the distance. Regions of low hydrogen density are indicated by L.

open spiral, with an average pitch angle of about 15°, and a greater separation between the major arms. He has worked primarily from velocity-longitude plots, and appears to have taken less account of the latitude widths and intensities of the various features. The main difference between the two pictures in the solar vicinity relates to the Carina arm: whether it passes through the Sun and joins to a major feature in Cygnus, or whether it can be joined to the Sagittarius arm, with the Cygnus feature disappearing into a small loop. Detailed comparative discussions of the two pictures can be expected in the literature shortly.

Comparisons between recombination-line velocities and H I data indicate that there is no substantial difference between the kinematics of H I and H II regions (Kerr *et al.*, 1968). The H II observations cannot lead to a spiral structure map on their own, primarily because the velocity dispersion of the individual H II regions tends to smear out the pattern. An important difference in the radial distribution is, however, observed: the 'giant' H II regions which are responsible for the recombination line and OH emission are almost wholly confined to a ring between 4 and 7 kpc from the galactic center, whereas the maximum density of H I is found outside this ring.

The hydrogen spiral arms are observed to be patchy on a scale of 500–1000 pc, and also they show a considerable amount of splitting and cross-linkage. An individual spiral arm is often centered a few tens of parsecs above or below the galactic plane, maintaining approximately the same displacement over a substantial distance. Many sections of spiral arms show a rolling motion, which might be related to a helical magnetic field of the type which has been postulated in the solar vicinity.

The central region is an especially interesting part of the large-scale structure problem. Here we find (i) a strong continuum source, Sgr A, which is partly non-thermal and partly thermal, (ii) several H II regions, observable in the continuum and in the recombination lines, and (iii) a widespread distribution of neutral hydrogen, which is largely moving outwards from the center. These features have been extensively reviewed (Burke, 1965; Downes and Maxwell, 1966; Kerr, 1967). I wish to discuss particularly the molecular cloud centered on $+40$ km/sec, in which OH and CH_2O can be seen in absorption, and NH_3 in emission. Occultation observations (Kerr and Sandqvist, 1968) give a diameter of 10 pc in OH, and a somewhat smaller value in CH_2O, and they also indicate that the object is rotating. The most surprising result is that no fine structure is found in very-high-resolution observations, in spite of the great breadth of the line (60 km/sec). We appear to be seeing a rather discrete object, whose existence in the central region raises interesting theoretical problems. Of course we do not know the precise distance of this cloud, but the close coincidence of its position with that of Sgr A suggests that it may well be quite close to the center.

2. Local Structure

To a first approximation, the local material can be regarded as consisting of a thin disk, about 120 pc between half-density points in H II, 200 in H I, and 450 in the synchrotron-emitting sources. There is, of course, considerable fine structure in all the components, and in the H I motions.

The prominent 'spurs' in the non-thermal continuum distribution are probably related to the local magnetic field, with an effective distance of about 100 pc (Bingham, 1967), although some authors have suggested a supernova type of origin. The local magnetic field strength and distribution can be studied through polarization observations to distances of a few hundred parsecs. Both the background polarization and the Faraday rotation work indicate a connection with the local spiral arm, while more precise studies show that the field is complicated, perhaps following a helical form.

3. H I Regions

3.1. Temperature and cloudiness

H I emission studies give a harmonic mean value of about 150 K for the spin temperature of the neutral hydrogen. For simplicity, it has commonly been assumed that a
constant value can be used throughout the Galaxy. As originally suggested by Clark
(1965), absorption observations clearly show that there are clouds of low temperature
(c. 50 K) and high density, moving around in a hotter and smoother medium at about
1000 K. Recent theoretical studies, which consider the ionization by low-energy
cosmic rays (Field *et al.*, 1969; Hjellming *et al.*, 1969), have also led to a two-compo-
nent model. On a larger scale, the two components can be considered to be inter-
mingled; the figure of 150 K can, therefore, be used for the mean temperature of the
neutral hydrogen in nearby spiral arms, but the value could well be different in interarm regions, or closer to the galactic center.

3.2. Dust

There appears to be a general correlation between H I and dust, but there is almost
an anticorrelation for the small dense clouds of dust, which often exhibit hydrogen
self-absorption, but produce very little 21-cm emission. Varsavsky (1968) and Heiles
(1969) have shown that most of the hydrogen must be molecular in these dense clouds,
while in addition the temperature must be low.

As described earlier, OH and CH_2O have both been found directly in several dust
clouds, which makes possible for the first time the measurement of velocities of dust
clouds, and thus a study of the kinematics of this subsystem of the Galaxy. In each
case, low excitation temperatures are indicated: 3–5 K for the OH, and < 1.8 K for
the CH_2O. The latter implies that an 'antimasering' phenomenon is involved, as the
excitation temperature would be expected to be at least 2.8 K, the radiation temperature
produced by the general microwave background.

The widespread OH and CH_2O absorption shows that these molecules are widely
distributed through the H I regions of the spiral arms. The small-scale association with
dust suggests that the general absorbing molecules may also be closely related to dust,
but we do not know how close the detailed association is. The OH absorption observations sometimes indicate the occurrence of masering, as the distributed OH sometimes goes into emission in the 1720-MHz line, while still appearing in absorption in
the other three lines. The presence of 1720 MHz emission implies a relatively high
density in the distributed OH. It is also interesting that the 1720-MHz line is expected
to be the first to go into emission as the optical depth increases, according to the
ultraviolet pumping theory of Litvak *et al.* (1966).

The interrelationship between hydrogen atoms, dust, and the hydrogenic molecules
is one of the most interesting new problems in radio astronomy today.

3.3. Magnetic field

Polarization and other evidence indicates that the general field is low, a few microgauss or less. Higher fields have been found by Verschuur (1969b) in a few clouds in 21-cm Zeeman effect observations, with an extreme value of $20\,\mu G$ in a cloud in the direction of the Crab nebula. As these are absorption observations, the high fields must refer to dense clouds, in which the field lines may be compressed. There is, therefore, no conflict with the evidence for a low general field.

3.4. INTERMEDIATE- AND HIGH-VELOCITY CLOUDS

Many studies have been made of the intermediate-velocity $(70 > |V| > 30$ km/sec) and high-velocity $(|V| > 70$ km/sec) clouds of neutral hydrogen, which have many interesting properties, including a high degree of concentration to a particular section of the sky. As their distance is not yet known, although many suggestions have been made about their origin and location (Oort, 1967; Verschuur, 1969a; Kerr and Sullivan, 1969), their relationship to galactic structure and evolution is still quite unclear. The IVC's are probably quite close, as they are often seen to be joined to the lower-velocity galactic disk hydrogen on the contour maps, but the distance of the HVC's is much less certain.

4. HII Regions

HII regions can be studied through the thermal continuum, the recombination lines, OH, and H_2O. The first two processes are fairly well understood, the last two are not. Observations of the recombination lines can give information on internal motions of the HII regions, and studies using each of the four types of emission can contribute to knowledge on the astrophysical processes taking place.

4.1. TEMPERATURE

Continuum observations give in principle a direct measure of the electron temperature. Mills and Shaver (1968) for example used a sufficiently high resolution to pick out the small opaque area in the center of the Orion nebula, and obtained a value of 7600 K for the brightness temperature, and thus the electron temperature. Recombination-line results are less direct in their interpretation, as there are believed to be slight departures from LTE. Temperatures obtained by both methods are almost always lower than the traditional optical value of 10000 K. The interrelationship between various methods of temperature measurement is a subject of considerable discussion, including the question of whether or not the various measures refer to the same positions in space.

4.2. DUST

Radio observations see the HII-region emission, unhampered by any effects of associated dust. By comparing detailed optical and radio isophotes for an HII region, it is possible to separate out the optical extinction, and thus to map out the dust distribution. This method has been applied for example by Ishida and Akabane (1968). It is clear that there is much more dust associated with HII regions than had been thought in earlier years, and this dust probably has an important role in star-formation processes.

4.3. COSMIC MASERS, OH AND H_2O

The strong OH- and H_2O-emission sources must be very small in physical size, as indicated by the VLB measurements for the OH and the rapid time variations for the H_2O. The radio-emitting H_2O in particular must be in high-density regions, because the transition takes place between highly excited levels with short lifetimes of the order of seconds. These considerations have led to many speculations that the source regions may be protostars, and the observed relationship of some of the sources with 'compact regions' of star formation supports this type of suggestion.

In considering the circumstances of the masers, we first ask what radiation is being amplified: it could be continuum radiation at the appropriate wavelength from discrete sources (such as the H II regions themselves), from the galactic background, or from the 2.8 K general microwave background. Various suggestions have been made for the pump source, which supplies the energy to invert the populations in the relevant levels. The protostar possibility suggests an infrared pump, although it is also possible that a hot internal object may be pumping a circumstellar dust cloud; in addition, an ultraviolet pump may well be important in other situations, such as that found in the general OH absorption.

Several quite distinct possibilities exist for the form of the maser. In a coherent process, the physical size of the sources may not be as small as their apparent size. Also, instead of individual small objects, there could well be a large volume of OH (or H_2O), in which apparently bright spots are produced by various possible combinations of rather accidental circumstances. It is possible also that the radiation is highly beamed, and we are only seeing the regions in which the beams are fortuitously in our direction.

5. Stars

Until recently, the only known radio-frequency radiation from stars, apart from the Sun, was the weakly-detectable and very sporadic bursts from a few flare stars. Radio emission is now also believed to have been detected from three other types of stars: neutron stars, infrared stars, and at least one X-ray source. The pulsar emission from neutron stars is the best known of these, with very precisely measured characteristics, but the other two are also increasing in importance. Molecular-line radiation can be expected to be detectable from many infrared sources, and many correlations are likely to be found between X-ray and radio continuum sources, as sensitivity increases.

Conclusion

As well as the information provided by radio astronomy on galactic structure and kinematics, many recent developments are contributing strongly to various branches of astrophysics. These studies are still young, and much more can be expected in the very near future.

Acknowledgment

The author's work has been supported by the U.S. National Science Foundation.

References

Bingham, R. G.: 1967, *Monthly Notices Roy. Astron. Soc.* **137**, 157.
Burke, B. F.: 1965, *Ann. Rev. Astron. Astrophys.* **3**, 275.
Clark, B. G.: 1965, *Astrophys. J.* **142**, 1398.
Downes, D. and Maxwell, A.: 1966, *Astrophys. J.* **146**, 653.
Field, G. B., Goldsmith, D. W., and Habing, H. J.: 1969, *Astrophys. J.* **155**, L149.
Heiles, C.: 1969, *Astrophys. J.* **156**, 493.
Hjellming, R. M., Gordon, C. P., and Gordon, K. J.: 1969, *Astron. Astrophys.* **2**, 202.
Ishida, K., and Akabane, K.: 1968, *Nature* **217**, 433.
Kerr, F. J.: 1967, in *Radio Astronomy and the Galactic System*, IAU Symposium No. 31 (ed. by H. van Woerden), Academic Press, London, p. 239.
Kerr, F. J.: 1969, *Ann. Rev. Astron. Astrophys.* **7**, 39.
Kerr, F. J.: 1970, in *The Spiral Structure of Our Galaxy*, IAU Symposium No. 38 (ed. by W. Becker and G. Contopoulos), D. Reidel, Dordrecht-Holland, p. 95.
Kerr, F. J. and Knapp, G. R.: 1970, *Australian J. Phys. Astrophys. Suppl.*, in press.
Kerr, F. J. and Sandqvist, A.: 1968, *Astrophys. Lett.* **2**, 195.
Kerr, F. J. and Sullivan III, W. T.: 1969, *Astrophys. J.* **158**, 115.
Kerr, F. J., Burke, B. F., Reifenstein, E. C., Wilson, T. L., and Mezger, P. G.: 1968, *Nature* **220**, 1210.
Litvak, M. M., McWhorter, A. L., Meeks, M. L., and Zeiger, H. J.: 1966, *Phys. Rev. Lett.* **17**, 821.
Mills, B. Y. and Shaver, P. A.: 1968, *Australian J. Phys.* **21**, 95.
Oort, J. H.: 1967, in *Radio Astronomy and the Galactic System*, IAU Symposium No. 31 (ed. by H. van Woerden), Academic Press, London, p. 279.
Varsavsky, C. M.: 1968, *Astrophys. J.* **153**, 627.
Verschuur, G. L.: 1969a, *Astrophys. J.* **156**, 771.
Verschuur, G. L.: 1969b, *Astrophys. J.* **156**, 861.
Weaver, H. F.: 1970, in *The Spiral Structure of Our Galaxy*, IAU Symposium No. 38 (ed. by W. Becker and G. Contopoulos), D. Reidel, Dordrecht-Holland, p. 126.

X-RAY SOURCES IN THE MILKY WAY

L. GRATTON

Università di Roma, Rome, Italy and
Dudley Observatory, Albany, N.Y., U.S.A.

1. Introduction

Non-solar X-ray Astronomy* was born in 1962 when by means of an experiment em-
ploying an 'Aerobee' rocket, Giacconi, Gursky, Paolini and Rossi (Giacconi *et al.*,
1962) discovered the first cosmic X-ray source. Since then many important results have
been obtained concerning both observations and their physical interpretation. There
are now a score of scientific groups actively engaged in experiments – mostly by means
of sounding rockets – and the number of papers on the subject is increasing very fast;
about 60 papers, among invited discourses and original contributions, have been
presented during the Rome IAU Symposium No. 37 on 'Non-Solar X- and Gamma-
Ray Astronomy' (1970), showing the wide interest of scientists in the subject.

Several excellent review papers exist; 1 will quote only Rossi's lecture at the Acca-
demia Nazionale dei Lincei (Roma) in 1968 (Rossi, 1969).

2. Instrumentation and Techniques

Since soft X-rays (1 Å $< \lambda <$ 100 Å) are absorbed by a small fraction of the atmosphere
(Figure 1), the observations must be made from above 100 km. Up to now, only sounding rockets have been used; satellite-borne experiments are in preparation and will
certainly yield very important results. For hard X-rays ($\lambda <$ 0.1 Å, energy greater than
100 keV) and for gamma-rays, balloons can be and are currently employed.

2.1. DETECTORS

The most frequently employed detectors are the *proportional counters.*

A typical counter consists of a metal box which is also the cathode; the box contains
a gas, generally a noble gas, like Ne, Ar, Kr or Xe, with a small proportion of a poli-
atomic gas to make the response more stable. The anode is a metallic wire crossing the
box.

One of the sides of the box is the window through which the X-photons are admitted – usually a very thin beryllium plate or a mylar sheet (Figure 2).

When a photon is absorbed by the counter, an atom of the gas is ionized; the photo-
electron usually generates a certain number of additional pairs of ion-electron which
in turn are accelerated by the field between cathode and anode and originate a secon-

---

* The adjective 'non-solar' will be consistently omitted, since this lecture concerns only results from
non-solar sources of X-rays.

L. N. Mavridis (ed.), Structure and Evolution of the Galaxy, 152–177. All Rights Reserved
Copyright © 1971 by D. Reidel Publishing Company, Dordrecht - Holland

dary emission. In this way each absorbed photon generates from 10^3 to 10^5 electrons which are collected by the anode and produce a current pulse; the pulse intensity, or more exactly the collected electric charge, is proportional to the number of initial electrons and to the counter gain and is, thus, a measure of the initial photon energy.

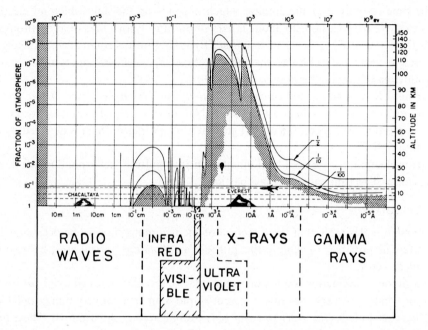

Fig. 1. Attenuation of electromagnetic radiation in the atmosphere. Solid curves indicate altitude (and corresponding pressure expressed as a fraction of 1 atm) at which a given attenuation occurs for radiation of a given wavelength.

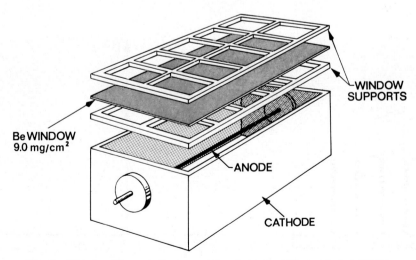

Fig. 2. Thin beryllium window proportional counter. (Giacconi *et al.*, 1968.)

The efficiency of the counter depends on the probability that a photon will pass through the window and later be absorbed by the gas. By varying the thickness and the material of the window and the pressure and chemical nature of the gas one can obtain different spectral responses of the counter. In the most favourable spectral region very high efficiencies may be obtained (up to 90%).

The time resolution of the counter which depends on the time which an electron takes to move from cathode to anode is very important. With a field of a few thousand Volts this time is of the order of one μsec. By taking into account the characteristic time of the electronics, one sees that photon fluxes up to 500000 photons sec^{-1} over the total area of the counter may be counted.

For hard X-rays and gamma-rays *scintillation counters* are also used. These consist of a scintillating crystal – the most useful are NaI and CsI crystals – and a photomultiplier. A photon arriving on the crystal generates a light pulse which is detected by the photomultiplier.

Various techniques are used to eliminate spurious pulses (background noise) due to other causes, like gamma-rays or fast particles.

2.2. COLLIMATION

Of course it is desirable that the direction of the incoming photons be determined as accurately as possible; to achieve this purpose a *collimating system* must be used before the detector.

The simplest collimator is a set of rectangular or circular tubes situated before the detector window. A very convenient system employs an array of very thin parallel and equally spaced blades (in order to cover only a small area of the window). These limit the photon beam only in the direction perpendicular to their planes and permit thus to obtain only one coordinate of a source. The acceptance angle ω_0 is given by $d/h = \mathrm{tg}\,\omega_0 = \omega_0$ (if $d \ll h$), where d is the distance between two blades and h their height (Figure 3).

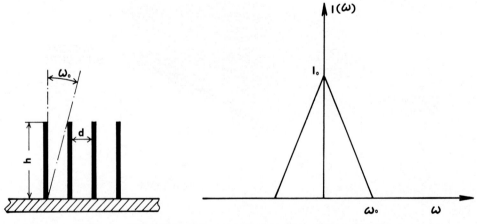

Fig. 3.   Calculation of the acceptance angle ω_0 of the collimator.

During the flight the rocket rotates uniformly around its axis; hence the collimator scans a zone of the celestial sphere. If there is an X-ray source inside the zone during the time in which its direction is within the angle of acceptance, the detector will count the photons from the source. Clearly the counting rate will be a maximum at the time at which the source is on the plane of the blades; thus the coordinate of the source is given by the time of maximum counting rate. To convert the time into an angular coordinate an optical device, called the *star sensor*, is used; the star sensor defines at each time the orientation of the collimator plane relative to some stars.

The accuracy of the coordinate determination is limited by the fact that the actual fluxes are very low and there exists a background X-radiation which cannot be eliminated.

Since only one coordinate is measured, it is convenient to employ two detectors or two banks of detectors whose collimator planes make an angle with the axis of rotation. By this means two coordinates are measured and the position of the source is determined. Note that if there are two sources the position may not be univoquely determined.

Another collimating system, called the *modulation collimator*, was invented by M. Oda. It consists of two or more parallel wire grids; the principle on which it works is shown in Figure 4. Since the response function is a multiple one, the data produced by

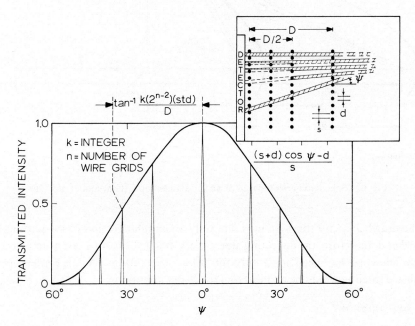

Fig. 4.   Response of a modulation collimator shown schematically in the upper right. The quantity *d* is the wire diameter, *s* is the wire to wire spacing, and *n* is the number of wire planes. In the four-grid modulation collimator used by the AS & E/MIT group to observe the Crab and Sco X-1 (40–42), $s = d = 0.005$ inches, $D = 24$ inches, which yielded a width at the base of each of the triangular transmission bands of 80 seconds of arc (Giacconi *et al.*, 1968.)

a modulation collimator are compatible with several positions of the source. This fact limits the applicability of the system in the case of a region crowded with many sources; on the other hand, the high accuracy which can be achieved with it made it possible to identify by this means some sources when nothing else was known.

Doubtless the most advanced collimating system is the *X-ray telescope*, which makes use of the property of X-rays to be reflected by a metal surface when the angle of incidence is nearly 90° (grazing incidence). Giacconi and Rossi (1960) suggested that two reflecting surfaces – paraboloid-hyperboloid – should be used (Figure 5), in order to obtain a geometrically corrected field of about 1°. The image formed by an X-ray telescope may be photographed directly or through some image converter.

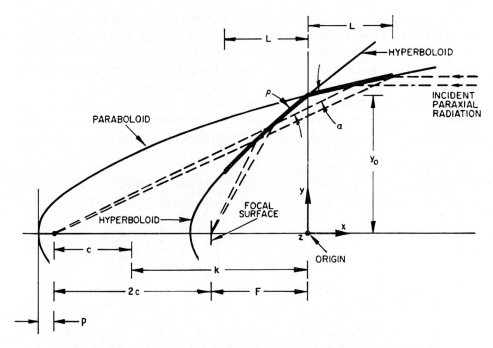

Fig. 5. Schematic cross section of an X-ray telescope. (Giacconi *et al.*, 1968.)

The drawback with this system is that only an annular section of the photon beam is used and, therefore, the collecting area is very small. For this reason up to now the X-ray telescope has been used only for solar observations. A large telescope in a stabilized satellite would be an ideal tool for X-ray astronomy.

2.3. SPECTROSCOPY

Initial information about the spectrum can be obtained by using detectors with windows of different materials and filled with different gases in order to obtain different spectral bands of maximum sensitivity. This is what is being done at present; the continuum spectrum of the sources is, thus, obtained at very low resolution. Note that since many X-ray sources vary in intensity, the observations at different wavelengths must be obtained simultaneously to be comparable.

Of course, if a large collecting area telescope were available and the observation time might be made very long, in principle, Bragg spectroscopy could be applied; this has been done only for solar work.

Gursky and Zehnpfennig (1966) suggested the use of a slitless spectrograph consisting of a grazing incidence telescope and a transmission grating. A system of this type is being tested at the AS & E group.

3. X-Ray Emission Processes

The interpretation of the observations of the cosmic X-ray sources requires first of all some knowledge of the emission processes.

In the case of the stars the spectrum is very near that of a black body, because the emitting region, the photosphere, is an optically thick gas near thermodynamic equilibrium. When the first X-ray source, the Crab nebula, was identified (it was assumed that the emission was due to the neutron star which, presumably) was formed by the outburst of the 1054 supernova which originated the nebula. If this assumption were true, the X-source would be a very dense body of small size and with a very high surface temperature of some 10^7 K.

We know now that the observed X-ray emission is due to the nebula itself; the neutron star is probably to be identified with the pulsar NP 0532 lying near the center of the nebula; the pulsar has its own X-ray emission pulsed (Bradt *et al.*, 1969; Fritz *et al.*, 1969a; Haymes *et al.*, 1969) with the same period as the optical and radio emission, and due presumably to the same process (Bertotti *et al.*, 1970). We will not discuss pulsars in this lecture.

In most of the X-sources it is likely that the emission takes place in more or less extended and relatively optically thin clouds of plasma. In this case the emission processes which come into consideration are the following (Ginzburg, 1967):

(a) thermal emission (bremsstrahlung) from a high temperature plasma ($T \simeq 10^7$ K);

(b) emission from a non-thermal plasma, in which electrons are excited to very high energies (relativistic electrons) by some generally unknown process. In this case two different radiative processes are known:

(b$_1$) synchrotron radiation, called also magnetobremsstrahlung;

(b$_2$) Compton effect.

Line emission by cosmically abundant atoms is, of course, possible. The Kα-lines and the K-edges of atoms like Na, Mg, Si, Ca, Fe, etc. fall into the observed region, but the resolving power of the spectral observations is not large enough to detect such features, if they exist (Fritz *et al.*, 1969b).

3.1. BREMSSTRAHLUNG

The process called *bremsstrahlung* – and also free-free transition – is due to a plasma electron which passes close to another charged particle (not another electron!). The theory of the process is well known; if the free electrons possess a maxwellian velocity distribution with temperature T, the power emitted per unit volume and unit frequency is

$$W(v) = 6.84 \times 10^{-38} \frac{n_e n_z Z^2}{\sqrt{T}} \exp\left(-\frac{hv}{kT}\right) \quad \text{erg cm}^{-3} (\text{c/sec})^{-1}, \quad (3.1)$$

where v is the frequency, n_e the electron and n_z the ion density, Z the ion charge; h and k are as usual Planck's and Boltzmann's constants. For a completely ionized hydrogen plasma, of course, $Z \simeq 1$, $n_e = n_z$.

If the plasma is optically thin, the total power emitted per unit frequency is obtained simply by multiplying $W(v)$ by the volume; if the plasma is optically thick, of course, one must integrate the corresponding transfer equation.

3.2. Synchrotron radiation

The theory of the *synchrotron radiation* is very well known for its application to the Crab nebula and the radiogalaxies. The case which is generally considered is that of a plasma in which the electrons have an energy spectrum given by a power law

$$dN = KE^{-\gamma} dE, \quad (3.2)$$

where dN is the number of electrons per unit volume having an energy between E and $E + dE$; K and γ are two constants. Equation (3.2) is assumed to hold in the range $E_1 \leqslant E \leqslant E_2$, where E_1 and E_2 are called the cut-off energies; outside these limits the energy spectrum drops to zero. Only the case in which $E \gg mc^2$ ($= 0.51$ MeV) is of interest (relativistic electrons).

The power radiated per unit frequency by an electron of energy $E \gg mc^2$ in a uniform magnetic field H at an angle θ with the velocity of the electron is given by

$$w(E, v) = \frac{\sqrt{3}}{4\pi} \frac{e^3}{mc^2} H \sin\theta \, F\left(\frac{v}{v_q}\right), \quad (3.3)$$

v_q being a characteristic frequency,

$$v_q = \frac{3}{4\pi} \frac{e}{mc} \left(\frac{E}{mc^2}\right)^2 H \sin\theta, \quad (3.4)$$

and F a function which can be reduced to the Bessel functions; $w(E, v)$ is equal to zero for $v = 0$, rises very steeply to a maximum at $v_{max} \simeq 0.3 v_q$ and then goes slowly to zero when $v \to \infty$. If E is given in MeV and H in Gauss and assuming $\pi/4$ as the average value of $\sin\theta$, one finds

$$v_q = 1.263 \times 10^7 E^2 H \quad \text{sec}^{-1}.$$

Table I gives some typical values of the energy for frequencies in the X-ray domain ($v_{max} \simeq 10^{18}$).

TABLE I

Energy and time of decay of relativistic electrons through
synchrotron radiation in the X-ray region.

H (Gauss)	E MeV	τ years	T K
10^{-3}	1.6×10^7	0.78	48
10^{-4}	5×10^7	25	15.2
10^{-5}	1.6×10^8	7.8×10^2	4.8
10^{-6}	5×10^8	2.5×10^4	1.52
10^{-7}	1.6×10^9	7.8×10^5	0.48
10^{-8}	5×10^9	2.5×10^7	0.15

By integrating Equation (3.3) over all frequencies one finds

$$w(E) = \frac{4}{9} \frac{e^4}{m^2 c^3} \left(\frac{E}{mc^2} \right)^2 H^2 , \tag{3.5}$$

where the average value of $\sin^2 \theta$ was assumed to be $\frac{2}{3}$. This is the total energy lost by
an electron per unit time through the synchrotron mechanism. The corresponding
time of decay – that is the time it takes the energy of the electron to reduce from E to
$E/2$ – is given by

$$\tau = \frac{9}{4} \frac{m^3 c^5}{e^4} \left(\frac{mc^2}{E} \right) \frac{1}{H^2}$$

$$= 12.5 \frac{1}{EH^2} \text{ years} . \tag{3.6}$$

The time of decay is also given in Table I (column 3); it may be seen that the life of
relativistic electrons emitting in the X-ray region is very short, unless the magnetic
field is very weak.

By combining Equations (3.2) and (3.3) it is found that the power emitted by a
plasma per unit volume and unit frequency in a chaotic magnetic field is given by

$$W(v) = K A_\gamma H^{(1+\gamma)/2} v^{(1-\gamma)/2} , \tag{3.7}$$

where A_γ is a constant whose value depends on γ.

Some typical values of the spectral index $\alpha = (\gamma - 1)/2$ and of the constant A_γ are
given in Table II (units are always in the cgs system).

Equation (3.7) holds for frequencies far from the cut-off frequencies, that are the
values of v_q for the cut-off energies.

If the plasma is optically thin the total power is simply obtained by integrating
$W(v)$ over the whole volume; when the plasma is optically thick, by integration of the
transfer equation in the homogeneous case one finds that the power of the emerging
radiation is proportional to $H^{-1/2} v^{5/2}$. More generally it is found that the plasma is
optically thin at high frequencies and may be thick at low frequencies; the general

shape of the spectrum is, then, qualitatively shown in Figure 6. Note that, if the plasma density changes, usually the variation has two effects; for instance, a density decrease will cause a general decrease of the emissivity and at the same time a shift of the maximum due to the decrease of the absorption; the latter may cause an increase of the power emitted at low frequencies.

TABLE II

Constants of the synchrotron
radiation formula

γ	α	A_γ
1.0	0.00	4.14×10^{-23}
1.5	0.25	1.08×10^{-18}
2.0	0.50	3.66×10^{-14}
2.5	0.75	1.44×10^{-9}
3.0	1.00	6.28×10^{-5}
3.5	1.25	2.94
4.0	1.50	1.46×10^{4}
4.5	1.75	7.55×10^{9}
5.0	2.00	4.09×10^{14}

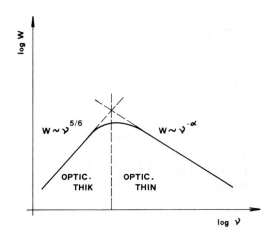

Fig. 6.   General shape of the spectrum of plasma.

Synchrotron radiation is polarized. When $\nu \gtrsim \nu_q$ then the intensity in the direction of polarization with the electric vector perpendicular to H is about 7.5 times greater than in the orthogonal direction.

3.3. COMPTON EFFECT

The scattering of low energy photons by relativistic electrons may produce X- and gamma-ray photons. This process is called the *inverse Compton effect*. This is a typical relativistic effect due to the change of the energy when passing from one frame of reference to another.

In the laboratory frame, which is the same as that of the Galaxy, the photon energy $h\nu_0$ is assumed to be very low and the electron energy very high ($E_0 \gg mc^2$). If the angle θ_0 between the motion of the electron and that of the incident photon is not too small, when transforming to the proper frame of reference of the electron – in which frame the electron energy is only the rest energy mc^2 – the energy of the photon becomes very large because its frequency is greatly increased by the Doppler effect. At the same time the angle between the motion of the electron and that of the photon becomes practically equal to π, for all not too small values of θ_0.

In other words, a relativistic electron moving in an isotropic field of soft photons, sees a well collimated beam of incoming hard photons. This is scattered in the frame of reference of the electron into a beam having a certain angular and energy distribution, according to the known laws of the Compton effect. Then going back to the laboratory frame, one finds that the photons which have changed their original direction (the majority) have a very high frequency.

In this way, in the laboratory frame of reference energy is transferred from the electron to the photons, thus obtaining a flux of hard X-ray or gamma-ray photons, from the original field of soft photons. One finds that the total power transferred to the photons by an electron of energy E moving in an isotropic radiation field whose energy density is ϱ_{rad} is given by

$$w(E) = \frac{32\pi}{9} \frac{e^4}{m^2 c^3} \left(\frac{E}{mc^2} \right)^2 \varrho_{rad}. \tag{3.8}$$

This equation may be compared with Equation (3.5); if one keeps in mind that the energy density of the magnetic field is $\varrho_{mag} = H^2/8\pi$, one sees that the energy lost by electrons by Compton effect is the same as that due to synchrotron radiation, if the energy densities of the radiation field and, respectively, of the magnetic field are the same. Also the time of decay will, of course, be the same. For instance, a black body radiation field at temperature T is equivalent to a magnetic field given by

$$H = \sqrt{8\pi a T^2} = 4.35 \times 10^{-7} T^2. \tag{3.9}$$

The corresponding temperatures are entered in the last column of Table I.

Also the spectrum of the scattered Compton radiation, in the case of electrons having an energy spectrum given by Equation (3.2), is very similar to the synchrotron spectrum. For instance, for black body radiation at temperature T it is found for the Compton spectrum

$$W(\nu) = KB_\gamma T^{(\gamma+5)/2} \nu^{(1-\gamma)/2} = KC_\gamma \varrho_{rad}^{(\gamma+5)/8} \nu^{(1-\gamma)/2}, \tag{3.10}$$

and the coefficients B_γ and C_γ are given in Table III.

TABLE III

Coefficients of the Compton radiation

γ	α	B_γ	C_γ
1.0	0.00	1.04×10^{-28}	4.06×10^{-18}
1.5	0.25	1.61×10^{-23}	4.79×10^{-12}
2.0	0.50	2.72×10^{-18}	6.18×10^{-6}
2.5	0.75	4.96×10^{-13}	8.60
3.0	1.00	9.55×10^{-8}	1.26×10^{7}
3.5	1.25	1.93×10^{-2}	1.95×10^{13}
4.0	1.50	4.04×10^{3}	3.11×10^{19}
4.5	1.75	8.78×10^{8}	5.16×10^{25}
5.0	2.00	1.96×10^{14}	8.77×10^{31}

4. Identification of X-Ray Sources with Optical Objects

Until present (July 1969) about 50 X-ray sources have been found with more or less certainty. They are named by the experimenters in different ways; for instance, the MIT and AS&E groups named the first sources which have been discovered by the constellation name, followed by X and by an ordinal number (e.g. Sco X-1, Tau X-1, Cyg X-2, ...), but later they began to name the more recently discovered sources by the letters GX followed by the degrees of galactic longitude and latitude (e.g. GX 3 + 1). The NRL group which is responsible for the discovery of a large number of sources uses the constellation name followed by XR and a number. Since this adds to the unavoidable confusion due to the low accuracy of the position of the sources, it would be desirable that the IAU assumed the task of naming the sources as it does for the variable stars.

Figure 7 shows the distribution of the sources in galactic coordinates; it is obvious by simple inspection that the majority of the sources are galactic.

The identification of a certain X-source with a possible optical object presents a very difficult problem because of the low precision of the position determined with the present techniques; many 'obvious' candidates have been found later to be wrong.

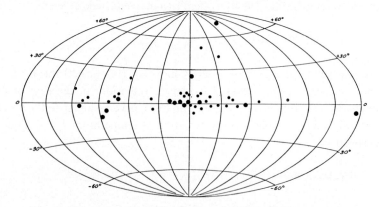

Fig. 7.   Apparent galactic distribution of the known X-ray sources.

At present only 8 sources may be considered as reasonably well identified with optical objects, namely:

Tau X-1, identified with the Crab nebula (Bowyer *et al.*, 1964); Sco X-1 (Sandage *et al.*, 1966), Cyg X-2 (Giacconi *et al.*, 1966), Cen XR-2 (Eggen *et al.*, 1968), GX 3+1 (Blanco *et al.*, 1968) identified with stellar objects having peculiar spectra and optical variations resembling those of old novae;

Vir XR-1 (Friedman and Byram, 1967; Bradt *et al.*, 1967) identified with the well known radiogalaxy M 87 (Virgo A);

Cas XR-1 (Byram *et al.*, 1966) and another source in Cassiopeia (Gorenstein, 1970) identified with the supernova remnants Cas A and SN Cas 1572 (Tycho's supernova).

Other possible identifications await confirmation.

As it is seen from these identifications the optical counterparts of X-ray sources belong to – at least – three different classes of objects, besides the pulsars; these are:

(a) supernova remnants: Tau X-1, Cas XR-1 and SN Cas 1572;

(b) peculiar stellar objects: Sco X-1, Cyg X-2, Cen XR-2, GX 3+1;

(c) radiogalaxies: M 87.

Clearly this refers to X sources sufficiently strong to be detected with present techniques.

When considering the distribution of the sources among the three classes, it was noted that the optical objects corresponding to (b) show a remarkable similarity with old novae; hence it is reasonable to compare the galactic distribution of the X sources with that of novae (Figure 8). Indeed the similarity in galactic distribution is remarkable; so much so if one compares it with the distribution of other possible candidates like high temperature stars, planetary nebulae, etc.

It is, therefore, tempting to conclude that the majority of the X-sources so far discovered belong to class (b) having as optical counterparts peculiar objects with spectra, light variations and galactic distribution similar to those of old novae. On the other side it would be wrong to identify these objects with old novae straight away, because not a single known nova has been as yet found to possess a detectable X-ray emission.

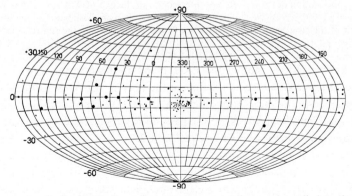

Fig. 8.   Apparent galactic distribution of novae. Large, medium and small dots denote maxima brighter than third magnitude, third to sixth magnitude, and fainter than sixth magnitude, respectively. (Gaposchkin, C. – Galactic Novae.)

5. Galactic Sources

5.1. Supernova remnants

Three galactic supernova remnants have been positively identified with X-sources; therefore there cannot be any doubt that at least some supernovae – possibly all – during a certain stage of their development are powerful X-ray emitters (Poveda and Woltjer, 1968).

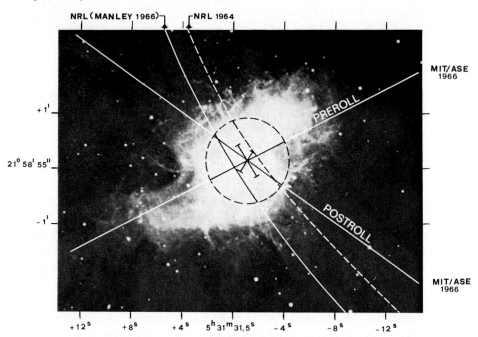

Fig. 9. Summary of results pertaining to the position of the X-ray source in the Crab Nebula. The dashed curved line is the line of position obtained from a lunar occultation experiment by Bowyer *et al.* (1964), and the solid curved line is the same result corrected by Manley and Ouellette (1965) for parallax due to the motion of the rocket. The intersection of the two MIT/AS & E lines of position is at (1950.0) $\alpha = 5^h 32^m 30^s$, $\delta = 21°59'.1$. The 100" diameter circle is an idealization of the source region. The origin of the coordinate system is the southwest component of the central double star.

The best known case is that of the Crab Nebula; the position and size of the X source coincide remarkably well with that of the plasma which emits the synchrotron radiation observed in the radio and optical spectrum (Oda *et al.*, 1967) (Figure 9). The high energy spectrum, from the X- to the gamma-ray region (Figure 10) can be represented by a power law $F(v) \sim v^{-\alpha}$ with $\alpha = 1.3$ or 1.2.

For these reasons it seems reasonable to consider also the X-ray spectrum as due to the synchrotron mechanism; on the other side (Figure 11) although a smooth curve may easily be drawn through the whole electromagnetic spectrum of the Crab Nebula – radio, optical, X and gamma – the slope is considerably different in the different spectral regions.

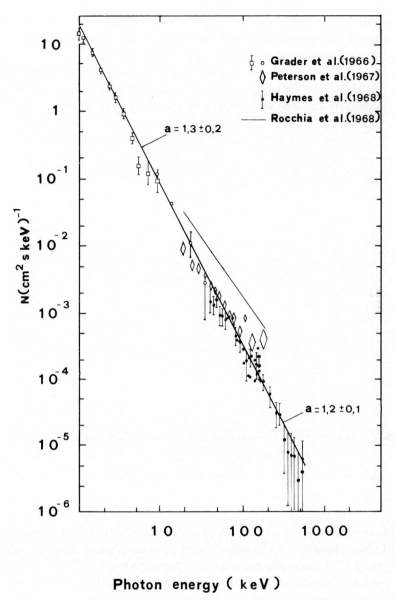

Fig. 10.   Spectrum of the Crab Nebula in the X- and gamma-ray region (N = constant $\times (h\nu)^{-(a+1)}$).

Woltjer (1958, 1970) has shown that the electrons emitting the optical synchrotron spectrum lose energy too fast and cannot, therefore, have been produced at the time of the outburst; the case is, of course, still worse for the X-ray emission. Some residual activity in the central part of the nebula can be traced from the light 'ripples' discovered by Baade and discussed recently by Scargle (1969); most probably it should in some way be associated with the pulsar.

There is also the possibility that the X-spectrum of the Crab Nebula is due to thermal bremsstrahlung (Sartori and Morrison, 1967); but in this case it is necessary to assume different temperatures in different points inside the nebula.

Very little can be said about Cas A and Tycho's supernova; the radio and X-ray spectra can be fitted by a single power law. The optical spectrum is much fainter; this might be due to strong interstellar absorption near the galactic plane.

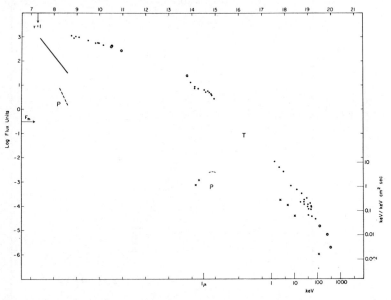

Fig. 11.   Spectrum of the Crab nebula in the radio, optical, X- and gamma-ray region.
(Woltjer, 1969.)

5.2. Sco x-1 sources; general properties of optical objects

The observations refer mainly to Sco X-1 (by far the strongest source yet discovered) and Cyg X-2; the observations of WX Cen, a variable star which is believed to be the optical counterpart of Cen XR-2, are in general agreement with those of Sco X-1. Very little is known about GX 3+1, which might be quite different from the other three objects. There is an observable radio emission from Sco X-1.

The three first objects are variable in the optical region and their variation is described as the superposition of three effects (Figures 12 and 13):

(a) rapid oscillations (flickering) of small amplitude ($\pm 0^{m}.1$) and characteristic time of some minutes (Hiltner and Mork, 1967; Sandage et al., 1969; Mark et al., 1969);

(b) a slower fluctuation with a characteristic time of some hours and an amplitude of about half stellar magnitude;

(c) short 'flares' of a few tenths of a magnitude lasting about ten minutes, which are usually seen when the magnitude due to (b) is the brightest.

The spectrum is a complex one and shows remarkable variations. In the case of Sco X-1 (Figures 14 and 15) the Balmer lines – up to H_{12} – and lines due to HeI, HeII,

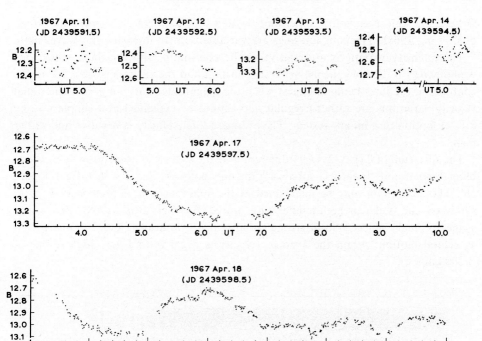

Fig. 12.   April and May, 1967, light-curves of Sco XR-1. Note that the magnitudes are given in *B* and the time in UT. (Hiltner *et al.*, 1967.)

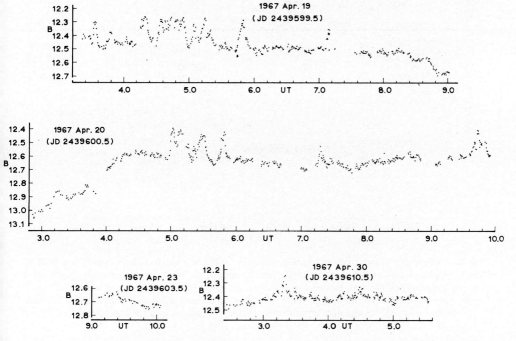

Fig. 13.   April and May, 1967, light-curves of Sco XR-1 (continued from Figure 12). Note that the magnitudes are in *B* and the time in UT. (Hiltner *et al.*, 1967.)

On etc. are seen in emission, superimposed to a high temperature continuum (West-phal *et al.*, 1968). The emission lines are variable both in intensity and wavelength.

The intensity variations are relative to the continuum and are those which one would expect if the lines themselves were constant and the variations were due to the continuum, the lines being strongest when the magnitude is at minimum. The radial velocity variations are rather irregular and cannot be explained in a simple way by orbital motion in a binary system. The average radial velocity is negative and rather high.

The spectrum of Cyg X-2 is very different and shows a set of absorption lines resembling those of an F or early G subdwarf; the only emission line is λ 4686 HeII (Figure 16). The continuum may be interpreted as the superposition of that of an F or G subdwarf and a flat or high temperature source (Peimbert *et al.*, 1968); this would suggest a binary object, but the strong radial velocity variations cannot be interpreted as orbital motion. Again the average velocity is negative and very high (order of −200 km/sec).

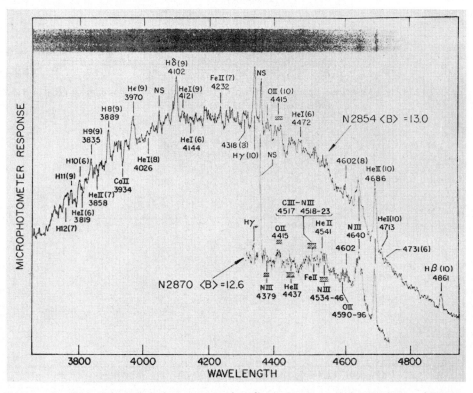

Fig. 14.   Computer-drawn microphotometer tracings for the two spectral plates N 2854 and N 2870. The microphotometer response is shown as plate transmission. No attempt at intensity rectification was made. Individual spectral features are identified. Numbers in parentheses show number of plates upon which the marked feature was detected. A photographic reproduction of plate N 2854, enlarged to the same scale as the microphotometer tracing, is shown above for comparison.
(Westphal *et al.*, 1968.)

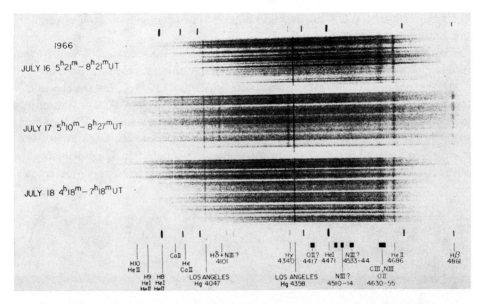

Fig. 15.   Three continuously trailed spectra in the blue spectral region. Time marks, made by closing the camera dark slide, are evident on the last two. Tentative line identifications are indicated. The comparison arc is He. The original dispersion was 85 Å/mm. (Sandage *et al.*, 1966.)

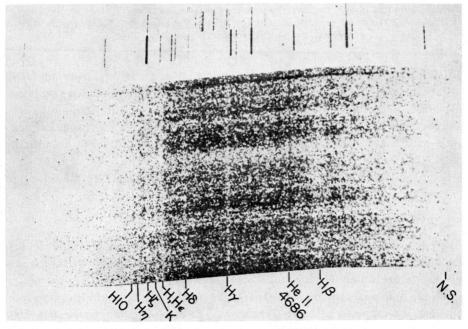

Fig. 16.   The spectrum of Cyg X-2. The upper and comparison spectra are, respectively, He + Ar + Ne and He + Hg. (Lynds, 1967.)

The spectrum of WX Cen (Eggen *et al.*, 1968) and its light variations are rather similar to those of Sco X-1 (Figure 17).

The interstellar Ca II H and K lines are well visible in the spectrum of Sco X-1 and should give a reliable estimate of the distance of the object; according to Wallerstein (1967) one obtains in this way a distance between 270 and 1000 pc. But the question of the distance is a matter of some controversy; the absorption of X-rays with energy less than 2 keV gives about 400 pc, in agreement with the estimate from the Ca II interstellar lines, but recently Sofia *et al.* (Sofia *et al.*, 1969; Gatewood and Sofia, 1968) from a new determination of the proper motion conclude that most probably the object is a member of the Scorpio-Centaurus association and considerably nearer to us.

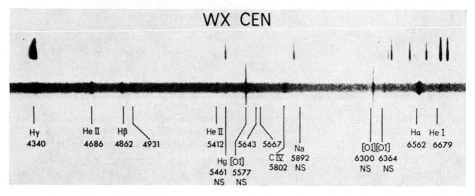

Fig 17.   Spectrum of WX Cen obtained with the Carnegie Institution's image-tube spectrograph attached to the Mount Stromlo 74-inch reflector. Original dispersion 180 Å/mm. The comparison spectrum is neon plus mercury. (Eggen *et al.*, 1968.)

The membership of Sco X-1 to a stellar association might be very important for an understanding of its nature. Unfortunately, the proper motion estimate is very uncertain and in view of the discrepancy with other estimates it would be premature to draw any conclusion. If there were any conclusion, the distribution of the X-sources in galactic coordinates does not suggest that a large proportion of them be members of stellar associations.

Sofia and Wilson (1968) have shown convincingly enough that Cyg X-2 belongs to an intermediate galactic population.

5.3. Sco x-1 sources; physical interpretation

Any physical model of sources of the type of Sco X-1 must account for the fact that the total energy emitted by these objects in the X- and gamma-ray spectral regions is by quite a few orders of magnitude larger than that corresponding to the optical spectrum. Assuming a distance of 500 pc for Sco X-1, by integrating the X-ray spectrum one finds a total power of 2×10^{37} erg sec^{-1}. The absolute photographic magnitude is $M_B = 3\overset{m}{.}6$, adopting an apparent magnitude $B = 1\overset{m}{.}30$ and an absorption of $0\overset{m}{.}9$ (Westphal *et al.*, 1968).

Hence in the hypothesis of a binary system formed by a star and an X-ray object – whichever the latter might be – even if the optical emission were due exclusively to the former, it would not be larger than 10^{33} or at most 10^{34} erg sec^{-1}; we would then be faced with the problem of the equilibrium of a star, an hemispherium of which would receive from an external source a quantity of energy considerably larger than that produced by its internal thermonuclear sources.

A similar conclusion may be drawn also in the case of Cyg X-2 and is essentially independent of the distance adopted.

On the other side the X-ray spectrum (Meekins et al., 1969) suggests as a source an optically thin plasma at a temperature of 6.5×10^7 K emitting with a bremsstrahlung mechanism (Figures 18a, b); a similar result, with a temperature of 6.0×10^7 K, is obtained also for Cyg X-2.

This has led several authors (Shklovsky, 1967; Burbidge and Prendergast, 1968) to consider a model formed by a close binary in which one of the components, the X-ray object, is a degenerated white dwarf or neutron star; matter is continuously transferred to the degenerated component from the other component which is still evolving

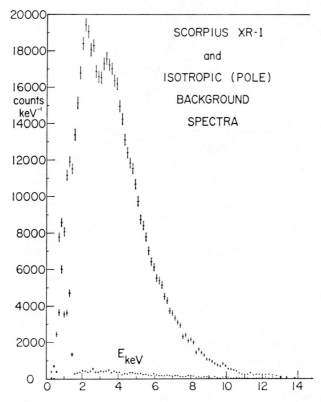

Fig. 18a.   Upper set of points shows the pulse-height distribution from Sco XR-1. Lower set shows the background (pole) distribution. Vertical lines indicate ± 1 standard deviation in counts.
(Meekins et al., 1969.)

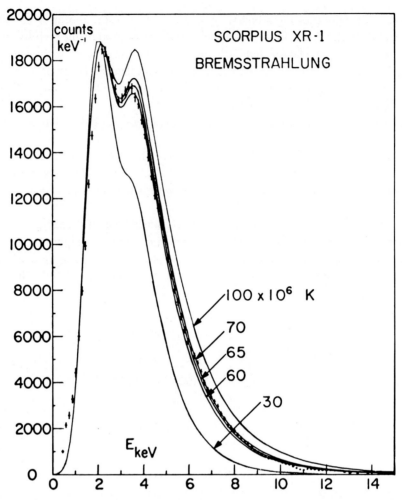

Fig. 18b. Pulse-height distribution from Sco XR-1 after removal of background and using a running mean of three channels, together with fitted bremsstrahlung source functions, of temperatures 60, 65, and 70 × 10⁶K. (Meekins *et al.*, 1969.)

and it was shown that the infall of the gas on the surface of the former may heat it to temperatures of the order of a few 10^7 K.

Leaving the actual model a bit more indeterminate, a comparison of the visual and infrared observations with the X-ray observed flux, according to Neugebauer *et al.* (1969) shows that the plasma although optically thin in the X-ray region must be optically thick to the visual and infrared radiation. On the assumptions that (a) the X-ray spectrum is due to bremsstrahlung from an hydrogen plasma at $T=5\times10^7$ K; (b) in the infrared the plasma is opaque and emits essentially like a black body at the same temperature, and (c) all the radiation – X, optical and infrared – is originated in a spherical plasma of radius R, it is possible to obtain a fairly consistent model.

From the observed X-ray flux one gets

$$n_e^2 R^3/d^2 = 10^{17} \text{ cm}^{-5}, \tag{5.1}$$

where n_e is the electron density of the plasma (assumed to be homogeneous) and d the distance of the object in cm. The near infrared flux may be approximated by the Rayleigh-Jeans formula, obtaining

$$R^2 T/d^2 = 10^{-17} \text{ K}. \tag{5.2}$$

With a distance $d = 500$ pc $= 1.5 \times 10^{21}$ cm, one finds $R \simeq 7 \times 10^8$ cm and $n_e = 2.7 \times \times 10^{16}$ cm^{-3}; the size of the object is that of a white dwarf and at that electron density the opacity to optical and infrared radiation is very large, so that the model is internally consistent.

Obviously this model is oversimplified, but it appears a rather reasonable one. It does not take into account the emission lines and their variations in intensity and wavelength.

An utterly different model has been proposed recently by Manley (Manley and Olbert, 1969). In this model an X-ray source is not a star or a binary system, but a system of plasmoids or of filamentary condensations of the interstellar matter formed by a sort of radiative instability; in turn Alfvén's hydromagnetic waves develop in these filaments causing large fluctuations of the magnetic field which accelerate the electrons up to relativistic energies. Interacting with the field the electrons emit X-ray through the synchrotron mechanism. The energy radiated is, ultimately, due to the dissipation of the interstellar magnetic fields.

For more details one is referred to the original paper by Manley and Olbert, who claim that this theory may account for all the observed properties of the X-ray sources of the type Sco X-1.

6. Extragalactic Sources

6.1. M 87

Vir XR-1 is the only X-source convincingly identified with an extragalactic object, although other objects have been suggested but not confirmed; the identification is with the radiogalaxy M 87 (Vir A).

In a sense, M 87 is somewhat exceptional among radiogalaxies, being the prototype of a rather small class of radiosources known as 'core-halo' sources, while the large majority of the radiosources connected with galaxies are double sources with the optical galaxy more or less in the middle. Typically M 87 is a spherical galaxy of very large mass, whose distinctive feature is a kind of luminous jet starting from the core and reaching a distance of about 25″; several knots are seen inside the jet and the light is polarized according to a complicated pattern (Figure 19). It may be assumed – with some reserve, however – that the radiosource coincides with the optical jet. The physical properties of the jet have been discussed recently by Felten (1968).

Assuming that the radio and X-spectrum are due to the jet and from the observed optical flux it is found that the entire electromagnetic spectrum of the jet can be re-

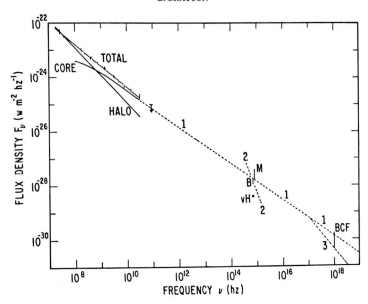

Fig. 19.   The electromagnetic spectrum of the M 87 jet. The three optical points are: B (Bless, 1962); M (Moroz, 1962); vH (van Houten, 1961). The point BCF is an X-ray measurement by Byram *et al.* (1966).

presented by the simple power law $F_\nu \sim \nu^{-2/3}$ (Adams *et al.*, 1969); but a recent negative result between 35 and 52 keV (McClintock *et al.*, 1969) suggests that the spectrum bends down in the region of hard X-rays. It is practically certain that the radio and optical emission of the jet is due to the synchrotron mechanism and it seems safe to assume that the same holds also for the X-rays; infrared and ultraviolet observations are, however, desirable.

According to Felten the relativistic electrons cannot be generated in the core and diffuse along the jet in a time longer than $L/c = 4000$ years – $L = 1200$ pc being the length of the jet – because the time of decay of the electrons computed from Equation (3.6) is too short, unless the magnetic field be unprobably weak. But if the velocity of diffusion would be of the order of c in a stationary plasma strong instabilities should arise. Hence the only possibility, if the relativistic electrons are created in the core, is that they diffuse in a plasma which is already moving with a velocity of the order of c. Another possibility is that the electrons are created in the jet itself by fast protons (cosmic rays!) through processes of the type (Burbidge, 1956)

$$p + p \rightarrow p + n + \pi^+,$$
$$p + p \rightarrow p + p + \pi^+ + \pi^-,$$

followed by the decay of pions

$$\pi^\pm \rightarrow \mu^\pm + \nu_\mu, \quad \mu^\pm \rightarrow e^\pm + 2\nu_e.$$

This secondary production of relativistic electrons from fast protons leads to a reasonable model of the jet with a plasma density of 200 cm^{-3}, an emission velocity of the matter of 10^4 km/sec and an age of the jet of 10^5 years; the total energy of the relativistic protons is 2×10^{57} ergs and the field corresponding to equipartition is 10^{-2} G. The proton energy must be at least 10^{13} eV if the synchrotron spectrum extends to the X-ray domain.

A difficulty with this model is that one would expect a strong flux of gamma photons, as it was shown by Gould and Burbidge (1967) for a similar model for the Crab nebula; these photons arriving at the Earth's atmosphere should originate observable Čerenkov light pulses due to showers produced by the photons themselves.

6.2. Diffuse X-ray Radiation

As it has been mentioned in passing, besides the individual X-sources, a diffuse background X-ray radiation has been discovered, whose characteristic is its remarkable isotropy. According to the most recent results (Gorenstein et al., 1969), the photon spectrum of this diffuse radiation follows a power law

$$\mathrm{d}N/\mathrm{d}E = 12 - 4E^{-1.7} \quad \text{photons } (\text{cm}^2 \text{ ster sec keV})^{-1},$$

for $1 \leqslant E \leqslant 40$ keV, and

$$\mathrm{d}N/\mathrm{d}E = 20E^{-2} \quad \text{photons } (\text{cm}^2 \text{ ster sec keV})^{-1}$$

for $E > 60$ keV. The corresponding energy density in space is 6×10^{-5} eV cm^{-3}, which can be compared with that corresponding to the 2.7 K black body radiation (0.3 eV cm^{-3}) and that due to the integrated light of the galaxies (10^{-2} eV cm^{-3}).

The isotropy of the radiation strongly suggests an extragalactic origin; two possibilities then exist: (a) the radiation is due to a mechanism acting in the intergalactic space, or (b) the radiation is due to the superposition of a large number of unresolved (extragalactic) sources.

The case (a) was discussed especially by Felten and Morrison (1966), who considered in a classical paper the interaction of a flux of intergalactic relativistic protons with the 2.7 K universal black body radiation; the X-ray diffuse radiation should be due to the inverse Compton effect arising from this interaction. The main difficulty with this theory is that the time of decay of the electrons is short, if compared, for instance, with the age of the universe, and, hence, one has to make some ad hoc assumption concerning the origin of the electrons.

Similar difficulties are encountered also in other theories based on case (a); therefore, a theory based on case (b) seems more promising (Setti and Rees, 1970). A very agreeable model assuming an evolutionary (big bang) cosmology was proposed by Bergamini, Londrillo and Setti (Bergamini et al., 1967).

It must be recalled that in evolutionary cosmologies the temperature of the black body radiation is inversely proportional to the radius of the universe and also, when we observe an object whose cosmological redshift is z, we see it as it was at a time at which the radius of the universe was $(1+z)^{-1}$ times the present radius. Consider, now, a very distant radiogalaxy, whose redshift z, if we could observe it, would be of the order of 3 or larger; the temperature of the black body radiation as measured at this radiogalaxy is then $1+z$ times greater than the temperature of 2.7 K measured now in our galaxy and the corresponding radiation density is $(1+z)^4$ times greater than the present radiation density. It follows (see section 3) that the Compton effect due to the relativistic electrons inside the radiogalaxy interacting with the universal black body radiation may compete favourably with the synchrotron mechanism.

In other words with increasing z or, the same identically, with increasing distance, we must expect that radiogalaxies transform gradually into strong X-ray sources and, ultimately, will radiate practically only in the X-ray spectrum through the inverse Compton effect mechanism. As a result at large distances a very large number of X-ray sources will appear and at the same time the number of radiosources will drop very quickly. It is remarkable that a sharp decrease in the counted numbers of radiosources was found by Ryle and his coworkers (Pooley and Ryle, 1968; Ryle, 1968) corresponding to the radiofluxes which should be due to sources with $z = 2$ or 3.

According to this theory the diffuse X-ray radiation is due to these distant sources; if it could be checked by direct observations, it might provide a strong argument in favour of evolutionary cosmology.

References

Adams, D. J., Cooke, B. A., Evans, K., and Pounds, K. A.: 1969, *Nature* **222**, 757.
Bergamini, R., Londrillo, P., and Setti, G.: 1967, *Nuovo Cimento* **52B**, 495.
Bertotti, B., Cavaliere, A., and Pacini, F.: 1969, in *Non-Solar X- and Gamma-Ray Astronomy*, IAU Symposium No. 37 (ed. by L. Gratton), D. Reidel, Dordrecht-Holland, p. 196.
Blanco, V., Kunkel, W., and Hiltner, W. A.: 1968, *Astrophys. J.* **152**, L137.
Bless, R. C.: 1962, *Astrophys. J.* **135**, 187.
Bowyer, S., Byram, E. T., Chubb, T. A., and Friedman, H.: 1964, *Science* **146**, 912.
Bradt, H., Mayer, W., Naranan, S., Rappaport, S., and Spada, G.: 1967, *Astrophys. J.* **150**, L199.
Bradt, H., Rappaport, S., Meyer, W., Nather, F. H., Warner, B., McFarlane, M., and Kristian, J.: 1969, *Nature* **222**, 728.
Burbidge, E. M., Lynds, C. R., and Stockton, A.: 1967, *Astrophys. J.* **150**, L95.
Burbidge, G. R.: 1956, *Astrophys. J.* **124**, 416.
Burbidge, G. R. and Prendergast, K. H.: 1968, *Astrophys. J.* **151**, L83.
Byram, E. T., Chubb, T. A., and Friedman, H.: 1966, *Science* **152**, 66.
Eggen, O. J., Freeman, K. C., and Sandage, A. R.: 1968, *Astrophys. J.* **154**, L27.
Felten, J. E.: 1968, *Astrophys. J.* **151**, 861.
Felten, J. E. and Morrison, P.: 1966, *Astrophys. J.* **146**, 686.
Friedman, H. and Byram, E. T.: 1967, *Science* **158**, 257.
Fritz, G., Henry, R. C., Meekins, J. F., Chubb, T. A., and Friedman, H.: 1969a, *Science* **164**, 709.
Fritz, G., Meekins, J. F., Henry, R. C., and Friedman, H.: 1969b, *Astrophys. J.* **156**, L33.
Gatewood, G. and Sofia, S.: 1968, *Astrophys. J.* **154**, L69.
Giacconi, R. and Rossi, B. B.: 1960, *J. Geophys. Res.* **65**, 773.
Giacconi, R., Gursky, H., Paolini, F. R., and Rossi, B. B.: 1962, *Phys. Rev. Lett.* **9**, 439.
Giacconi, R., Gorenstein, P., Gursky, H., Usher, P. D., Waters, G. R., Sandage, A. R., Osmer, P., and Peach, J. K.: 1966, *Astrophys. J.* **148**, L129.
Giacconi, R., Gursky, H., and Speybroek, van L.: 1968, *Ann. Rev. Astron. Astrophys.* **6**, 373.
Ginzburg, V. L.: 1967, in *High Energy Astrophysics, Vol. 1: Radiosources and Their Interpretation* (ed. by C. de Witt, E. Schatzman and P. Véron), Gordon and Breach, New York, p. 19.

Gorenstein, P.: 1970, in *Non-Solar X- and Gamma-Ray Astronomy*, IAU Symposium No. 37 (ed. by L. Gratton), D. Reidel, Dordrecht-Holland, p. 134.

Gorenstein, P., Kellogg, E. M., and Gursky, H.: 1969, *Astrophys. J.* **156**, 315.

Gould, R. J. and Burbidge, G. R.: 1967, in *Cosmic Rays II* (ed. by K. Sitte), Springer-Verlag, Berlin, Heidelberg, New York, p. 265.

Grader, R. J., Hill, R. W., Seward, F. D., and Toor, A.: 1966, *Science* **152**, 1499.

Gursky, H. and Zehnpfennig, T.: 1966, *Appl. Opt.* **5**, 875.

Haymes, R. C., Ellis, D. V., Fishman, G. J., Kurfess, J. D., and Tucker, W. H.: 1968, *Astrophys. J.* **151**, L9.

Haymes, R. C., Freeman, K. C., and Harnden, Jr., F. R.: 1969, *Astrophys. J.* **156**, L107.

Hiltner, W. A. and Mork, D. E.: 1967, *Astrophys. J.* **150**, L23, 851.

Houten, van, C. J.: 1963, *Bull. Astron. Inst. Neth.* **16**, 1.

Kraft, R. P. and Demoulin, M. H.: 1967, *Astrophys. J.* **150**, L183.

Lynds, C. R.: 1967, *Astrophys. J.* **149**, L41.

Manley, O. and Olbert, S.: 1969, *Astrophys. J.* **157**, 223.

Mark, H., Price, R. E., Rodriguez, R., Seward, F. D., Swift, C. D., and Hiltner, W. A.: 1969, *Astrophys. J.* **156**, L67.

McClintock, J. E., Lewin, W. H. G., Sullivan, R. J., and Clark, G. W.: 1969, *Nature* **223**, 162.

Meekins, J. F., Henry, R. C., Fritz, G., Friedman, H., and Byram, T. E.: 1969, *Astrophys. J.* **157**, 197.

Neugebauer, G., Oke, J. B., Becklin, E., and Garmire, G.: 1969, *Astrophys. J.* **155**, 1.

Oda, M., Bradt, H., Garmire, G., Spada, G., Sreekantan, B. V., Gursky, H., Giacconi, R., Gorenstein, P., and Waters, J. R.: 1967, *Astrophys. J.* **148**, L5.

Peimbert, M., Spinrad, H., Taylor, B. F., and Johnson, H. M.: 1968, *Astrophys. J.* **151**, L93.

Petersen, L. E., Jerde, R. L., Jacobson, A. S.: 1967, *Ann. Inst. Aeronaut. Astronaut. J.* **5**, 1921.

Pooley, G. G. and Ryle, M.: 1968, *Monthly Notices Roy. Astron. Soc.* **139**, 515.

Poveda, A. and Woltjer, L.: 1968, *Astron. J.* **73**, 65.

Rossi, B. B.: 1969, *Astronomia in raggi X*, Annuario della ES&T, Mondadori, Milano, p. 272.

Ryle, M.: 1968, *Ann. Rev. Astron. Astrophys.* **6**, 249.

Sandage, A. R., Osmer, P., Giacconi, R., Gorenstein, P., Gursky, H., Waters, J., Bradt, H., Garmire, G., Sreekantan, B. V., Oda, M., Osawa, K., and Jugaku, G.: 1966, *Astrophys. J.* **146**, 316.

Sandage, A. R., Westphal, J. A., and Kristian, J.: 1969, *Astrophys. J.* **156**, 927.

Sartori, L. and Morrison, P.: 1967, *Astrophys. J.* **150**, 385.

Scargle, J. D.: 1969, *Astrophys. J.* **156**, 401.

Setti, G. and Rees, M. J.: 1970, in *Non-Solar X- and Gamma-Ray Astronomy*, IAU Symposium No. 37 (ed. by L. Gratton), D. Reidel, Dordrecht-Holland, p. 352.

Shklovsky, I. S.: 1967, *Astrophys. J.* **148**, L1.

Sofia, S. and Wilson, R. E.: 1968, *Nature* **218**, 73.

Sofia, G., Eichhorn, H., and Gatewood, G.: 1969, *Astron. J.* **74**, 20.

Wallerstein, G.: 1967, *Astrophys. Lett.* **1**, 31.

Westphal, J. A., Sandage, A. R., and Kristian, J.: 1968, *Astrophys. J.* **154**, 139.

Woltjer, L.: 1958, *Bull. Astron. Inst. Neth.* **14**, 39.

Woltjer, L.: 1970, in *Non-Solar X- and Gamma-Ray Astronomy*, IAU Symposium No. 37 (ed. by L. Gratton), D. Reidel, Dordrecht-Holland, p. 208.

MOTIONS OF THE NEARBY STARS

SIR RICHARD WOOLLEY

Royal Greenwich Observatory, Herstmonceux Castle, Hailsham, Sussex, United Kingdom

1

The motions of the nearby stars can be studied in especial detail since, if the stars are sufficiently close to admit a determination of the trigonometrical parallax, all three rectangular components of the velocity relative to the Sun can be calculated from the parallax, proper motions and radial velocity. For example if the stars are within 20 pc of the Sun, as in Gliese's catalogue (1957), the parallaxes exceed 0".050 and are in good cases known within about 10%, and for stars as close as this the proper motions are in most cases reliable. Although a volume of radius 20 pc centred on the Sun is a small volume relative to the size of the Galaxy as a whole, the stars at present within this volume do not remain so throughout a galactic year. Many of them wander one or two kiloparsecs away from the Sun, so that the fraction of the Galaxy sampled in a survey of the nearby stars is much larger than the volume in which they find themselves temporarily, and is in fact from a statistical point of view a sampling of a reasonably large fraction of the outer parts of the Galaxy.

We are of course interested in the orbits of these stars, but there is a great difference between the study of stellar orbits in the Galaxy and a study of planetary orbits in the solar system. In the latter case the orbits can be seen (except in the case of the outermost planets) as a whole, and have been described many times during the history of astronomical observation: whereas the galactic orbits take hundreds of millions of years to describe, and all that has been observed is a tiny fraction of the orbit. The study of galactic orbits is necessarily a study of the statistics of short arcs.

It is therefore of some interest to enquire whether there are any *invariants* of the orbits which are connected with the short arcs that can be measured. While this cannot, perhaps, be fully decided in so far as the galactic attracting field is not completely known, some progress may be made by adopting some hypothetical or plausible models of the galactic field and determining the corresponding invariants.

The simplest case is associated with the name of Lindblad. If it is assumed that the galactic attraction is central and spherically symmetrical we can say that the potential ψ is $\psi(r)$, a function of r, the distance from the centre, alone. If we write

$$u = 1/r$$

and set

$$\psi(u) = \alpha u^2 + \tfrac{1}{3}\beta u^3 \tag{1.1}$$

as an approximation to the galactic potential, then the orbit is given exactly by

$$u = \bar{\omega}^{-1}\{1 + e\cos n(\phi - \phi_0)\}$$

where ϕ is the azimuthal angle and $\bar{\omega}$, e, and ϕ_0 are invariants of the orbit. The quantity n is given by

$$n^2 = \alpha\bar{\omega}/(\alpha\bar{\omega} + \beta).$$

The orbit can be usefully regarded as an epicycle on a point moving round the Galaxy with the circular velocity (which is only true if the square of the eccentricity e can be neglected).

However the Galaxy is highly flattened, and is very far from being spherically symmetrical, so that a higher approximation to the truth can be made by recognizing a potential term depending on θ the angular distance from the pole of the Galaxy.

If the potential takes a form investigated by Eddington (1915), Clark (1936), and Lynden-Bell (1962) namely

$$\psi = \psi(r) + G(\theta)/r^2,$$

then three independent integrals of the motion can be written down. They are

$$I_1 = \tfrac{1}{2}\{\dot{r}^2 + (r\dot{\theta})^2 + (r\sin\theta\dot{\phi})^2\} - \psi$$
$$I_2 = r^2\sin^2\theta\dot{\phi}$$
$$I_3 = \tfrac{1}{2}\{(r^2\sin\theta\dot{\phi})^2 + (r^2\dot{\theta})^2\} - G(\theta)$$

and it can be shown that the orbit is confined to a *box* whose dimensions can be specified and are in fact the invariants of the orbit (Woolley and Candy, 1968). The box is continuous round the Galaxy in the azimuthal angle ϕ but is limited in r, the distance from the centre of the Galaxy, and θ, the angular distance from the galactic pole. It is more convenient to specify the complement of θ, or $\zeta = \pi/2 - \theta$. With the special case $\psi(r) = \alpha u^2 + \tfrac{1}{3}\beta u^3$ the box limits are

$$\bar{\omega}/(1 + e) \leqslant 1/r \leqslant \bar{\omega}/(1 - e)$$

and

$$|\zeta| \leqslant i$$

where

$$2I_3 = \alpha\bar{\omega} + \beta$$
$$e^2 - 1 = 2I_1\bar{\omega}/\alpha$$

and i satisfies $I_3 = \tfrac{1}{2} I_2^2 \sec^2 i + g\ (i)$ with the convention $g\ (\zeta) = -G\ (\theta)$.

The invariants of the box orbit are $\bar{\omega}$, e and i, and they are determined as shown above from I_1, I_2, I_3, that is to say from short arcs of the stellar motions.

2

The force perpendicular to the galactic plane can be assessed by comparing the statistics of stellar velocity components perpendicular to the galactic plane and the distribution of the same classes of stars in parallel layers above and below the plane (Oort, 1932; Woolley 1957; Woolley and Stewart 1967). An elegant account of the method was given by Camm (1949). In all these papers the force perpendicular to the plane is taken to be independent of the distance from the centre of the Galaxy, so that there is only one integral of the motion

$$I = \tfrac{1}{2}Z^2 - \psi(z)$$

where Z is the velocity perpendicular to the plane. Jeans' relation for the stellar number density $v(Z, \psi(z))$ is

$$v(Z, \psi(z)) = F(Z^2 - 2\psi(z))$$

and if a Gaussian distribution of Z velocities is assumed

$$v(Z, \psi(z)) = \text{const} \times \exp\{-j^2(Z^2 - 2\psi(z))\}$$

where j is a constant. Integrating with respect to Z

$$v(z) = \text{const} \times \exp(2j^2\psi(z))$$

and writing $2j^2(\psi(z) - \psi(0)) = \psi$,

$$v(z)/v(0) = e^{-\psi}.$$

If the stars have the same mass, the density $\varrho(z)$ is given by

$$\varrho(z)/\varrho(0) = e^{-\psi}.$$

As the matter is taken to be stratified in parallel planes, Poisson's equation becomes

$$d^2\psi(z)/dz^2 = -4\pi\Gamma\varrho$$

and if $z = x/l$ where $l \equiv (8\pi\Gamma\varrho_0 j^2)^{-1/2}$

$$d^2\psi/dx^2 = e^{-\psi}$$

of which the solution, given by Camm, is

$$\psi = 2\ln\cosh(x/\sqrt{2}).$$

A suitable generalization of this where the matter is no longer stratified in parallel planes can be obtained by using Jeans' relation

$$v = F(I_1, I_2, I_3)$$

and replacing the supposition that the distribution of Z velocities is Gaussian by the more general supposition that it is an exponential function of a general quadratic function of the velocities $\dot{r}$, $r\dot{\theta}$ and $r\sin\theta\dot{\phi}$, that is to say by taking the velocity function to be

$$e^{-C\xi}$$

where C is a constant and

$$\xi = 2I_1 + 2k_1 I_2 + k_2 I_2^2 + 2k_3 I_3.$$

The constants k_1, k_2 and k_3 are characteristic of a particular distance from the galactic centre. This leads (Woolley and Candy, 1968, pp. 281–283) to an expression similar to Camm's (to order ζ^2) if we set

$$g(\zeta) = H \ln \cosh K\zeta$$

the constants H and K being determined from observation.

3

Given the form of $g(\zeta)$ it is possible to construct an *explicit* orbit for the star, that is to say to give explicit expressions for r, ζ and for ϕ, the azimuthal angle, as functions of the time. Without going into all the details here, we remark that the equation for the acceleration towards the centre in polar coordinates is

$$\ddot{r} - r\dot{\theta}^2 - r\sin\theta\dot{\phi}^2 = \partial\psi/\partial r. \tag{3.1}$$

Introduce an auxiliary angle λ defined by

$$\dot{\lambda} = u^2\sqrt{(2I_3)}. \tag{3.2}$$

Then

$$\ddot{r} = -2I_3 u^2\, d^2u/d\lambda^2.$$

Again

$$r\dot{\theta}^2 + r\sin^2\theta\dot{\phi}^2 = u^3\{(r^2\dot{\theta})^2 + (r^2\sin\theta\dot{\phi})^2\}$$
$$= u^3\{2I_3 - 2g(\zeta)\},$$

and substituting in Equation (3.1)

$$2I_3 u^2\frac{d^2u}{d\lambda^2} + u^3\{2I_3 - g(\zeta)\} = -\frac{d\psi(r)}{dr} - \frac{2g(\zeta)}{r^3},$$

so that

$$\frac{d^2u}{d\lambda^2} + u = -\frac{1}{2I^3 u^2}\frac{d\psi(r)}{dr}. \tag{3.3}$$

If $\psi(r) = \alpha u + \tfrac{1}{2}\beta u^2$ and $d\psi(r)/dr = -\alpha u^2 - \beta u^3$ the solution of (3.3) is

$$u = \bar{\omega}^{-1}\{1 + e\cos n(\lambda - \lambda_0)\}, \tag{3.4}$$

where

$$2I_3 = \alpha\bar{\omega} + \beta \quad\text{and}\quad n^2 = \alpha\bar{\omega}/(\alpha\bar{\omega} + \beta).$$

The explicit equation for λ in terms of the time t is

$$\frac{d\lambda}{\{1 + e\cos n(\lambda - \lambda_0)\}^2} = \frac{\sqrt{(\alpha\bar{\omega} + \beta)}}{\bar{\omega}^2}\,dt.$$

From (3.2) and (3.4) we have

$$U = \dot{r} = \sqrt{(\alpha/\bar{\omega})}\, e\sin n(\lambda - \lambda_0).$$

If we define a velocity Λ by

$$\Lambda = \sqrt{(\alpha/\bar{\omega})}\, e\cos n(\lambda - \lambda_0)$$

then

$$U^2 + \Lambda^2 = (\alpha/\bar{\omega})\, e^2 .$$

It can be shown that if T is the tangential velocity ($T^2 = (r\dot\theta)^2 + (r\sin\theta\dot\phi)^2$) and C the circular velocity at the star's place (i.e. at distance r and not necessarily $\bar{\omega}$ from the centre of the Galaxy) then, neglecting e^2,

$$\Lambda = \frac{2}{n}(T - C).$$

Hence if it is assumed that $n(e^2) = \text{const} \times e \exp(-k^2 e^2)$ there will be Gaussian distributions of the velocities U and $(T-C)$ (i.e. an ellipsoidal distribution of the velocities) and the ratio of the radial and tangential velocity dispersions can be used to determine the ratio β/α (and ultimately the ratio of the Oort's constants) (Woolley and Candy, 1968, pp. 280–281.)

4. The Gliese Catalogues

In 1957 W. Gliese published a 'Katalog der Sterne näher als 20 Parsek für 1950.0 (*Mitteilungen Heidelberg* Ser. A, No. 8) and he has followed this with 'a Catalogue of Nearby Stars', edition 1969 (Gliese, 1969). This catalogue specifies all the stars supposed to be within 20 pc from the Sun, and gives the parallaxes, proper motions and radial velocities where available. It combines these to give three rectangular components of the star's motion relative to the Sun, in kilometers per second, namely U directed *towards* the galactic centre, V in the galactic plane in the direction of galactic rotation, and W in the direction of the north galactic pole.

In addition a catalogue is being prepared at Greenwich containing additional stars and enumerating the box orbit parameters $\bar{\omega}$, e and i.

These are calculated as follows in practice. It is convenient to work in units such that the circular velocity (at the Sun's place) is unit velocity, and the distance from the Sun to the centre of the Galaxy is unit distance.

Then

$$\alpha + \beta = 1 .$$

Since the circular velocity C is given by

$$C^2 = \alpha u + \beta u^2$$

and the Oort's constants A and B are given by

$$2A = C/r - dC/dr, \qquad 2B = -C/r - dC/dr$$

we have, at $r = 1$,

$$-B/A = \alpha/(3\alpha + 4\beta).$$

In earlier work at the Royal Greenwich Observatory the ratio $-B/A$ was taken to be

$\frac{1}{2}$, giving $\alpha = \frac{4}{3}$, $\beta = -\frac{1}{3}$ but a careful examination of more recent data suggests

$$- B/A = 0.7877$$

giving

$$\alpha = 1.762\,45, \qquad \beta = -0.762\,45.$$

It is necessary to adopt a value for the circular velocity in kilometers per second, or $C = 250$ km/sec, and to adopt numerical values for $g(\zeta)$, in fact

$$g(\zeta) = H \ln \cosh K\zeta = 0.002\,77 \ln \cosh(56.1\,\zeta)$$

this derivation being in Woolley and Candy (1968). Further definite values of the solar velocity relative to the circular velocity must be adopted, namely

$$U_\odot = +10 \text{ km/sec}$$
$$V_\odot = +10 \text{ km/sec}$$
$$W_\odot = +7 \text{ km/sec}.$$

The Sun's velocity in fixed axes in galactic units is then

$$u = +0.040$$
$$v = +1.040$$
$$w = +0.028.$$

We have therefore

$$2I_1 = u^2 + v^2 + w^2 - (2\alpha + \beta) = -1.6785$$
$$2I_3 = v^2 + w^2 \qquad\qquad\qquad = 1.0824$$
$$\alpha\bar{\omega} + \beta = 2I_3 \qquad\qquad\qquad \bar{\omega} = 1.0467$$
$$e^2 - 1 = \frac{2}{\alpha} I_1 \bar{\omega} \qquad\qquad\qquad e = 0.0560.$$

Since the distance from the Sun to the centre of the Galaxy is about 10 kpc, the epi-centre of the Sun's galactic orbit is 467 pc outside its present position and its furthest distance from the centre of the Galaxy is another 560 pc outside that.

The box angle i is found by interpolation among tables of $g(\zeta)$ using

$$2I_3 = I_2 \sec^2 i + 2g(i)$$

or

$$v^2 + w^2 = v^2 \sec^2 i + 2g(i).$$

For the Sun, $i = 0.0091$ radians. The Sun therefore moves within 91 pc on either side of the galactic plane.

5. Statistics of the Invariants

The distribution of the parameters e and i with spectral type is shown in Tables I and II, which are compiled from the Royal Greenwich Observatory extension of the Gliese's catalogue. It is clear that both eccentricity and box angle increase systematically from

TABLE I

Distribution of eccentricity e with spectral type

Range in e	B	A	F0-F4	F5-F9	G	K	M	DA, DB	Other	Total
0 to 0.050	3	32	17	27	33	62	41			215
0.050 to 0.075	1	14	14	25	45	53	31		3	186
0.075 to 0.100		11	15	31	56	55	45			213
0.100 to 0.125		5	9	17	29	37	48	2		147
0.125 to 0.150			4	7	32	42	32	2		119
0.15 to 0.20			1	13	45	59	44	3		165
0.20 to 0.25			2	5	36	41	25	3		112
0.25 to 0.30				3	11	26	13			53
0.30 to 0.40				1	17	17	24			59
> 0.40				1	4	4	7	1		17
Total	4	62	62	130	308	396	310	11	3	1286
Median e		0.0481	0.0740	0.0876	0.1168	0.1183	0.1175			

Special classes included in above

| Range in e | MK Types II, III, and IV | | | | | | MK type VI | Me stars |
	F0-F4	F5-F9	G	K	M	All		
0 to 0.050	5	7	6	8	3	29	0	10
0.050 to 0.075	8	5	11	8	2	34	0	6
0.075 to 0.100	8	3	8	6		25	1	2
0.100 to 0.175	5	1	3	6		15	0	9
0.125 to 0.150		2	4	5		11	1	3
0.15 to 0.20		4	7	7		18	2	4
0.20 to 0.25		1	5	2		8	3	2
0.25 to 0.30			1	1		2	1	1
0.30 to 0.40			3	2		5	1	2
> 0.40							8	
Total	26	23	48	45	5	147	17	39
Median e	0.0753	0.0669	0.0953	0.1012		0.0812	0.3612	0.1045

Stars of MK types II to IV and Me stars are slightly *less* eccentric on the average than the other stars of the same spectral type.

Stars of MK type VI are markedly *more* eccentric than the other stars of the same spectral type.

early-type stars (spectral class A) to late-type stars (spectral classes G, K and M, which are not clearly differentiated one from another). It is attractive to interpret this progression as a matter of stellar age. To do this we can construct HR diagrams for stars of eccentricity 0 to 0.05 etc., which can be done accurately in those cases where V and $B-V$ are known from photoelectric photometry and the trigonometrical parallax is well determined. Three such HR diagrams are shown in Figures 1 to 3. From these diagrams it is clear that the stars of low eccentricity are *young* in terms of current theories of stellar evolution, while the stars of high eccentricity are *old*. Similar results are got if the stars are classified in terms of box angle i rather than eccentricity e.

TABLE II

Distribution in box angle

Range in i radians	No. of stars	Range in i radians	No. of stars
0 to 0.0049	258	0.0500 to 0.0599	45
0.0050 to 0.0099	268	0.0600 to 0.0699	24
0.0100 to 0.0149	179	0.0700 to 0.0799	21
0.0150 to 0.0199	137	0.0800 to 0.0899	10
0.0200 to 0.0299	154	0.0900 to 0.0999	7
0.0300 to 0.0399	85	0.1000 to 0.1499	22
0.0400 to 0.0499	52	> 0.1500	24

Distribution of box angle with spectral type

Spectral class	No. of stars	Median i	Maximum i	No. of stars with $i > 0.0500$
A, B	66	0.0052	0.0209	0
F0 to F4	62	0.0091	0.0542	1
F5 to F9	130	0.0104	0.3120	13
G	308	0.0130	0.2934	40
K	396	0.0143	0.8216	47
M	310	0.0156	0.4677	47

Special classes included in above

MK II, III, IV	147	0.0104	0.2263	5
MK VI	17	0.0569	0.8216	9
Me	39	0.0124	0.1150	5

Stars of MK types II to IV and Me stars have on the average slightly smaller box angles i than other stars of the same spectral type, but stars of MK type VI have larger box angles.

While the age is clear in the case of A-type stars and fairly clear in the case of late-type giants, the main sequence late-type stars may be either new or old, and the statistics must be regarded as the statistics of a mixture. Recently, however, O. C. Wilson has proposed a method of determining the relative ages of late-type main sequence stars by classifying the emissions in the H and K lines of calcium. The connexion between these reversals and the box parameters has been investigated by Wilson and Woolley (1970) and the box angles for those stars to which Wilson assigns calcium emission parameters of $+3$ or greater are very similar to those of the Gliese stars of spectral type A – that is to say, of young stars.

6. Deviation of the Vertex

A particular feature of the distribution of the velocities of the A-type stars, which is shared by that of the young late-type main sequence stars with Ca^+ emission $+3$ or greater, is the deviation of the vertex. It has been known, for the A stars, since the work of Strömberg (1946) and is shown in Figure 4 which shows the U and V components of A stars (filled circles) and stars classified by Wilson as $+3$ or greater (open circles). The positions in the diagram of Eggen's Sirius and Hyades groups are also shown.

This deviation can have one of the following explanations:

(a) the deviation is confined to certain stars only, specifically young stars, and is a consequence of the initial distribution of their velocities or their position at the time of their formation;

(b) the deviation is a manifestation of a peculiarity of the attraction in the neighbourhood of the Sun, which is non-central (in the galactic plane) on account of the attractions of spiral arms.

We may attempt to distinguish between the validity of these hypotheses by constructing a plot of the (U, V) velocities of stars which may be regarded as old. Figure 5 shows the appropriate diagram for stars to which Wilson assigns calcium emission 0

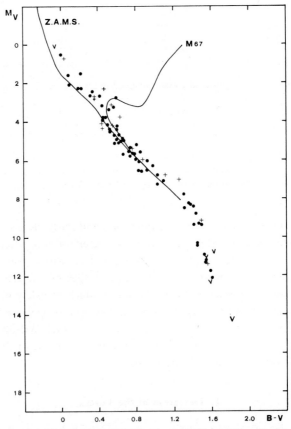

Fig. 1.   HR diagram for the nearby stars with $e < 0.075$. The symbols $+$, $\vee$ represent close double stars with combined photometry and variable stars respectively.

(crosses), −1 (open circles), or −2, −3, −4 (filled circles) and Figure 6 shows the diagram for evolved catalogue stars, that is stars of MK classes II to IV, the different symbols representing different spectral types. In neither of these diagrams is there a marked deviation of the vertex, and this seems to indicate that hypothesis (a) is to be preferred to hypothesis (b). The deviation of the vertex is then connected with the initial distribution of young stars, and is not something to do with spiral arms pulling all stellar orbits into some asymmetrical pattern.

A numerical analysis can be made as follows. The quadratic

$$AU^2 + 2FUV + BV^2$$

can be referred to its principal axes so that it becomes

$$\mathfrak{A}u^2 + \mathfrak{B}v^2$$

by the substitution

$$u = lU + mV$$
$$v = -mU + lV$$

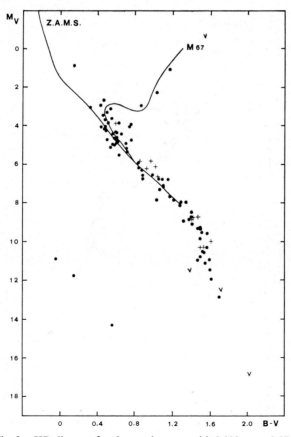

Fig. 2.   HR diagram for the nearby stars with $0.100 < e < 0.175$.

where $l = \cos\theta$ and $m = \sin\theta$ and

$$\tan 2\theta = -2F/(A - B).$$

Now if the velocity function is elliptical, i.e.

$$N(U, V) = \text{const} \times \exp\{-(AU^2 + 2FUV + BV^2)\}$$

the ratios of the mean values of U^2, UV and V^2 are given by

$$\overline{U^2} : \overline{UV} : \overline{V^2} :: B : -F : A$$

so that the deviation of the vertex θ can be found from

$$\tfrac{1}{2}\tan 2\theta = \frac{\sum UV}{\sum U^2 - \sum V^2}.$$

Once θ has been found, we can compute $l = \cos\theta$ and $m = \sin\theta$ and then find the ratio of the principal axes from

$$\frac{\mathfrak{A}}{\mathfrak{B}} = \frac{Al^2 - Bm^2}{Bl^2 - Am^2} = \frac{l^2 \sum U^2 - m^2 \sum V^2}{l^2 \sum V^2 - m^2 \sum U^2}.$$

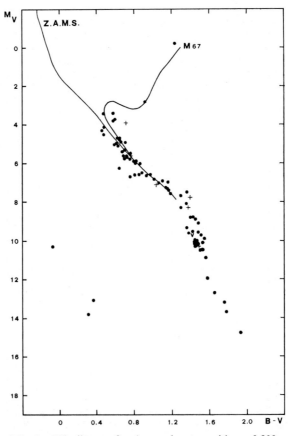

Fig. 3.   HR diagram for the nearby stars with $e > 0.200$.

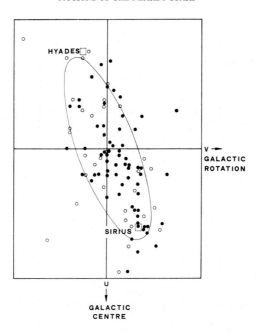

Fig. 4. Plot of the U, V velocity components of the nearby A stars (filled circles) and the nearby stars with Ca⁺ emission $\geq +3$ (open circles) showing the limiting velocity ellipse.

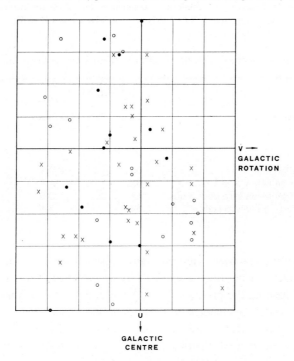

Fig. 5. Plot of the U, V velocity components of the nearby stars with Ca⁺ emission 0 (crosses), -1 (open circles) and -2, -3, -4 (filled circles).

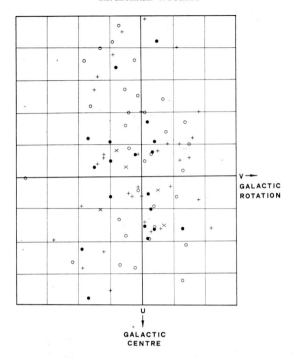

Fig. 6.   Plot of the U, V velocity components of the nearby stars of the spectral types F (filled circles), G (open circles), K (+), M (×), and MK classes II to IV.

Applied to the stars selected for illustration the results are as given in Table III.

TABLE III

Class of star	No. of stars	Deviation θ	Ratio of principal axes $(\mathfrak{A}/\mathfrak{B})^{1/2}$
Sp. Type A	66	$+22°.3$	3.25
Ca$^+$ emission $\geqslant +3$	33	$+19°.2$	2.85
Ca$^+$ emission zero and minus	65	$+ \ 3.0$	1.41
Late-type evolved stars	69	$+ \ 3.1$	1.82

In all cases a ceiling has been imposed, namely only those stars for which the eccentricity of the galactic orbit does not exceed $e = 0.15$ have been kept. This has been done partly to avoid loading the products $\sum U^2$, $\sum UV$ and $\sum V^2$ with very high entries and partly to avoid confusing changes in deviation due to stellar age with changes due to the eccentricity of the orbit – since the young stars are all stars with low eccentricity. A breakdown of the data for the old stars confirms the result that the low tilt is *not* a matter of eccentricity. Notice that the ratio of the principal axes is quite different for the young and old stars.

7

The initial conditions which give rise to the appearance of a particular velocity ellipse in the neighbourhood of the Sun can be investigated with the help of Bok's (1934) equations. They are of course very similar to equations used by Lindblad.

We consider motion in two dimensions only (that is, in the plane of the Galaxy).

If (x, y) are rectangular coordinates in fixed axes in the Galaxy $(R+\xi, \eta)$ are rectangular coordinates in axes rotating with a constant angular speed ω in the Galaxy; then at an instant when the $(R+\xi)$-axis makes an angle θ with the x-axis

$$x = (R + \xi) \cos \theta - \eta \sin \theta, \qquad y = (R + \xi) \sin \theta + \eta \cos \theta$$
$$\dot{x} = (\dot{\xi} - \eta\omega) \cos \theta - (\dot{\eta} + (\xi + R) \omega) \sin \theta$$
$$\dot{y} = (\dot{\eta} + (R + \xi) \omega) \cos \theta + (\dot{\xi} - \eta\omega) \sin \theta$$
$$\ddot{x} = [\ddot{\xi} - 2\dot{\eta}\omega - (\xi + R) \omega^2] \cos \theta - (\ddot{\eta} + 2\omega\dot{\xi} - \eta\omega^2) \sin \theta$$
$$\ddot{y} = (\ddot{\eta} + 2\dot{\xi}\omega + \eta\omega^2) \cos \theta + [\ddot{\xi} - 2\dot{\eta}\omega - (R + \xi) \omega^2] \sin \theta$$

and when $\cos\theta = 1$ and $\sin\theta = 0$

$$x = R + \xi \qquad\qquad\qquad y = \eta$$
$$\dot{x} = \dot{\xi} - \eta\omega \qquad\qquad\qquad \dot{y} = \dot{\eta} + (R + \xi) \omega$$
$$\ddot{x} = \ddot{\xi} - 2\dot{\eta}\omega - (R + \xi) \omega^2 \qquad \ddot{y} = \ddot{\eta} + 2\dot{\xi}\omega - \eta\omega^2 .$$

If r is the distance from the centre of the Galaxy, or $r^2 = x^2 + y^2 = (R+\xi)^2 + \eta^2$, the forces $F(x)$ and $F(y)$ directed to the centre of the Galaxy are related to the force $F(r)$ by

$$F(x) = \frac{R + \xi}{x} F(r), \qquad F(y) = \frac{n}{r} F(r),$$

and if we approximate to the galactic attraction by setting

$$F(r) = - \alpha r^{-2} - \beta r^{-3},$$

then, ignoring quantities of the second order,

$$F(x) = - \alpha R^{-2} - \beta R^{-3} + \xi(2\alpha R^{-3} + 3\beta R^{-4})$$
$$F(y) = - \eta(\alpha R^{-3} + \beta R^{-4}).$$

Now

$$\ddot{x} = F(x) \quad \text{and} \quad \ddot{y} = F(y).$$

Choose ω so that

$$R\omega^2 = \alpha R^{-2} + \beta R^{-3}. \tag{7.1}$$

Then

$$\ddot{\xi} - 2\dot{\eta}\omega - \xi(3\omega^2 + \beta R^{-4}) = 0$$
$$\ddot{\eta} + 2\dot{\xi}\omega \qquad\qquad\qquad = 0. \tag{7.2}$$

These are Bok's equations for the case $F(r) = \alpha r^{-2} + \beta r^{-3}$. These equations admit of a solution, namely

$$\xi = A \cos(pt + \gamma) + \xi_0 \qquad \eta = - B \sin(pt + \gamma) + \lambda t + \eta_0 \tag{7.3}$$

if

$$-p^2 A + 2\omega pB - A(3\omega^2 + \beta R^{-4}) = 0 \tag{7.4}$$

$$2\omega\lambda + \xi_0(3\omega^2 + \beta R^{-4}) = 0 \tag{7.5}$$

$$pB - 2\omega A = 0. \tag{7.6}$$

Eliminating A/B from (7.4) and (7.6) we have

$$p^2 = \omega^2 - \beta R^{-4}. \tag{7.7}$$

Given R, ω is fixed by (7.1) so that p is fixed by (7.7). Since A/B is fixed by (7.6) and λ/ξ_0 by (7.5), we are left with four disposable constants

$$A, \gamma, \xi_0, \eta_0$$

to be determined by the initial conditions, i.e. the values of

$$\xi, \eta, \dot{\xi}, \dot{\eta} \quad \text{at} \quad t = 0.$$

8

Consider a set of stars at $\xi=0$ and $\eta=0$ at $t=0$ (i.e. the stars now in the solar neighbourhood) and consider a set of these whose velocities lie on an ellipse (centred on the circular velocity). If

$$\xi = 0 \quad \text{and} \quad \eta = 0 \quad \text{at} \quad t = 0,$$

$$0 = A \cos\gamma + \xi_0 \qquad\qquad 0 = -B \sin\gamma + \eta_0$$

$$\dot{\xi}_0 = -pA \sin\gamma \qquad\qquad \dot{\eta}_0 = -pB \cos\gamma + \lambda.$$

Hence

$$\dot{\xi}_0 = -p\frac{A}{B}\eta_0 = -\frac{p^2}{2\omega}\eta_0$$

$$\dot{\eta}_0 = p\frac{B}{A}\xi_0 + \lambda = \xi_0\left(\frac{pB}{A} + \frac{\lambda}{\xi_0}\right)$$

$$= (p^2/2\omega)\xi_0 \qquad\qquad \text{(using Equations (7.5), (7.6), (7.7)).}$$

Hence if the velocities $\dot{\xi}_0$ and $\dot{\eta}_0$ lie on an ellipse, so do ξ_0 and η_0 (which are in fact the coordinates of the epicentre relative to the local standard of rest).

Let this ellipse be

$$a\xi_0^2 + 2f\xi_0\eta_0 + b\eta_0^2 = C^2. \tag{8.1}$$

But

$$\xi = \xi_0 + A\cos(pt + \gamma)$$

$$= \xi_0 + A\cos pt \cos\gamma - A\sin pt \sin\gamma$$

$$= \xi_0(1 - \cos pt) - \frac{A}{B}\eta_0 \sin pt$$

$$\eta = \eta_0 + \lambda t - B \sin(pt + \gamma)$$

$$= \eta_0 + \lambda t - B[\sin pt \cos \gamma + \cos pt \sin \gamma]$$

$$= \eta_0(1 - \cos pt) + \xi_0\left(\frac{\lambda t}{\xi_0} + \frac{B}{A}\sin pt\right)$$

so that ξ, η are given by the relations

$$\xi = h\xi_0 + k\eta_0$$

$$\eta = l\xi_0 + h\eta_0$$

where at any given time h, l and k are constants given by

$$h = 1 - \cos pt \qquad l = \frac{\lambda}{\xi_0}t + \frac{B}{A}\sin pt = \frac{1}{\xi_0}t + \frac{2\omega}{p}\sin pt,$$

$$k = -\frac{A}{B}\sin pt = -\frac{p}{2\omega}\sin pt.$$

Solving for ξ_0, η_0 in terms of ξ and η, we obtain

$$\xi_0 = \frac{h\xi - k\eta}{h^2 - kl}$$

$$\eta_0 = \frac{h\eta - l\xi}{h^2 - kl}.$$

Accordingly, if ξ_0, η_0 satisfy (8.1), ξ, η satisfy

$$a(h\xi - k\eta)^2 + 2f(h\xi - k\eta)(h\eta - l\xi) + b(h\eta - l\xi)^2 = C^2(h^2 - kl)^2,$$

which can be written as

$$\mathfrak{A}\xi^2 + 2\mathfrak{J}\xi\eta + \mathfrak{B}\eta^2 = C^2(h^2 - kl)^2 \qquad (8.2)$$

where

$$\mathfrak{A} = ah^2 - 2f\,lh + bl^2$$

$$2\mathfrak{J} = -2ahk + 2f(h^2 + kl) - 2bhl$$

$$\mathfrak{B} = ak^2 - 2f\,kh + bh^2.$$

Accordingly (ξ, η) lies on an ellipse also centred on the local standard of rest and given by (8.2). It also follows that every star whose velocity at $t=0$ lies within the original velocity ellipse, so that its epicentre (ξ_0, η_0) lies within (8.1), lies inside the ellipse (8.2) at any time t. (It lies on an ellipse similar to (8.2) but with a smaller value of C^2).

9

As in Section 4 we shall work in units in which $R=1$ and $\omega=1$. These imply $\alpha+\beta=1$.

In these units

$$p^2 = \alpha$$
$$A/B = \sqrt{\alpha/2}$$
$$\lambda = -\frac{3+\beta}{2}\,\xi_0.$$

In investigations currently being pursued at the Royal Greenwich Observatory we use $\alpha = 1.76245$. With this value of α

$$p = 1.327\,57$$
$$B/A = 1.506\,51$$
$$\lambda = -1.118\,78\,\xi_0$$
$$\dot{\xi}_0 = -1.881\,22\,\eta_0$$
$$\dot{\eta}_0 = +0.881\,22\,\xi_0.$$

The ellipse which contains most of the velocities of the A-type stars and young K stars (with strong Ca^+ reversals) is

$$2.297\ U^2 + 6.320\ UV + 8.704\ V^2 = 900 \quad (km/sec)^2 \tag{9.1}$$

with the convention that U is directed away from the centre of the Galaxy, i.e. $U = +\dot{\xi}$, and V is positive in the direction in which rotation takes place. Then

$$8.704\xi_0^2 - 6.320\xi_0\eta_0 + 2.296\eta_0^2 = 0.018\,544 \tag{9.2}$$

is the figure describing the limiting epicentres.

At $t = -3$ and $t = -4$ we have the following limiting ellipse (see Table IV).

TABLE IV

	$t = -3$	$t = -4$
$\mathfrak{A}$	117.2	92.6
$2\mathfrak{J}$	-23.46	$+11.32$
$\mathfrak{B}$	3.296	1.54
$C^2(h^2 - kl)^2$	0.4625	0.2055
Tilt	$-5°.8$	$+3°.5$
Ratio of principal axes	7.49	7.76
Length of semi-axis major	0.47	0.41

At $pt = -2\pi$, or $t = -4.74$, the ellipse degenerates to a straight line along the η-axis whose half length is 0.244 units.

10

For ellipses as large as this it is not really accurate to ignore the square of the eccentricity, which is implied in using the Bok's equations; but the errors involved can be ascertained in any particular case by running a computer programme. The analytical and approximate solution gives a qualitative idea of the events; and the young stars now seen in the solar neighbourhood must have occupied areas very like the ellipses given by the approximate theory.

To see the significance of these areas, we enquire what are the figures corresponding to those velocities which we do *not* see in the young stars in the solar neighbourhood, and we examine the velocity ellipse

$$2.296 \, U^2 + 6.320 \, UV + 8.704 \, V^2 = 900 \quad (\text{km/sec})^2 \tag{10.1}$$

that is an ellipse similar to (9.1) but tilted in the opposite sense.

The ellipse for $t = -3$ corresponding to (10.1) is

$$22.85 \, \xi^2 - 16.38 \, \xi\eta + 13.27 \, \eta^2 = 0.4625 \tag{10.2}$$

and it is shown in Figure 7 plotted together with the ellipse

$$117.2 \, \xi^2 - 23.46 \, \xi\eta + 3.30 \, \eta^2 = 0.4625 \tag{10.3}$$

which corresponds to (9.1). The shaded area is *forbidden* in the sense that if any stars had been found at $t = -3$ in this area, and had got to the solar neighbourhood at $t = 0$, then their velocities would have been outside (9.1) (but inside (10.1) instead). The semi-axis minor of (10.3) is 0.063 units or 630 parsec. This means that the star forming arm had a width of 1260 parsec, which seems reasonable.

A direct computation was made by Miss R. Johnston of the positions of each of the stars in Figure 4, at $t = -4$, assuming only the force field

$$F(R) = -1.762\,45R^{-2} + 0.762\,45R^{-3}$$

and the points are shown in Figure 8, in axes fixed in the Galaxy at $t = -4$ and the origin being the position of the present centre of rest carried back to that time – i.e. the point occupied at $t = -4$ by an object now in the solar neighbourhood and having the circular velocity, or $U = 0$ and $V = 0$. This figure shows the points occupied in space, at that time, of objects which are all *now* in the solar neighbourhood, but are distributed in *velocity* space as we see in Figure 4.

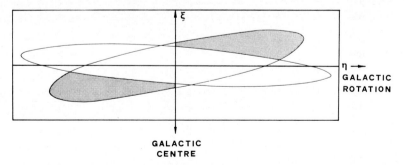

Fig. 7.   Figure showing the ellipse occupied by the stars, now in the solar neighbourhood with velocities inside the limiting ellipse, at $t = -3$. Hatched area is the 'forbidden' region.

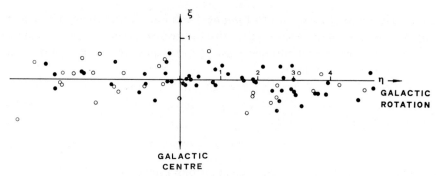

Fig. 8. Figure showing the ξ, η distribution of the stars shown in Figure 4 at $t = -4$ assuming a specific force field. Open circles correspond to the nearby stars with Ca⁺ emission $\geqslant +3$ while filled circles correspond to the Gliese A and B stars.

11

Figure 8 suggests that the young Gliese stars may have been formed in a region resembling a spiral arm at $t = -3$ or $t = -4$, or that is to say between 1.0 and 1.3×10^8 years ago. We may test this by enquiring how the stars compare with those in the Pleiades and Hyades clusters, whose ages Sandage (1957) placed at 2×10^7 and 4×10^8 years respectively.

This could be readily examined if we were able to draw a satisfactory HR diagram for the Gliese A stars. Unfortunately the only means of doing this is by constructing absolute magnitudes of the Gliese stars from their trigonometrical parallaxes, some of which are poorly determined, and it seems better to refer to spectral types, of which the counts are as follows (see Table V).

These figures indicate strongly that the early-type Gliese stars are much younger than the Hyades, and support the age of 1 to 1.5×10^8 years for the majority of the early-type Gliese stars.

TABLE V

Spectral type	Number of stars		
	Pleiades	Gliese stars	Hyades
B5	–	1	–
B6	3	–	–
B7	3	1	–
B8	5	1	–
B9	3	1	–
A0	1	13	–
A1	4	2	–
A2	1	12	1
A3	4	8	0
A4	2	8	0
A5	2	4	3
A6	2	1	3
A7	2	10	3
A8	1	0	3
A9	?	2	4

Acknowledgement

Miss R. Johnston has helped me very considerably with the preparation of this lecture.

References

Bok, B. J.: 1934, *Harvard Coll. Observ. Circ.* 384.
Camm, G. L.: 1949, *Monthly Notices Roy. Astron. Soc.* **110**, 309.
Clark, G. L.: 1936, *Monthly Notices Roy. Astron. Soc.* **97**, 182.
Eddington, A. S.: 1915, *Monthly Notices Roy. Astron. Soc.* **76**, 37.
Gliese, W.: 1957, *Astron. Rechen-Inst. Heidelberg, Mitt.* A., No. 8.
Gliese, W.: 1969, *Veröff. Astron. Rechen-Inst. Heidelberg*, No. 22.
Lynden-Bell, D.: 1962, *Monthly Notices Roy. Astron. Soc.* **124**, 95.
Oort, J. H.: 1932, *Bull. Astron. Inst. Neth.* **6**, 262.
Sandage, A.: 1957, *Astrophys. J.* **126**, 326.
Strömberg, G.: 1946, *Astrophys. J.* **104**, 12.
Wilson, O. and Woolley, R.: 1970, *Monthly Notices Roy. Astron. Soc.* **148**, 463.
Woolley, R.: 1957, *Monthly Notices Roy. Astron. Soc.* **117**, 198.
Woolley, R. and Candy, M. P.: 1968, *Monthly Notices Roy. Astron. Soc.* **141**, 277.
Woolley, R. and Stewart, J. M.: 1967, *Monthly Notices Roy. Astron. Soc.* **136**, 329.

RECENT DEVELOPMENTS IN GALACTIC DYNAMICS

G. CONTOPOULOS

University of Thessaloniki, Thessaloniki, Greece

1. Introduction

Galactic dynamics has been influenced considerably in recent years by plasma dynamics. For instance, Lin's gravitational theory of spiral structure is using methods that were first developed for a plasma. The same is true for many studies of collisionless and collisional relaxation of stellar systems, problems of stability, etc.

A stellar system sometimes is called a gravitational plasma. In fact, there is a basic similarity between a stellar system and a plasma; in both cases we have inverse square forces. Such forces have particular 'long range' effects, that do not appear in the case of 'short range' forces between molecules or atoms in a gas. The fact that in a plasma we have both attractive and repulsive forces is of smaller importance. However, this difference, combined with the fact that in a plasma we usually have almost equal numbers of positive and negative charges, makes the problems of plasma easier. In fact, the effective forces in a plasma at distances large with respect to the Debye radius are weakened by *shielding*. On the other hand, in a stellar system the Debye radius is equal to the radius of the system and no shielding appears.

There are three kinds of problems common in plasma dynamics and stellar dynamics:

(1) First we have problems of orbits in *given* fields, e.g. the orbits of charged particles in a given magnetic field, like a mirror machine or the van Allen belts, or the orbits of stars in galaxies. Such are most of the problems of celestial mechanics.

(2) Then come the problems of collisionless dynamics. When collisions (encounters) are unimportant, namely in cases of long relaxation times, we study collective effects, like oscillations, instabilities, collisionless mixing, etc.

(3) Finally we have problems of collisional dynamics, like the time of relaxation, and the long-time evolution of a plasma or a stellar system.

2. Collisionless Dynamics

The basic equation of collisionless dynamics is the collisionless Boltzmann's equation (known as Vlasov's equation by plasma physicists)

$$\frac{\partial f}{\partial t} + \mathbf{v} \cdot \frac{\partial f}{\partial \mathbf{x}} - \frac{\partial V}{\partial \mathbf{x}} \cdot \frac{\partial f}{\partial \mathbf{v}} = 0, \tag{2.1a}$$

or

$$\frac{\partial f}{\partial t} + \mathbf{v} \cdot \frac{\partial f}{\partial \mathbf{x}} + \frac{e}{m}\left[\mathbf{E} + \frac{\mathbf{v} \times \mathbf{B}}{c}\right] \cdot \frac{\partial f}{\partial \mathbf{v}} = 0, \tag{2.1b}$$

L. N. Mavridis (ed.), Structure and Evolution of the Galaxy, 198–207. All Rights Reserved where f is the distribution function, V the gravitational potential, and $\mathbf{E}$, $\mathbf{B}$ are the intensities of the electric and magnetic field respectively.

The density ϱ is found by integrating f over all velocities; in a similar way we find the charge and current densities. Then we can derive the gravitational potential V, or the electric field $\mathbf{E}$, by means of Poisson's equation, and the magnetic field B can be found from Maxwell's equations. In self-consistent problems V, $\mathbf{E}$, $\mathbf{B}$ are the same with those entering Equations (2.1.) Therefore Equations (2.1) are not linear in f. On the other hand, if V, $\mathbf{E}$, $\mathbf{B}$ are *given*, Equations (2.1) are linear. One can also consider intermediate cases where an exterior field is added to the field of the moving particles. Our discussion will be restricted to gravitational self-consistent problems. Then Poisson's equation, relating V and f, can be written

$$\nabla^2 V = 4\pi G \int f \, d\mathbf{v}. \tag{2.2}$$

In many cases it is possible to find simple solutions of Equations (2.1) and (2.2). For example, one can find simple axisymmetric models of galaxies. Then in order to find approximate self-consistent solutions 'near' a given solution we can write

$$V = V_0 + V_1, \quad f = f_0 + f_1, \tag{2.3}$$

where V_0, f_0 is the 'unperturbed solution' and V_1, f_1 are small quantities, with respect to V_0, f_0. We insert these values in Equations (2.1), omitting all terms higher than the first in V_1, f_1, and find the 'linearized collisionless Boltzmann's equation' (or 'linearized Vlasov's equation')

$$\frac{\partial f_1}{\partial t} + \mathbf{v} \cdot \frac{\partial f_1}{\partial \mathbf{x}} - \frac{\partial V_0}{\partial \mathbf{x}} \cdot \frac{\partial f_1}{\partial \mathbf{v}} = \frac{\partial V_1}{\partial \mathbf{x}} \cdot \frac{\partial f_0}{\partial \mathbf{v}}. \tag{2.4}$$

The corresponding Poisson's equation is now

$$\nabla^2 V_1 = 4\pi G \int f_1 \, d\mathbf{v}. \tag{2.5}$$

The solution of Equations (2.4) and (2.5) is one of the main problems of collisionless dynamics.

If we have perturbations of a flat axisymmetric galaxy, defined by the potential $V_0 = V_0(r)$, we can write in polar coordinates

$$\frac{\partial f_1}{\partial t} + \dot{r} \frac{\partial f_1}{\partial r} + \frac{J_0}{r^2} \frac{\partial f_1}{\partial \vartheta} = \frac{\partial f_0}{\partial E_0} \left(\dot{r} \frac{\partial V_1}{\partial r} + \frac{J_0}{r^2} \frac{\partial V_1}{\partial \vartheta} \right) + \frac{\partial f_0}{\partial J_0} \frac{\partial V_1}{\partial \vartheta}, \tag{2.6}$$

where

$$\dot{r} = \frac{dr}{dt}, \quad J_0 = r^2 \frac{d\vartheta}{dt}, \quad E_0 = \tfrac{1}{2} \left(\dot{r}^2 + \frac{J_0^2}{r^2} \right) + V_0(r); \tag{2.7}$$

E_0 can be used as a variable instead of $\dot{r}$.

The solution of the linearized equation can be found by a variant of the general method used by Landau in the case of the perturbations of a homogeneous plasma. We perform a Fourier analysis in ϑ (instead of Fourier transforms in both coordinates) and a Laplace transform in time.

To exemplify our procedure let us consider only two-armed perturbations

$$V_1 = V_1 \exp(-2i\vartheta), \quad f_1 = f_1 \exp(-2i\vartheta), \tag{2.8}$$

and let V_1^*, f_1^* be the Laplace transforms of V_1, f_1 namely

$$V_1^* = \int_0^\infty V_1 \exp(-i\omega t)\,\mathrm{d}t, \quad f_1^* = \int_0^\infty f_1 \exp(-i\omega t)\,\mathrm{d}t. \tag{2.9}$$

Then after some operations we find

$$i\left(\omega - 2\frac{J_0}{r^2}\right) f_1^* + \dot{r}\frac{\mathrm{d}f_1^*}{\mathrm{d}r} = \frac{\partial f_0}{\partial E_0}\dot{r}\frac{\mathrm{d}V_1^*}{\mathrm{d}r}$$
$$- 2iV_1^*\left(\frac{\partial f_0}{\partial E_0}\frac{J_0}{r^2} + \frac{\partial f_0}{\partial J_0}\right) + f_{10} = Q_1, \tag{2.10}$$

where f_{10} is the initial value of f_1 (for $t=0$).

If we consider V_1^* as known in Equation (2.10) we can solve this linear differential equation for f_1^* easily in the form

$$f_1^* = \int P_\omega(f_0, V_1^*)\,\mathrm{d}\tau + \phi_{10}, \tag{2.11}$$

where P_ω is an operator linear in f_0 and in V_1^*, depending also on ω; τ is the time along unperturbed orbits (orbits in the field V_0), and ϕ_{10} is an integral containing f_{10} as a factor. Away from resonances we have explicitly

$$f_1^* = \frac{1}{2i\sin(\omega\tau_0 - 2\vartheta_0)}\int_{-\tau_0}^{\tau_0} Q[r(\tau)]\exp\{i[\omega\tau - 2\vartheta(\tau)]\}\,\mathrm{d}\tau,$$

where $2\tau_0$ is the period of an unperturbed orbit, corresponding to the given values of E_0 and J_0; ϑ_0 is the angle between pericentron and apocentron, and $r(\tau)$, $\vartheta(\tau)$ give the motion along the orbit.

If we integrate f_1^* over all velocities we find the surface density σ_1^*. If then we replace V_1^* by a solution of Poisson's equation, written in the form

$$\nabla^2 V_1^* = 4\pi G\,\delta(z)\,\sigma_1^*, \tag{2.12}$$

where $\delta(z)$ is Dirac's function, we find the relation

$$r\sigma_1^*(r) = \int K_\omega(r, r')\,r'\sigma_1^*(r')\,\mathrm{d}\;' + s_{10}(r), \tag{2.13}$$

where $K_\omega(r, r')$ is a function of r, r' depending also on ω. Equation (2.13) is an integral equation similar to a non-homogeneous Fredholm's equation of the second kind (the difference is that the present equation is not linear in the parameter ω, and it may be singular).

If the eigenvalues of the integral equation are known, then its solution can be found in the form

$$r\sigma_1^*(r) = s_{10}(r) + \int G_\omega(r, r') s_{10}(r') \, dr', \qquad (2.14)$$

where $G_\omega(r, r')$ is called the 'solving kernel'. If the eigenvalues of the integral equation are isolated, then $G_\omega(r, r')$ has these eigenvalues as poles.

The integral equation (2.13) can be simplified considerably in particular cases, e.g., in the case of a homogeneous plasma the corresponding integral equation can be solved explicitly, and the eigenvalues are given by a simple 'dispersion relation'.

The problem of a galaxy is more complicated because a galaxy is far from homogeneous. However Lin and Shu made this problem tractable in the case of a tightly wound spiral. In this case the radial wavelength λ is small and higher order terms of λ can be omitted. Thus Lin and Shu derived a 'dispersion relation'

$$D \equiv 1 - \frac{2\pi G \sigma_0}{|k| \langle \dot{r}^2 \rangle} \left[1 - \frac{\nu\pi}{\sin\nu\pi} \, \mathfrak{G}_\nu(\chi_*) \right] = 0, \qquad (2.15)$$

with

$$\mathfrak{G}_\nu(\chi_*) = \frac{1}{2\pi} \int\limits_{-\pi}^{\pi} \cos\nu\gamma \, \exp\left[- \chi_*(1 + \cos\gamma)\right] d\gamma, \qquad (2.16)$$

where σ_0 is the unperturbed surface density, $\langle \dot{r}^2 \rangle^{1/2}$ is the dispersion of the unperturbed velocities, k is the wave number $(k = 2\pi/\lambda)$, ν is the relative frequency

$$\nu = (\omega - 2\Omega)/\kappa, \qquad (2.17)$$

with Ω the angular velocity of the galaxy at distance r from the center, and κ the corresponding 'epicyclic frequency', and

$$\chi_* = k^2 \langle \dot{r}^2 \rangle/\kappa^2. \qquad (2.18)$$

It is assumed that the unperturbed distribution function at any point $r = r_0$ is given by a 'Schwarzschild' formula

$$f_0 = \frac{\Omega(r_0) \sigma_0(r_0)}{\kappa(r_0) \pi \langle \,\,^2 \rangle(r_0)} \exp\left[- (E_0 - E_{00})/\langle \dot{r}^2 \rangle(r_0)\right], \qquad (2.19)$$

with

$$J_0 = r_0^2 \Omega(r_0), \quad \text{and} \quad E_{00} = \tfrac{1}{2} J_0^2/r_0^2 + V_0(r_0), \qquad (2.20)$$

and the solution of Poisson's equation is written in the lowest order approximation

$$\sigma_1^* = - |k| V_1^*/2\pi G. \qquad (2.21)$$

The 'dispersion relation' $D=0$ relates ω (through v) to k and r. If ω is assumed known, we find k as a function of r, and the form of the spiral arms is

$$\vartheta' = \vartheta - \Omega_s t = \tfrac{1}{2} \int k \, dr + \text{const}\,(+\pi). \tag{2.22}$$

The spiral arms rotate as rigid bodies with angular velocity $\Omega_s = \omega/2$.

Thus Lin and Shu solved the problem of modes (eigenvalues). If we compare the spiral arms (2.22) for various ω with the spiral arms of our Galaxy, we can find the approximate value of ω. Lin thus derived values of Ω_s near 13 km/sec per kpc.

In a similar way one can treat the initial value problem. We assume a two-armed initial perturbation

$$f_{1;0} = f_{1;0}^* \exp(-2i\vartheta), \tag{2.23}$$

where

$$f_{1;0}^* = f_0 a \exp(i\phi). \tag{2.24}$$

Let us assume, further, that a, ϕ depend only on r and not on E_0, or J_0. Then we find, following the same method as Lin and Shu,

$$\sigma_1^* = \frac{\pi\sigma_0 \, a \, \exp\left[i(\phi - \pi/2)\right] \, \mathfrak{G}_v(\chi_{*;0})}{D\kappa \sin v\pi}, \tag{2.25}$$

where D is the function (2.15), and $\chi_{*;0} = (\phi')^2 \langle \dot{r}^2 \rangle / \kappa^2$. Thus $\chi_{*;0}$ is the initial value of χ_*, because ϕ' is the initial value of k.

If we invert now the Laplace transform for σ_1^* we find

$$\boldsymbol{\sigma}_1 = \frac{1}{2\pi} \int\limits_{-\infty + i\omega_I}^{\infty + i\omega_I} \sigma_1^* \exp(i\omega t) \, d\omega, \tag{2.26}$$

where the integration takes place along a line $\omega = i\omega_I$ below the singularities of the integrand. Thus

$$\sigma_1 = \sigma_0 a \exp\left[i\left(\phi - \frac{\pi}{2} - 2\vartheta\right)\right] \frac{1}{2\pi\kappa}$$

$$\times \int\limits_{-\infty + i\omega_I}^{\infty + i\omega_I} \frac{1}{\sin v\pi} \frac{\mathfrak{G}_v(\chi_{*;0})}{D} \exp(i\omega t) \, d\omega. \tag{2.27}$$

If we assume that the singularities of the integrand are only poles, and move the line of integration above the singularities (in the same way as in the case of Landau), we find that after a transition period only the contributions from the poles remain, namely

$$\sigma_1 = \sigma_0 a \exp\left[i(\phi - 2\vartheta)\right] \sum_i \frac{\mathfrak{G}_{v_i}(\chi_{*;0})}{\sin v_i \pi \, (\partial D / \partial v)_{v = v_i}} \exp(i\omega_i t). \tag{2.28}$$

assuming always that $\sin v_i \pi \neq 0$, i.e. that we are far from resonances.

Therefore after a long time we have only oscillations with frequencies ω_i, where ω_i satisfy Lin's dispersion relation (2.13). Thus practically any initial perturbation excites Lin's modes.

This does not mean that the frequencies ω_i are constant in space and time. In fact, let us consider one root of Equation (2.15), $\omega = \omega(k, r)$, and assume that it is unique. At time $t = 0$, k is a given function of r, namely $k(r) = \phi'(r)$, hence

$$\omega = \omega(k(r), r), \tag{2.29}$$

i.e. ω is a function of r. If we assume that Equation (2.28) is valid for all $t \geqslant 0$, then, at time $t = \Delta t$ the function k becomes

$$k(r) = \phi'(r) + \omega'(r)\,\Delta t. \tag{2.30}$$

Therefore in the particular case that ω is independent of r (i.e. if the initial perturbation is such that $k(r) = \phi'(r)$ satisfies Lin's dispersion relation (2.15) for a fixed ω), then the form of the wave is presented, and ω remains constant in time also. In the present case, however, $k(r)$ changes in time, thus ω also changes in time. Hence Lin's modes imply, in general, a change of ω in time and space. In particular, if the initial perturbation is a superposition of slightly different spirals, we have a group of waves, moving inwards with group velocity $d\omega/dk$. Toomre studied in detail the properties of such groups of waves. The problem is now what happens to these groups of waves as they reach the inner Lindblad resonance. According to Toomre they are probably damped there. Lin, on the other hand, considers the possibility that these waves are changed near the inner Lindblad resonance into waves of large wavelength, which move outwards until they reach the outer edges of the Galaxy, and thus a quasi-stationary configuration is established.

We must stress that the effects taking place near the inner Lindblad resonance are basically non-linear. Such phenomena cannot be described approximately by a linear theory, like those developed up to now. Among other non-linear effects, we have found that near the inner Lindblad resonance a density distribution with a roughly quadruple symmetry is established. The non-linear theory of spiral structure is most promising today.

A few more remarks should be made here concerning the linear theory of spiral structure.

(1) The difference between the integral equation (2.13) and the dispersion relation (2.15), is not only one of degree of difficulty. The basic difference is that Equation (2.13) gives specific eigenvalues ω and corresponding eigenfunctions $r\sigma_1^*(r)$, while the dispersion relation (2.15) gives, for every r a relation between ω and k (k is the derivative of the phase of $r\sigma_1^*(r)$). Thus the dispersion relation allows modes ω for every $r\sigma_1^*(r)$, while the integral equation restricts both ω and $r\sigma_1^*(r)$. The difference is due to the fact that (because of the approximations introduced by Lin) the density σ_1^* is always in phase with the potential V_1^*. Thus one restriction is eliminated. However the difficulties avoided by this method are encountered at the boundaries of the system,

where the dispersion relation cases hold. On the other hand, the integral equation incorporates the boundary conditions, and gives solutions valid everywhere.* For this reason Lin's theory is sometimes called 'local', in the sense that it excludes a treatment of the boundary conditions. On the other hand, Kalnajs has recently found numerical solutions of the integral equation. In the case of a model of M31 he found rather open spiral waves, rotating at an angular velocity $\Omega_s = \omega/2$ of the order of 30 km/sec per kpc.

(2) Near resonances a linear theory is valid only for short times. In fact, f_1 contains the denominator $\sin v\pi$, and if this becomes zero, or very near zero, the f_1 contains a secular term, or a small divisor term, and it becomes large with respect to f_0. Thus the basic assumption of the linearization is not valid. However, even under these restrictions a linear theory near resonances can give interesting results. For example, while it is known that away from resonances the density is in phase with the imposed spiral potential, we have found that near the inner Lindblad resonance the response to a slightly growing imposed wave is out of phase and always trailing. Thus even if we impose a leading wave the response tends to become trailing. We have here a clear preference of trailing spiral waves, a fact for which no satisfactory explanation had been given before.**

3. Collisional Dynamics

In collisional dynamics we start with a Boltzmann-type equation

$$\frac{\partial f}{\partial t} + \mathbf{v} \cdot \frac{\partial f}{\partial \mathbf{x}} - \frac{\partial V}{\partial \mathbf{x}} \cdot \frac{\partial f}{\partial \mathbf{v}} = \left(\frac{\partial f}{\partial t}\right)_{\text{coll}}, \tag{3.1}$$

and emphasis is put on the collision term. The classical Boltzmann collision term has the form

$$\left(\frac{\partial f}{\partial t}\right)_{\text{coll}} = n_0 \int \left[f(\mathbf{v}') f(\mathbf{v}_1') - f(\mathbf{v}) f(\mathbf{v}_1) \right] |\mathbf{v} - \mathbf{v}_1| \, d\sigma \, d\mathbf{v}_1 , \tag{3.2}$$

and is due to 'collisions' (or encounters) of a given particle (of velocity $\mathbf{v}$) with the rest of the particles (of velocities $\mathbf{v}_1$); the velocities after the collision are $\mathbf{v}'$ and $\mathbf{v}_1'$ respectively; n_0 is the particle density, and $d\sigma = b \, db \, d\varepsilon$, where b is the impact parameter and ε the angle of the scattering plane (the plane defined by $\mathbf{v} - \mathbf{v}_1$ and $\mathbf{v}' - \mathbf{v}_1'$) with a fixed plane through the axis $\mathbf{v} - \mathbf{v}_1$.

If we set $\mathbf{v}' = \mathbf{v} + \Delta\mathbf{v}$, $\mathbf{v}_1' = \mathbf{v}_1 + \Delta\mathbf{v}_1$ and expand $f(\mathbf{v}')$, $f(\mathbf{v}_1')$, retaining only terms up to the second in $\Delta\mathbf{v}$, $\Delta\mathbf{v}_1$, we find a Fokker-Planck's equation, of the form

$$\left(\frac{\partial f}{\partial t}\right)_{\text{coll}} = -\frac{\partial}{\partial \mathbf{v}} \cdot (f F) + \frac{1}{2} \frac{\partial^2}{\partial \mathbf{v} \, \partial \mathbf{v}} : (f D), \tag{3.3}$$

---

* Unless non-linear effects become important. The problem of the relative importance of non-linear effects versus boundary conditions in a linear theory of our Galaxy is open.
** Except for the trailing 'wavelets', provided by local (in the usual sense of a small region) theories of spiral structure.

where F, D are functions of $\mathbf{v}$; F is the coefficient of dynamical friction, while D is the dispersion coefficient.

Boltzmann's equation and Fokker-Planck's equation are not always appropriate for problems involving long range forces, like inverse square forces.* Thus if we calculate the coefficients F, D in the usual way, by considering all 'collisions' as binary effects (two-body problems) we find some divergent integrals. We are obliged to introduce somewhat artificial cut-offs in order to find finite results. This method was used by Chandrasekhar and many others in calculating the time of relaxation, the effect of dynamical friction, and the long range evolution of stellar systems. It is interesting to note that certain quantities, like the relaxation time, do not change considerably, even if we vary substantially the cut-off used. Numerical experiments give results in rough agreement with theory.

However there are results that are not reproduced by Boltzmann's equation. For example extensive numerical n-body experiments indicate that stellar systems do not tend towards a Maxwellian distribution. They have no equilibrium state of maximum entropy.

A stellar system develops a central nucleus of high density and a halo, while stars escape gradually out of the system. The central density increases continuously, while the escape of stars reduces the size of the system until it becomes a double star, or (more rarely) a stable multiple star.

The nearest analog to Boltzmann's H-theorem is a tendency of a stellar system towards a homologous system. A homologous system evolves by changing its dimensions and mass but remaining similar to itself. A spherical homologous model was found, as a special solution of Fokker-Planck's equation, by Hénon. Hénon found also examples where non-homologous systems tend to become homologous. If this is the general case, then we may be able to formulate a variant of the H-theorem for stellar systems.

Another form of the collision term is provided by the BBGKY theory:

$$\left(\frac{\partial f}{\partial t}\right)_{\text{coll}} = \frac{n_0}{m} \int \frac{\partial \phi_1}{\partial \mathbf{x}} \cdot \frac{\partial P}{\partial \mathbf{v}} \, d\mathbf{x}_1 \, d\mathbf{v}_1, \tag{3.4}$$

where $\phi_1 = -Gm^2/|\mathbf{x}-\mathbf{x}_1|$ is the potential between the particles $(\mathbf{x}, \mathbf{v})$ and $(\mathbf{x}_1, \mathbf{v}_1)$ and $P = P(\mathbf{x}, \mathbf{x}_1, \mathbf{v}, \mathbf{v}_1)$ is the 'correlation function'; if $f_2(\mathbf{x}, \mathbf{x}_1, \mathbf{v}, \mathbf{v}_1)$ is the 2-particles distribution function, P is defined as

$$P = f_2(\mathbf{x}, \mathbf{x}_1, \mathbf{v}, \mathbf{v}_1) - f(\mathbf{x}, \mathbf{v}) f(\mathbf{x}_1, \mathbf{v}_1). \tag{3.5}$$

However, P is not known, and its calculation is an extremely difficult problem. Many efforts have been made in this direction by people working in plasma physics. A particular form of the collision term, appropriate for a plasma, has been provided by Balescu and Lenard, and in some applications it is used instead of Boltzmann's collision term.

---

* An interesting discussion of the assumptions underlying the derivation of the Boltzmann's collision term is given by H. Grad: 1967, 'Principles of Kinetic Theory of Cases, in *Handbuch der Physik* **12**.

In the case of stellar systems very little has been done from the point of view of the BBGKY theory. Prigogine and his associates have attacked the gravitational problem using similar methods, developed in the non-equilibrium theory of statistical mechanics. The gravitational problem, however, is particularly difficult and no specific results have been derived yet. Prigogine and his associates have stressed the non-markovian character of the n-body problem, the appearance of 'memory' effects (memory of the initial conditions), and the fact that the entropy of a stellar system does not necessarily increase. In fact, there are numerical experiments showing spectacular memory effects, where the evolution depends critically on the initial conditions.

It is hoped that numerical n-body experiments will give enough experience for a renewed attack of the problems of collisional dynamics.

This new branch of experimental astronomy has had an impressive initial success, but at the present moment we are faced with difficulties that must be surmounted. During the 1967 Paris Colloquium on the Gravitational n-Body Problem, a comparison study was made by Lecar of various calculations of the same 25-body problem (same initial conditions) run by several investigators. Each investigator used a different machine or a different method of integration. The results diverged considerably after a short time, not only as regards the individual orbits (this was expected), but also as regards some average quantities. It seems that only a few average quantities are reliably calculated by n-body experiments. Sometimes it may even be necessary to have many runs of the same problem in different machines and/or by different methods, in order to find useful averages.

In order to discuss recent results and problems of the gravitational n-body problem a second IAU Colloquium is organized in Cambridge, England, during August 1970.

References

We give here only a few basic references related to the subjects we discussed above.
(1) The theory of Lin and his associates can be found in:
Lin, C. C. and Shu, F. H.: 1964, *Astrophys. J.* **140**, 646.
Lin, C. C. and Shu, F. H.: 1966, *Proc. Natl. Acad. Sci. U.S.* **55**, 229.
Lin, C. C., Yuan, C., and Shu, F. H.: 1969, *Astrophys. J.* **155**, 721.
    Further developments of the theory and applications are found in the Proceedings of IAU Symposium No. 38, *The Spiral Structure of our Galaxy* (ed. by W. Becker and G. Contopoulos), D. Reidel, Dordrecht, Holland, 1970 (papers by Contopoulos; Marochnik; Kalnajs; Shu; Hunter; Lynden-Bell; Toomre; Berry and Vandervoort; Vandervoort; Barbanis; Freeman; de Vaucouleurs and Freeman; Miller, Prendergast, and Quirk; Hohl; Lin; Yuan; Burton and Shane; W. W. Roberts, etc.).
    (2) A general discussion of cooperative phenomena in stellar dynamics is found in the article by D. Lynden-Bell: 1967, 'Cooperative Phenomena in Stellar Dynamics', in *Relativity Theory and Astrophysics. 2. Galactic Structure* (ed. by J. Ehlers), Am. Math. Soc., Providence, R.I., p. 131.
    (3) Books dealing with the kinetic theory of plasmas (Vlasov's equation, Landau damping, waves, instabilities, BBGKY theory, etc.):
Montgomery, D. C. and Tidman, D. A.: 1964, *Plasma Kinetic Theory*, McGraw-Hill, New York, London.
Stix, T. W.: 1962, *The Theory of Plasma Waves*, McGraw-Hill, New York, London,
and further books on plasma dynamics.
    A recent work about stability problems in plasma is:

Coppi, B.: 1969, 'Plasma Collective Modes Involving Geometry and Velocity Space', to be published in *La Revista del Nuovo Cimento*.

(4) The methods used by Prigogine and his associates may be found in the books:

Prigogine, I.: 1962, *Non-Equilibrium Statistical Mechanics*, Interscience, New York, London.

Balescu, R.: 1963, *Statistical Mechanics of Charged Particles*, Interscience, New York, London.

(5) Proceedings of conferences on numerical experiments:

IAU Colloquium, *Gravitational n-Body Problem*, Paris 1968; *Bull. Astron.* **3**, Nos. 2–3, 1968.

Symposium on Computer Simulation of Plasma and Many-Body Problems, 1967, NASA special publication SP-153.

(6) Detailed references on recent work in galactic dynamics (orbit theory, collisionless dynamics, spiral structure, collisional dynamics, etc.) can be found in:

Contopoulos, G.: 1970, Report for IAU Commission 33, 'Structure and Dynamics of the Galactic System', *Contr. astr. Dep. Univ. Thessaloniki*, No. 53.

PRESTELLAR EVOLUTION

P. LEDOUX

Institut d'Astrophysique de l'Université de Liège, Cointe-Sclessin, Belgium

A symposium on a very similar subject was held in Liège at the beginning of July 1969 and although, for reasons of time and competence, it is quite out of the question for me to try to summarize its proceedings here systematically, I shall certainly refer often to results presented there.

The general impression during this meeting, reinforced by many of the previous lectures in the present course, is certainly that we are entering a new era in which, thanks to radio, microwave and infrared astronomy, the direct observational information on the medium from which we believe that stars are born and on the initial phases of their lives will become increasingly abundant and precise. Perhaps with time, we shall even learn something observationally on the difficult problems raised by the conservation of angular momentum or magnetic flux and on the stages during which the protostar manages to get rid of most of it.

As Prof. Herbig put it in his introductory talk in Liège, "now the time has come to match up theory and observations in a more satisfactory manner". However as he proceeded to review the relevant evidence, the inferences and conclusions that he drew from it were so often and so strongly at variance with the opinions of many other people that one must conclude that we are still far from a one-to-one relationship between observations and theories.

1. General Interstellar Medium

Let us try to summarize briefly some properties of the interstellar medium in which we believe that stars form. As far as the interstellar matter in bulk is concerned, we can deduce an upper limit to its total mass or its density $\bar{\varrho}_0$ in the galactic plane by subtracting from the total density of the matter responsible for the gravity at right angle to the galactic plane that part (about, 4×10^{-24} g cm^{-3}) due to known stars. This yields in the solar neighbourhood, according to Spitzer (1968)

$$\bar{\varrho}_0 \leqslant 6.0 \times 10^{-24} \text{ g cm}^{-3}$$

or about 2.6 H atoms per cm^3 if a He/H ratio of 10% by number is assumed. Of course, unknown components, like black dwarfs, could still account for part of this density.

A large part of this material must be concentrated in clouds distributed along the spiral arms, which according to the latest information, both radio and optical (cf. for instance lectures of Prof. Kerr and Prof. Schmidt-Kaler), seem to have considerable extension at right angle to the galactic plane. Perhaps we should think of them more as 'ribbons' wound orthogonally to the galactic plane than as flat spiral bands in the latter.

L. N. Mavridis (ed.), Structure and Evolution of the Galaxy, 208–235. All Rights Reserved
Copyright © 1971 by D. Reidel Publishing Company, Dordrecht-Holland

The observed density of neutral H, $n(\text{H\textsc{i}})$, in the Orion arm is estimated at 1 to 2 atoms per cm^3; in the Perseus arm, it may amount to 2 to 3 atoms per cm^3, while the average value of $n(\text{H\textsc{i}})$ in the galactic plane is about 0.5/cm^3. The effective thickness of this gas is about 200 pc with $n(\text{H\textsc{i}})$ falling perhaps to 0.1 to 0.3 per cm^3 at a height of several hundred parsecs above the galactic plane. Of course, this still does not account for an important fraction of the total interstellar density $\bar{\varrho}_0$, which might correspond to ionized H (H\textsc{ii}), He and molecular hydrogen H$_2$, the heavier elements representing a very small percentage anyway and the solid matter in grains accounting at most probably for an average density of the order of 1% of that of the gas.

Detailed studies of particular regions at intermediate latitudes also show that H\textsc{i} fills rather sharply bounded structures which, in some cases, appear as fairly uniform on several hundred parsecs with only a few percent of the mass in smaller clouds and cloudlets a few parsecs in size (Heiles, 1967).

Usually one associates with this neutral hydrogen a temperature of some 100 to 125 K but perhaps its real meaning is not yet completely elucidated. Simonson (1970) finds, for the dark cloud in the direction of θ Ophiuchi, a density as high as 2.3×10^{-23} g cm^{-3} (10–15 H/cm^3), a low spin temperature varying from 26 to 40 K and a large overabundance of dust, its total mass 62 $M_\odot$ being about one tenth of the total mass of H\textsc{i} estimated at 675 $M_\odot$ and filling a volume with dimensions of the order of 10 pc.

The velocity dispersion of clouds in the galactic plane is of the order of 6 to 10 km/sec while the velocity dispersion within a single cloud may be of the order of 2 to 3 km/sec in some cases. In particular, comparison of internal velocity dispersion in 21 cm emission and absorption lines seems to indicate that there are two types of H-clouds representing about the same total fraction of the mass: relatively rapidly moving clouds of higher temperature and slowly moving ones at a lower temperature with perhaps a preponderance of small masses among the first ($< 30\ M_\odot$) and large masses among the second ($> 300\ M_\odot$).

The maintenance of the cloud motions and, to some extent, their origin are usually attributed to the expanding H\textsc{ii} regions with temperatures of the order of 10 000 K around bright young stars and perhaps to the effects of supernovae explosions.

It might be wrong however to think essentially of only these two types of regions: neutral H in clouds at some 100 K and ionized H at some 10 000 K. In H\textsc{i} regions, the interstellar clouds occupy not more probably than 7% of the available volume, the intercloud medium having a density much smaller, the general properties of these two phases depending critically on the available heating and cooling mechanisms. Field et al. (1969) have recently reviewed the situation taking into account the ionization by soft cosmic rays with an appropriately extrapolated energy density. The cooling is due to collisional excitation of H and He at high temperatures and of trace elements C$^+$, Si$^+$ etc. at low temperatures. These latter cooling agents are however depleted by large factors, 10 to 30 due to accretion on grains. They find that three phases are possible for the gas, respectively at $T \approx 10^4$ K, $T \approx 5 \times 10^3$ K and $T \approx 300$ K, the middle one being thermally unstable with an *e*-folding time as small as 10^6 years so that it should not practically exist in nature. Thus previous pictures (Field, 1962; Pikelner, 1967) in which the intercloud medium was identified with this intermediate phase must be rejected, the previously unknown high temperature phase being adopted instead.

This leads to a picture in which more than 90% of the mass of H$\textsc{i}$ is concentrated in dense clouds ($n(\text{H}\textsc{i}) \approx 14$–$7$ cm^{-3}, $T \approx 110$–80 K as z varies from 0 to 180 pc, in good agreement with average values ≈ 10 cm^{-3}, 100 K) in pressure equilibrium ($nT \approx 1200$ cm^{-3} deg at $z=0$ to 700 cm^{-3} deg at $z \approx 180$ pc) with the intercloud medium ($n(\text{H}\textsc{i}) \approx 0.1$) which is about 15% ionized. Since the agitation of the clouds is maintained by active H$\textsc{ii}$ regions and supernovae explosions at some 10 km/sec, which is close to the sound velocity in the intercloud medium, the scale heights Z are about the same for these two phases and of the order of $Z=180$ pc if the pressure of a magnetic field having about the equipartition value ($B=(12np)^{1/2} \approx 2.5\ \mu$G at $z=0$) is included.

Lower depletion factors of cooling agents might sometimes occur and account for unusually cold and dense clouds (for instance a reduction of this factor to 10 yields $n=30$ cm^{-3}, $T=40$ K at $z=0$).

The sizes of clouds in this model are limited on the small side by thermal conduction to a small fraction of $M_\odot$ and on the large side by gravitational instability (10^3–$10^4\ M_\odot$).

Note that at any place where the gas density is less than 15% of the local total density of matter, no clouds would form because the whole gas could then be accommodated in the hot phase.

This model leads to reasonable agreement with the observations. The clouds occupy some 4 to 7% of the total available volume with, on the average, a density 125 times that in the intercloud medium. The pulsar data on the average interstellar density of free electrons n_e agree with that in the model ($n_e \approx 0.016$ cm^{-3}) within a factor 2. Data on free-free absorption of low-frequency radio-waves (clouds) can also be accommodated with n_e in clouds an order of magnitude higher than the classical value 4×10^{-4}.

As far as the 21 cm line profile is concerned, the profile for the intercloud medium agrees well with profiles observed in many cases above $b \approx 50°$ where one might expect regions free of clouds. In other directions (especially anticentre), the clouds contribute 80 K to the brilliance temperature and the intercloud medium 35 K, giving a total of 115 K in good agreement with the observed value of 120 K. Absorption at 21 cm with a strong preference for cold clouds is also accounted for satisfactorily.

A somewhat similar picture with thermal instability leading to separation of two phases, hot intercloud and cold clouds, has also apparently been developed by J. H. Hunter (1969a, b).

In this model, the clouds themselves are formed by thermal instability filtering through a somewhat chaotic medium agitated by the formation of H$\textsc{ii}$ regions and the explosions of supernovae. This can very well lead to a population of clouds of all sizes reaching, through collision interactions, some kind of stationary state with a balance between the number of clouds formed and destroyed in each interval of mass (cf. Field and Saslaw, 1965). In particular, the depletion of the upper range of masses (10^3 to 10^4 $M_\odot$) by gravitational collapse is compensated by the formation of new large clouds by coalescence of smaller clouds during collisions, the kinetic energy dissipated probably being got rid of in the end as infrared radiation.

This general picture can be adapted to take into account the effects of a general magnetic field of the order suggested above and which might also play its rôle in the formation and the containment of clouds according to Parker's ideas (Parker, 1968; Lerche and Parker, 1968).

Once collapse of a large cloud occurs, it may, according to the usual views, be followed by fragmentation yielding a mass spectrum to be compared to that in galactic clusters and associations (cf. for instance Nakano, 1966). This process of star formation does not use up quickly the whole gas because, as already suggested by Oort (1954; cf. also Field and Saslaw, 1965), one or a few massive stars would tend to form at the centre of the cloud as OB stars and blow a good deal of the collapsing material away. Certainly on this picture, stars tend to form in groups in the spiral arms and there is indeed ample observational confirmation of this at least for large classes of stars (cf. for instance recent discussion by Hartmann, 1970a).

2. Individual Objects

Apart from these fairly general aspects, there exists also information, sometimes fairly detailed, on individual objects which, at first sight at least, seem to belong to the phases of star formation or early developments.

Those which have been known for the longest time are the globules (Bok and Reilly, 1947; Bok, 1948). An analysis of their distribution (Sim, 1968) shows a preferential concentration in those OB star systems with a large proportion of dust embedded stars. Their masses are of the order of $M_\odot$ (≈ 0.16–3 $M_\odot$) and their radii range from less than 0.1 pc to about 1 pc. Their association in many cases with young stars, their fairly regular shapes, and the order of magnitude of their masses have usually been considered as confirming Bok's suggestion that they are protostars.

Herbig (1970) however took a strong stand against this view arguing that they are practically absent in the two best-studied younger associations, Orion and NGC 2264, and that their velocity dispersion is different from that in the clouds. But Penston (1969) has studied in detail some globules visible against the Orion bright nebula and found the same general characteristic as above. Furthermore, his measures of the varying obscuration across them reveal a mass distribution which is identical to the projected density of an isothermal sphere at a very low temperature of the order of 10 K. This suggests certainly very strongly that the material has already relaxed inside these globules towards some kind of equilibrium against gravity. In fact, at the surface, GM/R and $\frac{1}{2}c^2$ are just about of the same order. It is difficult to see what else could happen to these globules but continuing contraction.

One should however note that, unless their obscuration properties are very different from those of ordinary interstellar grains, they could only give birth to relatively small mass stars. On the other hand, the circumstellar dust in which so many of the OB stars are embedded may correspond (Reddish, 1967) to as much as 50 $M_\odot$ around one single star if it contains a normal proportion of H and He. If the circumstellar cloud consists only of refractory grains, this mass may be reduced to about 0.5 $M_\odot$ and Reddish has noted the absence of Hα emission in about $\frac{3}{4}$ of these dust embedded OB star systems.

The whole question of the relationship between young stars and dust, which is also of course very well marked in the case of the T Tauri stars, is somewhat controversial. Is dust a primary factor in the formation of stars? Its shielding properties against outside stellar radiation were advocated a long time ago as giving rise to an extra compressing force which could help in the condensation of lumps of interstellar material. At present, Reddish and Wickramasinghe for instance (1968, 1969; Reddish, 1970a) argue that cloud collapse and star formation are directly related to the large drop in pressure accompanying the freezing of solid hydrogen on very cold grains (graphite or silicates) at temperatures close to 3 K. This could also explain the absence of Hα emission in some OB star systems, as reported above. But Werner and Salpeter (1970) find that the grain temperature in a cloud could hardly fall below about 4.5 K (cf. also Field, 1969).

Is the formation of dust itself, on the contrary, associated with the general phenomena of condensation giving birth to stars? One knows (Larimer, 1967) that finely-divided metal and silicate-rich grains can condense directly out of the gas at temperatures up to 1600 K. It is usually admitted also that graphite can form in the atmospheres of cool stars. In this way, dust could perhaps be considered as a by-product of stellar formation and some phases of stellar evolution, a point of view defended for instance by Herbig (1970) and, to some extent, by Hartmann (1970b) at the Liège meeting.

In the last few years, radioastronomy and infrared astronomy have revealed other objects which one tends naturally to associate with the formation of stars or their early life.

First of all, we might recall the cloudlets discovered by Heiles (1967) from intensified emission at 21 cm which have dimensions and masses respectively of the order of 1 pc and 1 $M_\odot$, in some ways, the radio counterparts of the optical globules.

The observations of Mezger and his colleagues at 15.4 GHz (cf. especially Schraml and Mezger, 1970, and Mezger, 1970) have established the existence of compact H II regions with densities $10^2 \leqslant n_H \leqslant 10^4$ cm^{-3}, radii $0.5 \leqslant R \leqslant 4$ pc and masses from around $M_\odot$ to a few times 10 $M_\odot$, embedded in usual extensive and tenuous H II regions. Since these condensations are not visible in Hα, a large source of obscuration accounting for several magnitudes in the visible must subsist in these regions. In many cases, the exciting stars are not visible either so that the obscuration must reach at least 10 magnitudes in the ultraviolet. How is it then that H II regions have developed around these obscured regions? On the other hand, how can the interstellar grains survive in the resulting hot gas with kinetic temperatures of the order of 10000 K? The answers to these questions will certainly bring interesting information on the optical and structural properties of the grains themselves. However, here we are more interested in whatever conclusions we can draw concerning the meaning of these compact H II regions with respect to star formation.

First of all, let us note that a recent survey at 15.4 GHz (Mezger, 1970) with an angular resolution of 2′ has shown that most nearby H II regions (within about 3 kpc) contain such compact components that the phenomenon seems to be fairly general. Mezger considers that we are witnessing here very early stages of the formation of Blaauw subgroups in OB-stars associations, each of these compact regions needing for their ionization one or more O stars which are most often completely obscured. Could these regions be hot cocoons with obscuring material hiding newly formed stars in their interiors? Or did these lumps of matter exist before the formation of the O stars which have caused H II regions to grow around them and in them? Are they protostars evolving on a longer time-scale?

This last possibility is certainly very likely in the case of the superdense H II condensations of very small dimensions (1.6×10^{-2} pc), high densities (10^5–10^6 at/cm^3) and small masses ($M(\text{H II}) \approx 2 \times 10^{-3} M_\odot$) revealed by the recent aperture synthesis observations of Webster and Altenhoff reported by Mezger at the Liège meeting. These could very well be the external collapsing layers of a protostar ionized by the ultraviolet radiation of surrounding newborn O stars. The same considerations apply probably to the small (2×10^{14} cm, 10^{-4} to $10^{-2} M_\odot$), dense (10^{10}–10^{13} at/cm^3) OH emission regions which have been known for a few years (Shklovsky, 1967; Mezger et al., 1967; Mezger and Robinson, 1968; Litvak, 1969), although we should remember here the words of caution of Prof. Kerr concerning the current interpretation.

In discussing OH emission associated with a dust cloud, Heiles (1969) derives from an analysis of the ratio of the intensities of the two emission lines at 1665 MHz and 1667 MHz, an excitation temperature of only 4 to 5 K. He favours a solution in which this corresponds really to the kinetic temperature of atoms in a dense cloud ($n = 2$ at/cm^3) which would then be unstable against contraction. The width of the OH lines could indeed correspond, in that case, to contraction velocities of the order of 1 km/sec leading to a complete collapse in about 10^6 years.

The recent discoveries of lines of H_2O (Knowles et al., 1969), NH_3 (Cheung et al., 1968, 1969), H_2CO (Palmer et al., 1969) also promise to yield new and important information on interstellar clouds and their physical characteristics.

What about infrared objects? Prof. Elsässer has already reviewed most of the evidence concerning infrared stars. Many of them are probably already well formed stars with part of their radiation thermalized at a few hundred degrees by circumstellar dust (or cold bodies anyway) and reradiated in the infrared. Their flat distribution, their associations with dark clouds, strong H II regions, sometimes OH emission, as well as the typical infrared excess of T Tauri stars etc., all suggest that again we have to deal with fairly young objects whose continued study is likely to yield new interesting information on the early phases of a star's life and perhaps on the formation of associated planetary systems.

Small infrared nebulosities with high densities ($n \geqslant 10^6$), such as those discovered by Ney and Allen (1969) in the Orion nebula should also find their place in this inventory of significant manifestations of stars in the making or at least in their early life which can be summarized in Table I adapted from Reddish (1970a).

TABLE I

Object	Number density in a.m.u./cm³
Region in between spiral arms?	
	— 0.01
Intercloud medium in a spiral arm	
	— 0.1
	— 1.0
Cloudlets	
Clouds	
	— 10^2
Compact H II regions	
Circumstellar clouds	— 10^4
Globules	
OH emission sources	
Superdense H II condensations	— 10^6
Small infrared nebulae	
C 109 Hα emission nebulae	— 10^8
Ambient nebulae	
of OH sources	— 10^{10}
OH sources	

Certainly, it reveals an extreme variety of conditions with atomic number densities varying from values perhaps as low as $10^{-2}/cm^3$ for the most dilute interstellar medium to values perhaps as large as 10^{13}–10^{14} in some of the densest OH emission regions. The latter is of the order of what is considered usually (cf. for instance Hattori *et al.*, 1969) as characteristic of the phase at which a lump of matter may start its life as an opaque protostar.

As we have already recalled, in many cases, stars form in groups sometimes fairly numerous. However this does not necessarily imply that they result from the fragmentation of a large collapsing cloud. Up to now, there is no evidence that any of the prestellar objects discussed above belong to families presenting the kind of relative systematic motions (contraction or expansion) which one would expect in such cases. Furthermore, as noted by Reddish (1970b), the volume number densities of OB stars in 66 OB clusters and associations studied don't seem to show any dependence on age. This suggests that these groups are not expanding from denser concentration of stars at a significant rate in the lifetime of an OB star. Thus it would seem that star formation occurs in the whole range of densities from the most dense clusters with smoothed-out mass densities $\bar{\varrho}$ of the order of 10^{-18} g cm^{-3} to the least dense associations with $\bar{\varrho} \approx 10^{-24}$ g cm^{-3}, with OB stars forming in the field.

On the other hand, the star rings, in favour of which Prof. Schmidt-Kaler made such a good case in his lectures and the distribution of which correlates rather well with the spiral arms, seem to open a new interesting line of evidence on star formation. It suggests strongly that the condensation of stars has taken place in ellipsoidal shells produced in some way (ejection of stellar matter, sweeping of interstellar matter, probably both) by a central object long since vanished in most cases and which might have been one of those single WR stars surrounded indeed by important expanding shells and which might represent an early stage of the process. Even if it can only be a secondary step in star formation, its continued study no doubt will be fruitful. It might indeed indicate that, in many cases, very large mass stars tend to form which then become vibrationally unstable when nuclear reactions start in their interiors leading to the slow ejection of a large part of their initial mass.

3. Gravitational and Thermal Instability

Granted that there are already hot stars and supernovae which cause ionization and agitation in the interstellar medium, one can conceive mechanisms which, on the average, maintain practically a steady distribution of clouds, as already outlined in reporting the recent investigation of Field et al. (1969). In this particular picture, gravitational instability comes in only to cause the collapse of the largest clouds and later in the process of fragmentation. Even so, these last steps are not so easily amenable to a really satisfactory treatment.

Furthermore, there must have been, at some time or other, a first generation of stars to be born in a medium not yet conditioned by the presence of pre-existing stars. Of course, we don't know much about the physical conditions that might have prevailed in this primeval galactic nebula. It must have had some net angular momentum and have been agitated by currents associated with its own birth and it may have been pervaded by general magnetic fields. Did it contain dust already? What was the density of cosmic rays at that time, if there were any, and what was their influence on the nebular gas? These are all difficult questions opening vast fields of speculation. However, one might feel entitled to start with conditions as simple as possible: some kind of very extended uniform medium with no motions and no magnetic fields, and ask whether anything can ever happen to it, granted that very slight natural fluctuations must be present.

This kind of question was formulated very early after the discovery of universal gravitation and Newton himself, in a letter to Bentley in 1692, expresses his opinion on it in a way which illustrates very well the fundamental aspects of the problem. He considers that, in the case of a finite uniform medium, under the effect of gravitation (in the absence of any other forces real or relative), the whole mass should concentrate in the end around its centre of mass to form a single body in which pressure and elasticity will balance gravity. On the other hand, if the medium is infinite, then it seems natural to him that the mass should subdivide, according to the small fluctuations present, in an infinite number of distinct masses.

Apparently however, one had to wait until 1902, before the first attempt to formulate the problem mathematically was made by Jeans (1902), yielding his celebrated criterion of gravitational instability. However, anyone who has studied Jeans' paper must have encountered conceptual difficulties and as Spitzer puts it in his latest book (1968), one might even be tempted to say that "it does not represent a solution of any physical problem". Perhaps, I wouldn't go as far as this but certainly, strictly speaking, Jeans' criterion applies only to an infinite uniform and static medium.

As is well known, Jeans' criterion can be derived easily from the general equations of conservation which can be written quite generally in Eulerian notations and with standard symbols (cf. for instance Ledoux, 1965)

$$\frac{\partial \varrho}{\partial t} + \mathbf{v} \cdot \text{grad} \, \varrho + \varrho \, \text{div} \, \mathbf{v} = 0 \tag{1}$$

$$\frac{\partial \mathbf{v}}{\partial t} + (\mathbf{v} \cdot \text{grad}) \, \mathbf{v} = - \, \text{grad} \, \Phi - \frac{1}{\varrho} \, \text{grad} \, p \tag{2}$$

$$\frac{\partial p}{\partial t} + \mathbf{v} \cdot \text{grad} \, p - \frac{\Gamma_1 p}{\varrho} \left(\frac{\partial \varrho}{\partial t} + \mathbf{v} \cdot \text{grad} \, \varrho \right) =$$

$$= (\Gamma_3 - 1) \varrho \left[- \mathscr{L} - \frac{1}{\varrho} \, \text{div} \, \mathbf{F} \right], \tag{3}$$

where we have introduced Field's generalized heat loss function $\mathscr{L}$ per unit mass and time instead of the thermal rate of nuclear energy generation in the stellar case.

We must add Poisson's equation

$$\nabla^2 \Phi = 4\pi G \varrho . \tag{4}$$

The general linearized perturbed equations become

$$\frac{\partial \varrho'}{\partial t} + \mathbf{v}' \cdot \text{grad} \, \varrho + \mathbf{v} \cdot \text{grad} \, \varrho' + \varrho' \, \text{div} \, \mathbf{v} + \varrho \, \text{div} \, \mathbf{v}' = 0 \tag{5}$$

$$\frac{\partial \mathbf{v}'}{\partial t} + (\mathbf{v}' \cdot \text{grad}) \, \mathbf{v} + (\mathbf{v} \cdot \text{grad}) \, \mathbf{v}' = - \, \text{grad} \, \Phi' + \frac{\varrho'}{\varrho^2} \, \text{grad} \, p - \frac{1}{\varrho} \, \text{grad} \, p' \tag{6}$$

$$\frac{\partial p'}{\partial t} + \mathbf{v}' \cdot \mathrm{grad}\, p + \mathbf{v} \cdot \mathrm{grad}\, p' + \left(\frac{\Gamma_1 p}{\varrho}\right)' \varrho\, \mathrm{div}\, \mathbf{v}$$

$$-\frac{\Gamma_1 p}{\varrho}\left(\frac{\partial \varrho'}{\partial t} + \mathbf{v}'\, \mathrm{grad}\, \varrho + \mathbf{v} \cdot \mathrm{grad}\, \varrho'\right)$$

$$= (\Gamma_3 - 1)\varrho\left[-\mathscr{L}' - \frac{1}{\varrho}\mathrm{div}\, \mathbf{F}' + \frac{\varrho'}{\varrho^2}\mathrm{div}\, \mathbf{F}\right]$$

$$+ \left[(\Gamma_3 - 1)\varrho\right]'\left[-\mathscr{L} - \frac{1}{\varrho}\mathrm{div}\, \mathbf{F}\right] \tag{7}$$

$$\nabla^2 \Phi' = 4\pi G \varrho', \tag{8}$$

where the prime denotes Eulerian perturbations.

If we assume, as Jeans, that the initial state is static $(\mathbf{v}=0)$, uniform $(p=c^t,\ \varrho=c^t)$ and infinite $(\mathrm{grad}\,\Phi=0)$ and if we neglect the dissipation terms on the left of (3) and (7), we obtain

$$\frac{\partial \varrho'}{\partial t} + \varrho\, \mathrm{div}\, \mathbf{v}' = 0 \tag{5'}$$

$$\frac{\partial \mathbf{v}'}{\partial t} = -\mathrm{grad}\, \Phi' - \frac{1}{\varrho}\mathrm{grad}\, p' \tag{6'}$$

$$\frac{\partial p'}{\partial t} - \frac{\Gamma_1 p}{\varrho}\frac{\partial \varrho'}{\partial t} = 0 \tag{7'}$$

$$\nabla^2 \Phi' = 4\pi G \varrho', \tag{8'}$$

where all the coefficients are constant. A conceptual difficulty arises here because Poisson's equation for an infinite medium is not really significant but its perturbed form (8') is correct since, by integration, it yields the correct gravitational forces acting in the perturbed medium.

Time and space can be separated, writing that all perturbations are proportional, say, to

$$e^{st} \cdot e^{i\mathbf{k}\cdot\mathbf{r}}.$$

Taking the divergence of (6') and using (5'), (7') and (8') to eliminate respectively $\mathrm{div}\,\mathbf{v}'$, $\nabla^2\Phi'$ and p' one gets the dispersion relation

$$s^2 = 4\pi G \varrho - k^2 \frac{\Gamma_1 p}{\varrho} = 4\pi G \varrho - k^2 c^2, \tag{9}$$

where c is the Laplacian sound velocity and $k^2 = k_x^2 + k_y^2 + k_z^2$. There is thus a critical wave-number

$$k_J = \sqrt{(4\pi G \varrho)}/c,$$

such that for $k > k_J$, s^2 is negative and the perturbation gives rise to some kind of generalized acoustic waves which in this 'adiabatic' approximation are undamped and the medium is gravitationally stable. On the other hand, if $k < k_J$, then s^2 is positive and the medium is gravitationally unstable, the perturbation considered giving rise to exponentially increasing condensations centered on the regions of initial maximum compression.

If, for simplicity, we assume $k_x = k_y = k_z = k/\sqrt{3}$ this gives for the critical dimensions d_J of the condensations above which continuous contraction occurs

$$d_J = \frac{\pi}{(k_x)_J} = \frac{\pi c\sqrt{3}}{\sqrt{4\pi G\varrho}} = \frac{\pi\sqrt{3\Gamma_1 p}}{\sqrt{4\pi G\varrho}} = \sqrt{\frac{3\Gamma_1\pi RT}{4G\bar{\mu}\varrho}}, \tag{10}$$

where R is the gas constant and $\bar{\mu}$, the mean molecular weight. The e-folding time for a condensation of dimensions $d > d_J$ is given by

$$\tau = \frac{1}{\sqrt{4\pi G\varrho\left[1 - (d_J/d)^2\right]}}, \tag{11}$$

which decreases continually as d increases and approaches fairly rapidly the free-fall time

$$\tau_f = \frac{1}{\sqrt{4\pi G\varrho}}. \tag{12}$$

Taking $\Gamma_1 = 1$ (isothermal perturbation) and introducing the numerical values, the critical Jeans' length becomes

$$d_J = 7.6 \times 10^7 \sqrt{\frac{T}{\bar{\mu}\varrho}} \quad \text{cm} \tag{13}$$

corresponding to a critical mass

$$M_J = 4.4 \times 10^{23} \frac{(T/\bar{\mu})^{2/3}}{\varrho^{1/2}}. \tag{14}$$

For values characteristic of the intercloud medium considered above this gives

$$M_J \approx 5 \times 10^8 \ M_\odot.$$

On the other hand, for conditions in the average cloud

$$M_J \approx 10^4 \ M_\odot.$$

One may note that a very similar result is obtained (factor $\frac{1}{2}$ to $\frac{1}{3}$) for the mass of a sphere, such that its gravitational potential at the surface is just equal to the thermal kinetic energy per unit mass $(GM/R \approx \frac{1}{2}\bar{c}^2)$.

The case of an infinite isothermal nebula stratified in a given direction (say z) under its own gravitation (Ledoux, 1951) is perhaps a little more satisfactory. If ϱ_0 is the density in the symmetry plane ($z = 0$), ϱ tending towards zero for z tending towards $\pm\infty$, the critical length for perturbations along the (x, y) directions (say $e^{i(k_x x + k_y y)}$) is again given by an expression similar to (10) with $\varrho = \varrho_0$ and affected by a slightly different numerical factor. In particular, if there is some axis of symmetry normal to the plane $z=0$, the instability could give rise to the condensation of cylindrical slabs with an horizontal thickness of this order.

Furthermore, as Simon (1965) has shown, there is, in this case, a wavelength of maximum instability about 2.2 to 2.8 times the critical length depending on whether $\Gamma_1 = 1$ or $\frac{5}{3}$. Assuming initial conditions $\varrho_0 \approx 2 \times 10^{-23}$ g/cm^3, $T = 10^4$ K this would give cylindrical slabs at distances of some 2.5 kpc contracting with an e-folding time of 2 to 3×10^7 years.

One must note however that there is no instability for perturbations $(e^{ik_z z})$ producing condensations and rarefactions in planes parallel to the symmetry plane $z=0$. This is due to the fact that the medium is in equilibrium in the z-direction. More generally, gravitational instability never occurs in a configuration which has reached a state of true static equilibrium. In fact, it has no meaning for such configurations and the Jeans' length is always greater than its global dimensions. This is the basis of many of the difficulties in the way of a clear-cut treatment of gravitational instability and, at times, it has created confusions. To get significant results, the initial unperturbed state has to be far from equilibrium at least in some sense and, in the previous examples, this was realized by letting the medium extend to infinity at least in some directions. This is a simple way to define a state far from equilibrium but which is still tractable although one might not be completely satisfied with it.

Another possibility is to introduce motions, simple ones as far as possible. Perhaps the simplest of all is obtained by making the isothermal stratified nebula considered above rotate with a constant angular velocity Ω, the centrifugal force in the plane being balanced by a gravitational field. One may even keep it infinite, in the x, y directions, if one introduces a distribution of mass (stars for instance) which does not participate in the perturbation and balance the centrifugal force in the initial state (Simon, 1967).

In that case, the rotation decreases the instability although the critical length and that of maximum instability are not very much affected. But, as the rotation increases, the e-folding times increase for all unstable modes until finally the latter vanish for a value of $\Omega^2/2\pi G\varrho_0$ which varies from about 0.4 to 0.3 as Γ_1 varies from 1 to $\frac{5}{3}$.

These numerical results are in general agreement with the analytical discussion of the marginal case by Goldreich and Lynden-Bell (1965, cf. also Savranov, 1960) who considered a finite disk and differential rotation, the latter increasing further the stability if Ω increases with the distance to the axis, it being understood that, in any case, it remains sufficiently small for the gravity to prevail at all points.

These types of instability may be significant for the formation of spiral arms especially if, as mentioned above, they extend vertically to fairly large heights on both sides of the galaxy plane. And after all, in many cases this may be the first step towards star formation.

At this very large scale, it seems now unlikely that a general magnetic field could play a decisive rôle. But other factors such as the thermal effects which are also neglected in Jeans' criterion can exercise a significant influence. They have been the object of a detailed discussion by Field (1965) in which he took more or less the opposite point of view, neglecting gravitational effects. This was justified for many of the cases to which he wanted to apply his criterion of thermal instability but there must be interesting cases where the two types of effects can interfere and it should not be too difficult to extend the linear analysis of Field to cover such cases.

Field's equations can be obtained immediately from Equations (5) to (7) by dropping the terms in Φ and Φ'. Since he retains the acceleration terms $\partial v'/\partial t$, his discussion covers also the damping or amplification of acoustic waves which are somewhat irrelevant to our present considerations.

If we neglect this acceleration term but keep the gravity and assume again the initial state to be static, uniform and infinite, the equations reduce, after introducing the factor $\exp(st + i\mathbf{k}\cdot\mathbf{r})$ to

$$\varrho' + i\mathbf{k}\cdot\delta\mathbf{r} = 0 \tag{5''}$$

$$4\pi G\varrho' - \frac{1}{\varrho}k^2 p' = 0, \tag{6''}$$

where Φ' has already been eliminated by (8),

$$s\left(p' - \frac{\Gamma_1 p}{\varrho}\varrho'\right) + (\Gamma_3 - 1)\varrho\left[\left(\frac{\partial \mathscr{L}}{\partial \varrho}\right)_T \varrho' + \left(\frac{\partial \mathscr{L}}{\partial T}\right)_\varrho T' + \frac{Kk^2 T'}{\varrho}\right] = 0, \tag{7''}$$

where, with Field, we assume that $\mathscr{L}$ can be defined as a function of ϱ and T. Furthermore, the conduction flux $\mathbf{F}$ is supposed to be given by

$$\mathbf{F} = -K \operatorname{grad} T, \tag{15}$$

where the thermal conductivity coefficient K is treated as a constant, independent of the physical conditions. This might not be always a very good approximation but again it should be fairly easy to improve on this. For a perfect gas, we may add the following relation

$$\frac{p'}{p} - \frac{\varrho'}{\varrho} - \frac{T'}{T} = 0. \tag{16}$$

Eliminating ϱ', p' and T' from (6''), (7'') and (16) yields the dispersion relation

$$s\left[1 - \frac{4\pi G\varrho}{c^2 k^2}\right] = \frac{(\Gamma_3 - 1)T}{c^2}\left[\frac{\varrho}{T}\left(\frac{\partial \mathscr{L}}{\partial \varrho}\right)_T - \left(\frac{\partial \mathscr{L}}{\partial T}\right)_\varrho - K\frac{k^2}{\varrho}\right]$$
$$+ \frac{4\pi G\varrho}{c^2 k^2}\frac{\bar{\mu}}{R}(\Gamma_3 - 1)\left[\left(\frac{\partial \mathscr{L}}{\partial T}\right)_\varrho + K\frac{k^2}{\varrho}\right]. \tag{17}$$

We notice that if Jeans' condition for stability is well satisfied ($k^2 \gg 4\pi G\varrho/c^2$), the relation above reduces essentially to

$$s = \frac{\Gamma_3 - 1}{c^2}T\left[\frac{\varrho}{T}\left(\frac{\partial \mathscr{L}}{\partial \varrho}\right)_T - \left(\frac{\partial \mathscr{L}}{\partial T}\right)_\varrho - K\frac{k^2}{\varrho}\right], \tag{18}$$

and the condition for thermal instability $(s > 0)$ can be written

$$\frac{\varrho}{T}\left(\frac{\partial \mathscr{L}}{\partial \varrho}\right)_T - \left(\frac{\partial \mathscr{L}}{\partial T}\right)_\varrho - K\frac{k^2}{\varrho} > 0. \tag{19}$$

This is Field's condition which is thus significant for perturbations of wavelengths much smaller than Jeans' critical length, i.e. when, on the scale considered, gravity is inefficient.

In the approximation used here $(K = ct)$, the conduction always favours stability. If its influence is negligible, then the condition for thermal instability reduces simply to

$$\left(\frac{\partial \mathscr{L}}{\partial T}\right)_\varrho - \frac{\varrho}{T}\left(\frac{\partial \mathscr{L}}{\partial \varrho}\right)_T = \left(\frac{\partial \mathscr{L}}{\partial T}\right)_p < 0. \tag{20}$$

It demands that, in a modification at constant pressure, the heat-loss function should increase when the temperature decreases, which is intuitively satisfactory. Let us note also that if $(\partial \mathscr{L}/\partial T)_\varrho$ is positive, this condition can also be written, for a medium in thermal equilibrium $(\mathscr{L} = 0)$ and using the definition of the implicit derivative,

$$\left(\frac{d \ln T}{d \ln \varrho}\right)_\mathscr{L} < -1. \tag{21}$$

On the other hand, because of the stabilizing effect of conduction, s tends always to become negative in going through zero for k large enough (say $k > k_K$) and Field finds that, for given physical parameters, there is a mode of maximum instability which corresponds to a value of k, say k_M, which varies from about $(\frac{1}{3})$ to $(\frac{1}{2}) k_K$ as the conductivity K becomes more and more important.

From the general formula (17) above, the condition becomes more complicated as one gets to values of k small enough to approach Jeans' critical value, the last term being then able to exercise some influence while the coefficient of s tends to decrease. Of course, the condition should then be generalized to include the inertial terms as well.

The applications depend of course mainly on the detailed mechanisms of heat gains and losses which enter the evaluation of $\mathscr{L}$ and which are affected by the prevailing physical conditions, the abundances of various elements, the fraction of matter in grains and the nature of the latter.

We have already recalled, at the beginning of the first section, the recent results of Field *et al.* (1969) for the general interstellar medium which, due to thermal instability (cf. Figure 1 and condition (21)) should comprise two phases, one at high $(\approx 10^4 \, \text{K})$ and the other at low $(\approx 10^2 \, \text{K})$ temperature in pressure equilibrium with each other.

If this process can work early in the history of the Galaxy, it could account for the formation of clouds in spiral arms if the latter are already there due to some kind of gravitational instability. Or, the clouds formed in a disk layer of some 30 pc initial thickness could then gather under dynamical influences, following the scheme of Prendergast and Miller summarized earlier in this course by Mrs. Burbidge in spiral arms. But, of course, we don't know whether all the physical factors required are present and in the right proportions at the time. Possibly, other thermally stable

phases like one at $T \approx 10^6$ K with bremsstrahlung as the chief cooling process (Field
et al., 1969) might also have to be considered in these early stages.

Anyway, the process certainly seems capable of yielding a whole range of clouds
at present and, in the resulting steady state, their densities are larger than the density
of the intercloud medium by a large factor of the order of 100. If thermal instability
can lead to this kind of density contrast, it is very satisfactory because the usual
gravitational instability might still, in realistic cases, come up against a serious
difficulty related to the notion of Roche limit.

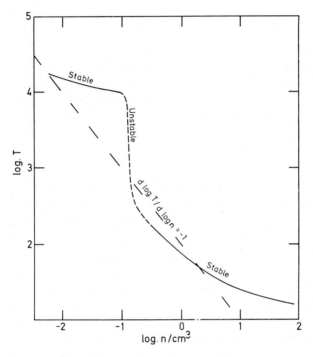

Fig. 1.   Possible phases in the interstellar medium according to Field *et al.* (1969) and their stability.

If a body M_2 whose cohesion is due to gravity alone is submitted to the gravitational
field of another one M_1, it remains stable against the tidal effect of the latter only if
its density ϱ_2 is some ten times the density of M_1 spread in a sphere of radius equal
to the distance d_{12} separating the centres of mass of M_1 and M_2. Consider a local
perturbation in a large but *finite* and more or less spherical system at a distance r
from its centre. According to what was said above, if the gravitational forces are the
only acting ones, this perturbation could persist and possibly grow in concentration
only if the density perturbation is finite and such that it raises the local density by a
factor 10 with respect to the density of the unperturbed central region of radius r.
Of course, this kind of tidal effect could also affect adversely the development of
thermal instability itself and a systematic investigation of the effect might be worthwhile.

4. Evolution of Clouds and the Formation of Protostars

Anyway, interstellar clouds exist with finite increment of density with respect to the surrounding medium. A considerable amount of work has been done starting with various distributions of physical conditions and velocity in these clouds and attempting to follow their subsequent history.

Most of this work has been done for spherical configurations, but even in absence of anisotropic forces like those resulting from rotation or magnetic field, large initial deviations from spherical symmetry seem likely especially when their formation is due to thermal instability. Even when the gravitational field takes over, an initial eccentricity tends to be amplified by the non-spherical gravitational field (Lin *et al.*, 1965) at least as long as the rôle of pressure is negligible. If sufficiently pronounced, this may possibly have some importance for the further fragmentation of the cloud. However, once the pressure gradient tends to balance gravity, the configuration in quasi-static equilibrium should again, in absence of rotation or magnetic field, approach the spherical shape as it corresponds to the minimum of the potential energy.

One of the best known cases relates to the instability of a finite, spherical, isothermal cloud of temperature T_0 and mass M_0 kept in static equilibrium by an external pressure p_a applied on its surface by the ambient medium. Ebert (1957) and McCrea (1957) have shown that there are no equilibrium states if

$$p_a > 17.6 \frac{RT_0^4}{4\pi G^3 \bar{\mu}^4 M_0^2}. \tag{22}$$

In that case, the isothermal sphere starts contracting, the pressure gradient becoming more and more negligible as compared to gravity as the contraction proceeds.

Inversely, if p_a is given, there exists a critical mass

$$M_c = \frac{4.2(RT_0)^2}{\sqrt{4\pi \bar{\mu}^2 G^{3/2} p_a^{1/2}}} \tag{23}$$

above which all such configurations must collapse. For instance, for a small H I region, in isothermal equilibrium at temperature T_I embedded in a H II region $(p_a = n_{II} k T_{II})$, one gets

$$M_c \approx 3 \times 10^{34} \frac{T_I^2}{(n_{II} T_{II})^{1/2}}.$$

With $T_{II} \approx 10^4\,\mathrm{K}$, its value varies from $10^4\,M_\odot$ for $T_I = 100$ and $n_{II} = 10^{-2}$ to $1\,M_\odot$ for $T_I \approx 10$ and $n_{II} = 10^2$.

The first case is not so different from the situation described by Field *et al.* (1969), the somewhat larger value of n_{II} ($\approx 10^{-1}$) reducing M_c to some $5 \times 10^3\,M_\odot$. Of course, for this to apply, the cloud should strictly have relaxed fairly close to true hydrostatic isothermal equilibrium which, in this case, is not too likely. But, anyway, the critical mass above is not so very different from Jeans' critical mass.

On the other hand, as we have seen, according to Penston (1969), globules are very close to isothermal spheres of very low temperature $T \approx 10$ K and high central density ($n \approx 10^5$ to 10^6/cm^3). For instance, in a H II region, they would be on the verge of instability if $n_{II} \approx 10^2$. In such cases, collapse could be started by an accidental elevation of pressure in the surroundings, for instance due to the passage of a shock wave or the formation or extension of a H II region. However one should not forget that ionization and shock fronts could also destroy some of these small clouds.

In general, following up the evolution of a cloud in detail is a complicated task and the efforts have been concentrated on two approaches. One, fairly general, attempts to determine only roughly this evolution in terms of some mean density and mean temperature of the cloud in a domain of ϱ and T as large as possible. In the other, one tries to follow as precisely as possible, through numerical integration, the evolution of a few typical clouds with various initial conditions.

The first method is best represented in the work of Hayashi and his colleagues (Hayashi and Nakano, 1965; Hattori *et al.*, 1969). The idea is to compute the total cooling and heating rates, Λ and Γ respectively, as functions of ϱ and T. Once these functions are known, one can define various characteristic times.

$t_c = 3R\varrho T/2\bar{\mu}(\Lambda - \Gamma)$ defines, when $\Lambda > \Gamma$, the time it takes for the temperature of the cloud to decrease by a factor e (cooling time);

$t_h = 3R\varrho T/2\bar{\mu}(\Gamma - \Lambda)$ has a similar meaning for the heating time ($\Gamma > \Lambda$);

$t_f = (32\pi G\varrho/3)^{-1/2}$ is some kind of free-fall time ($kT/\bar{\mu}m_H \ll GM/R$) during which the temperature increases by a factor e due to compression. Hayashi supposes that the corresponding contraction is adiabatic ($\varrho \sim T^{3/2}$), which is not a bad approximation when this time-scale is significant;

$t_e = (3M/4\pi\varrho)^{1/3} (\bar{\mu}/8RT)^{1/2}$ is an expansion time ($kT/\bar{\mu}m_H \gg GM/R$) during which the temperature decreases (adiabatically) by a factor e.

In the ($\log\varrho$, $\log T$) plane (cf. Figure 2) the line $t_e = t_f$ separates the regions where

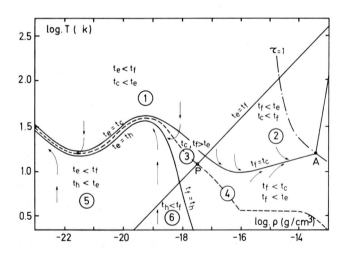

Fig. 2.   Evolutionary tracks, according to Hattori *et al.* (1969), in the ($\log T$, $\log\varrho$) plane (explanations in the text) for $M = 1\ M_\odot$.

the pressure effects predominate on the gravity (above, $kT/\bar{\mu}m_H \gg GM/R$) from those where the gravity is dominant (below, $kT/\bar{\mu}m_H \ll GM/R$).

Of course, the quantity $\Lambda - \Gamma$ is nothing but Field's heat loss function $\mathscr{L}$ and $\mathscr{L}=0$ is the condition for thermal balance. The corresponding line is contained between the two curves $t_c = t_e$ and $t_h = t_e$ or $t_c = t_f$ and $t_h = t_f$ also drawn on the diagram. The condition (21) for thermal instability should apply along the line $\mathscr{L}=0$ in the region of the diagram where the influence of gravity is practically negligible (above the line $t_e = t_f$).

The solid curves divide the diagram in six regions corresponding to the various inequalities between the different time-scales. The optical depth of the particular cloud considered becomes equal to one on the curve $\tau = 1$, the phases above and to the right of this curve being opaque while below and to the left they are treated as completely transparent. In this diagram, the dominant process corresponding to the shortest time-scale is immediately apparent in the various regions and one can draw schematically the evolutionary path of a cloud starting with given initial conditions (ϱ, T) at any point. For instance, in region (1), the cloud is expanding but the cooling is so efficient that the temperature decreases much more rapidly than the density and the representative point goes down nearly vertically until it approaches the curve $t_e = t_c$ and turns to the left, the cloud expanding along the line $t_e = t_h$. The arrows in the different regions of the diagram illustrate the possible evolutionary paths.

The main conclusion is that if a cloud is born with a density smaller than that corresponding to point P (about 2×10^{-18} g/cm³ for the cloud $M = 1\ M_\odot$) it finally expands nearly along the line $t_e = t_h$ whatever its initial temperature and returns to the interstellar medium. On the contrary, if it is born with a density higher than 2×10^{-18} g/cm³, it finally contracts nearly along the curve $t_f = t_c$ to the point A where it enters the opaque phase of its contraction at about $\varrho \approx 4 \times 10^{-14}$ g/cm³ and $T \approx 15$ K.

In the last version of their work (Hattori et al., 1969), the authors discuss carefully all the cooling processes for a chemical composition

$$(n_H + 2n_{H_2}): n(C): n(O): n(Si) = 10^6: 400: 890: 32$$

admitting that all the H_2 molecules are of the para-type with $n_{H_2} = 0.1\ n_H$ in the stage transparent to radiation. The cooling processes considered are thermal emission from interstellar grains $(n_g = 10^{-13}\ n$, n being the total number density in atomic mass units) and line emission by H_2 molecules, C^+ and Si^+ ions and C and O atoms. The heating processes are the absorption of star-light by grains and by C and Si atoms and the ionization of H by cosmic ray particles with an energy density at 10 MeV equal to one tenth that adopted by Hayakawa et al. (1961). Self-absorption of line-emission photons is also taken into account.

They have treated the problem for four masses, 10^4, 10^2, 1 and $10^{-2}\ M_\odot$. For the first two masses the results are qualitatively very similar to those for 1 $M_\odot$, the critical density (point P) below which the cloud returns to the interstellar medium decreasing respectively to values 10^{-24} and 10^{-20} g/cm³ while the position of the opaque phases (points A) is relatively little affected.

On the other hand, the case of a very small cloud ($M \approx 0.01\ M_\odot$) is quite different, the point A lying very close to the curve $t_e = t_f$ typical of gravitational equilibrium. To be able to contract, the cloud must be born with a density greater than about 6×10^{-15} g/cm^3. It can then get very quickly on the line $t_e = t_f$ along which it contracts slowly while radiating according to the Kelvin-Helmholtz scheme, a refinement of a previous result of Gaustad (1963) who found a somewhat larger critical mass of the order 0.1 $M_\odot$.

They have also studied the sensitivity of the evolution to changes in some of the uncertain parameters like the abundance of H$_2$ or the density of cosmic rays and found that this mainly affects the large mass clouds ($M \gtrsim 10^4\ M_\odot$): very little really as far as the variations of H$_2$ is concerned while a decrease of the intensity of cosmic rays by a factor 10^2 reduces the critical density for contraction to some 6×10^{-25} g/cm^3 i.e. about 10 times smaller than before.

In all this, the cloud has been considered in itself, the surrounding medium being neglected. But the influence of the latter may be significant in some cases and a more general approach taking the boundary conditions into account would be interesting.

Hayashi and his co-workers have extended their discussion to the evolution in the opaque stages past the point A, starting from:

(a) the conservation of energy

$$T \frac{dS}{dt} = -\frac{1}{\varrho} \operatorname{div} \mathbf{F} = -\frac{dL(r)}{dm(r)}$$

with the definition of the entropy S for a perfect gas

$$T \frac{dS}{dt} = \frac{RT}{\bar{\mu}} \frac{d}{dt} \ln \frac{T^{1/(\gamma-1)}}{\varrho},$$

where, at the beginning at least, γ is set equal to $\frac{7}{5}$, all H being in molecular form;

(b) the equation of motion which, at the beginning again just beyond point A, corresponds essentially to free-fall and can be written in terms of the mean density

$$\frac{1}{\varrho} \frac{d\varrho}{dt} \approx 2(6\pi G\varrho)^{1/2};$$

(c) the definition of $L(r)$ associated with the diffusion of photons

$$L(r) = -4\pi r^2 \frac{4acT_R^3}{3\kappa\varrho} \frac{dT_R}{dr}.$$

Replacing everywhere the variables by their mean values and the space derivatives by ratios across the whole configuration ($dL(r)/dm(r) \approx L/M$), they get an equation for the evolution in the ($\log T, \log \varrho$) plane which can be written

$$\frac{1}{\gamma-1} \frac{d \log T}{d \log \varrho} = 1 - \Phi \tag{24}$$

with

$$\Phi = \frac{4ac}{3R} \frac{\bar{\mu}}{M^{2/3}} \left(\frac{\pi}{6G}\right)^{1/2} \frac{T^3}{\kappa\varrho^{11/6}} \left(\frac{T_R}{T}\right)^4,$$

where the radiation temperature T_R soon becomes comparable to the gas temperature. The integration of Equation (24) gives the evolutionary path starting from A. As Φ decreases, this path approaches asymptotically the adiabat as illustrated in Figure 3 for three masses. The solid curves correspond to $n_g = 10^{-13}(n_H + 2n_{H_2})$ and the dashed curves, to $n_g = 10^{-14}(n_H + 2n_{H_2})$. The dots represent the points where, on each track, the star becomes opaque (points A). One sees that the evolution of the protostar is rather insensitive to the mass. This facilitates the study of the subsequent evolution. In particular, if one of the large cloud subdivides after point A, in fragments which are themselves opaque, the latter will follow the same adiabat as the parent cloud and this differs only slightly from the track of a protostar born directly, in the transparent stage, with the same mass as the fragment.

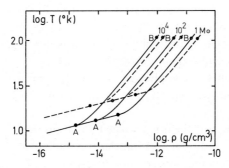

Fig. 3.   Evolutionary tracks in the $(\log T, \log \varrho)$ plane beyond point A, according to Hattori et al. (1969) for three masses.

A general objection could be raised concerning the use throughout this discussion of only average values of density and temperature while there must be some kind of stratification. Thus, the representative points of various parts of the star might fall, in the $(\log T, \log \varrho)$ plane, in regions corresponding to different behaviour and evolutionary paths so that, in some cases, for instance, an envelope could expand and return to the interstellar medium while a central core could contract and evolve to form a star.

Of course, the type of work needed to cover these aspects is of the second detailed type mentioned above. At the Liège symposium, Wright (1970; cf. also Disney et al., 1968, 1969) presented such an hydrodynamical program supposed to give the detailed evolution of a cloud contracting from $\varrho \approx 10^{-23}$ g/cm^3 to the point where thermonuclear reactions, possibly in a fragment, come into play. He noted that a first phase of free fall with rapid cooling by radiation is followed by a phase during which radiative cooling and compressional heating more or less balance at a constant temperature of the order of 10 K, subsidiary fluctuations tending to grow, leading possibly to fragmentation. Somewhat similar computations were made already by C. Hunter (1962, 1964) and J. H. Hunter Jr. (1966, 1967, 1969a, b) and a detailed comparison between them would certainly be of interest.

Wright notes also that in clouds of some 500 $M_\odot$ or smaller, heating by cosmic rays in the external layers can expand the latter sufficiently to lead to their escape, the separation occurring at about half the radius.

Some way beyond the point A, where the cloud becomes opaque, in the vicinity of the end of the trajectories in Figure 3 (say point B), this type of detailed calculations becomes necessary anyway to evaluate the exact distribution of physical conditions inside the protostar when it reaches a true quasi-static equilibrium corresponding to phases of slow contraction towards the main sequence.

From fairly simple hydrostatic considerations (Hayashi, 1961; cf. also Ledoux, 1967) one knows that there are no strict equilibrium configurations in the Hertzsprung-Russell diagram to the right of a nearly vertical line corresponding to effective temperature of the order of 2500 to 3000 K (Hayashi forbidden region). This line, let us call it H, corresponds to purely convective stars along which, following Hayashi, all the stars contract at least for some time on their way towards the main sequence. Since the points A and B are well to the right in this forbidden region, the corresponding configurations must collapse rapidly towards the line H.

These phases are further complicated by the dynamical instabilities due to the dissociation of H_2 and the ionization of H (Cameron, 1962) which must occur on the way.

Hayashi (1965) again was the first to sketch the transition from point B to the first state of quasi-static equilibrium. Noting that the radiative losses must probably be very small during the collapse, conservation of energy may be applied and this coupled with the condition that the star must end up anyway on the line H, enables one to compute a first approximation for its radius and its luminosity when it arrives on this line, roughly $R \approx 100\,R_\odot$ and $L \approx 300\,L_\odot$.

Since then, the detailed evolution across the forbidden region has been computed numerically taking explicitly into account the complete dynamical equations of stellar structure with various degrees of approximation by Hayashi and Nakano (1965) for $M=1\,M_\odot$, by Bodenheimer (1968) for $M=1$ and 12 $M_\odot$, by Nakano et al. (1968, 1970) for masses in the range 1–$10^4\,M_\odot$ and recently quite extensively by Narita et al. (1970); cf. also Hayashi et al. (1969).

Let us describe very summarily the latest results which were presented by Hayashi (1969) at the Liège meeting. First, one has to transfer the results obtained at the points A and B previously reached (Figure 3) on a Hertzsprung-Russell diagram (Figure 4) for some of the masses considered. This requires adopting some type of distribution of the physical conditions inside the protostar so as to be able to deduce surface values corresponding to the average conditions $\bar\varrho \approx 2 \times 10^{-11}$ g/cm^3, $\bar T \approx 100$ K at B. As all computations show (McNally, 1964; J. H. Hunter, 1967; Penston, 1966; Wright, 1970; McNally, 1970) the density, in spherical condensation, always tends to acquire a well-marked peak at the centre. This led Hayashi et al. to take, for the initial density distribution at point B, that corresponding to a polytrope $n=4$, and, for the temperature, a distribution $T=100\,(\varrho/\varrho_c)^{1/5}$ K which allow one to evaluate $T_{\rm eff}$ and L.

Assuming further a velocity field corresponding to free fall from infinity, the complete hydrodynamical set of equations is then integrated numerically, taking even into account the anisotropy of the radiation in the extended, tenuous envelope of the protostar. The Böhm-Vitense (1958) version of the mixing-length theory of convection is used. The chemical composition adopted is $X=0.70$, $Y=0.28$, $Z=0.02$, all H being treated as being molecular in the initial phases. Of course, dissociation and ionization are taken into account as well as the best available results for the opacity in the difficult ranges of ϱ and T concerned here.

Let us consider the mass $M=1\,M_\odot$ and set the time $t=0$ at the point B when practically free fall starts. At $t=0.5$ year, hydrostatic equilibrium is just established at the centre and the pressure reaction gives rise to a shock wave behind which the temperature rises sharply and the infalling material is brought to a standstill. The shock reaches the external layers ($\tau\approx30$) at $t=2.2$ years (point C in Figure 4) and gives rise to an intense flare-up to a peak luminosity of the order of $5\times10^3\,L_\odot$ in about 10 days, the time of diffusion of the photons through the layer of optical thickness $\tau\approx30$. The luminosity then decreases to point D ($t=4.7$ years) and rises again to about $10^3\,L_\odot$ under the influence of the convection which gets established in the external layers.

The luminosity then keeps nearly constant at this value while the effective temperature increases as the envelope contracts. The representative point crosses the H line represented by the dotted line on Figure 4 characterizing a wholly convective star.

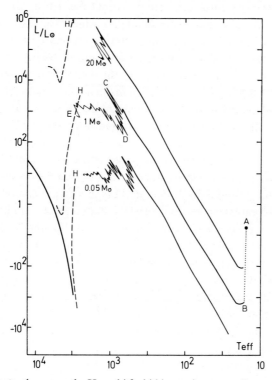

Fig. 4. Evolutionary tracks across the Hayashi forbidden region, according to Hattori *et al.* (1969).

Finally, at $t = 14.7$ years, quasi-static equilibrium is established nearly everywhere and the star reaches the point E, where the radius of the photosphere is about $110\, R_\odot$ and the convective envelope contains about 30% of the mass. The inner radiative region is still highly centrally condensed corresponding to an average effective polytropic index of about 4.6, the central temperature and density being respectively equal to $1.2 \times 10^5\, \mathrm{K}$ and $3.5 \times 10^{-3}\, \mathrm{g/cm^3}$.

Varying of the initial conditions (all atomic H; initial polytropic index $n = 1.5$; larger initial radius) does not seem to affect the final state when quasi-static equilibrium is reached, nor the peak luminosity.

Results for masses equal to 0.5 and 20 $M_\odot$ are also given and correspond to very similar evolution paths except that the final luminosity depends very sensitively on the mass (8 $L_\odot$ and $5 \times 10^5\, L_\odot$ respectively).

Hayashi does not refer explicitly, as in previous papers, to more than one 'bounce' at the centre nor, correspondingly to more than one shock going through the star. Wright (1970), considers that the behaviour of the protostar may be very different depending on whether the 'bounce' occurs after the hydrogen is already completely ionized or before in which case he thinks that dispersal of part or all of the star could occur.

Recently, Larson (1969, 1970), has also computed the collapse phase of protostars of various masses in the range 1–5 $M_\odot$ starting with different initial conditions in very tenuous and transparent stages. He finds that the density distribution becomes extremely peaked at the centre in a very small core which reaches equilibrium at stellar conditions of T and ϱ before most of the material has had time to collapse very far. This core grows as the surrounding material falls into it. Nearly up to the time when all the protostellar material has been accreted, the embryostar is completely obscured by the dust in the infalling layers which transform the radiation coming from the centre into infrared thermal emission of the grains. For $M = M_\odot$, the final star is almost on the Hayashi track, but with a small radius of about 2 $R_\odot$ and a luminosity as low as 1.3 $L_\odot$. For masses greater than about 2 $M_\odot$, he finds that the radius never becomes large enough for a convective Hayashi phase to exist at all.

It is difficult at present to point out exactly the source of these very divergent results. Hayashi thinks that it might be found in Larson's treatment of the shock wave, the reabsorption of radiation by the infalling matter being minimized by the way the fitting is made between regions on both sides of the shock. This way the star can radiate much more energy during the collapse than in Hayashi's solution. This could explain why only very low luminosities are reached at the end of the collapse since, as we have recalled above, qualitatively the high luminosities along the Hayashi track are essentially due to the quasi-conservation of the star energy during the collapse. On the other hand, the exact rôle of the extended dust layers subsisting in the envelope for a long time in Larson's solution should also be cleared up.

In Hayashi's solution, it remains to study the evolution from the last stage of the collapse (point E) as described above and which does not correspond yet to a wholly convective star, to the Hayashi track H itself. The discussion of the likely redistribution of entropy in the star and the computations of Bodenheimer (1966) and Sengbush (1968) suggest that indeed the star should move rapidly to this track at a luminosity of the same order as the one obtaining at the end of the collapse.

Of course, as a result of the collapse, the luminosity oscillates fairly rapidly around point E, but one expects these oscillations to damp out as the star moves on the line H. Nevertheless Graham (1970) reported a somewhat disturbing result at the Liège symposium. Using the general hydrodynamical equations adapted to purely radial motions including a fairly reasonable treatment of turbulent convection, he finds that, if the time-dependent terms are neglected except the usual $T dS/dt$ term in the conservation of energy, any initial solution started high up around the Hayashi track converges to it in about 20 years for $M = M_\odot$. He keeps evolving this model for another 30 years so as to get fairly steady conditions and then reintroduces all the time dependent terms. The star starts oscillating somewhat irregularly with an increasing amplitude. A rough evaluation of the work done by the pressure fluctuation seems to relate the incipient instability to the ionization zones of H and HeI. Could it be that these highly evolved protostars are vibrationally unstable?

As far as the comparison with observation is concerned, the slowest phase in all this evolution, as Wright (1970) remarks, corresponds to the stage of a cold cloud at some 10–15 K during which radiative cooling more or less balance compressional heating (phases $t_f = t_e$ in Hayashi's notation). According to Wright's computations, this may last for about 10^7 years and should then be the easiest stage to observe.

There seems to be little evidence in favour of this although Mebold (1969) has recently suggested that the frequency of cold clouds might have been considerably underestimated (cf. also Verschuur, 1969).

On the other hand, the high luminosity Hayashi phases last a short time and as such there is little chance of ever observing them. However, as Hayashi remarks, these protostars should still be surrounded by extensive dust clouds with a fairly long free-fall time (of the order of 10^3 years for $M \approx M_\odot$) which could make them look like infrared objects by converting the radiation of the central star. Nevertheless this could not increase the life-time of the star close to its peak luminosity and one would have to appeal to fairly large masses to account for recently discovered infrared objects like R Mon or FU Ori which fall in the region of the Hayashi track at very high luminosities ($\approx 10^3 L_\odot$).

On his side, Larson has worked out a simple approximation of the radiative transfer problem in the infalling dust clouds which surrounds his stars and finds that the emitted spectrum could account for some aspects of the infrared objects and the T Tauri stars.

5. Fragmentation, Rotation, Magnetic Fields

We have referred to fragmentation only occasionally but it is certainly a very important question since, on the whole, it seems easier to realize the conditions for the contraction of a large cloud than for a small one. However it still remains highly controversial. As the main cloud contracts, it is easy to imagine the formation of a whole hierarchy

of fragments of smaller and smaller dimensions and masses (Hoyle, 1953) corresponding to the decrease of the Jeans' length as the general mean density increases. However,
the ultimate fate of these fragments is not clear since as Layzer (1963, 1964, 1967) has
often insisted, it is quite possible that these fragments will be destroyed by shear and
turbulence or that they will be overtaken and crushed by the collapse of the cloud
as a whole.

Rotation might perhaps assist in this respect since the increase of the centrifugal
force (as $1/R^3$) is always able at some time to stop the effect of gravity (which increases
only as $1/R^2$) in the equatorial region which ceases to contract while condensation
parallel to the axis of rotation can continue. Fragmentation in the resulting disk could
then occur for, if the centrifugal force associated with orbital motion balances the
overall gravity of the cloud, the centrifugal force associated with the spin of the
fragment is smaller than its self-gravitation. However as shown by Mestel (1965;
cf. also Spitzer, 1968) it is impossible to form realistic stars in this manner without an
appreciable loss of angular momentum at some stage. The most direct way would be
to concentrate this angular momentum in the orbital motion of detached parts of the
initial mass by forming either a double star or a planetary system (Cameron, 1969).
A magnetic field could also help by providing magnetic breaking through the magnetic
lines of force anchored in surrounding material.

Magnetic fields, in their own right, might also have to be taken into account
although, of late, the impression has been growing that they play only a subsidiary
role. Anyway, one knows (cf. Mestel, 1965; Strittmatter, 1966, Spitzer, 1968) that a
large scale magnetic field can prevent the condensation of masses smaller than

$$M_c = \left(\frac{5}{9\pi^2}\frac{F^2}{G}\right)^{1/2} = \left(\frac{5}{G}\right)^{3/2}\left(\frac{H}{\varrho^{2/3}}\right)^3 \bigg/ 48\pi^2 = \frac{4\pi}{3}\varrho\left[\left(\frac{5H^2}{16\pi^2G\varrho^2}\right)^{1/2}\right]^3,$$

where $F = \pi R^2 H$ is the total flux and where, in the last expression, the critical 'magnetic
Jeans' length' of Chandrasekhar and Fermi (1953; cf. also Chandrasekhar, 1961) has
been isolated. In typical interstellar conditions, this leads again to values of M_c of
the order of 10^3 to $10^5\ M_\odot$. For an isotropic contraction M_c does not change since
both the magnetic and gravitational potential energy vary as R^{-1} and, in that case,
no fragmentation is possible. However contraction along the lines of force lowers
the value of M_c, H remaining constant while ϱ varies as $1/d$ if d is the decreasing
thickness parallel to H and subcondensations can occur. It would seem that, in this
way, one could form stars with masses as low as $1\ M_\odot$ or even smaller. However,
their total magnetic energy, although still smaller than their gravitational energy,
would be much larger than anything suggested by the observations.

In conclusion, it appears that, in any case, in the course of contraction to form a
star, the material must somehow be able to lose a large part of its angular momentum
and magnetic energy.

Perhaps, the simultaneous presence of these two factors and their interactions may
help to satisfy this requirement. But this certainly involves complex questions reviewed
at the Liège meeting by Mestel (1970) like the formation of multiple components with

orbital motion, the diffusion of magnetic fields during special phases of the prestellar evolution, the transfer of angular momentum through magnetic torques and (or) turbulence, the pushing of magnetic flux tubes outside convective regions, the snapping of lines of force and their reconnection outside the star itself, anyone of which is likely to take long efforts before it receives a realistic and reliable solution.

On the other hand, many of the objects which we tend to assimilate with early phases of prestellar evolution have properties, fairly regular round shapes, narrow lines etc., which do not especially suggest the presence of large rotation or magnetic field although they already correspond sometimes to large degrees of condensation with respect to the interstellar medium. Does it mean that they have already managed to lose angular momentum and magnetic energy in large proportions? On the other hand, if interstellar gas clouds are transitory objects, maybe, as suggested by McNally (1970), there never is a gross angular momentum problem during collapse from gas cloud to protostar. But we know stars, like the fast rotating Be stars that definitively have difficulties with their angular momentum and other young stars or infrared objects (cf. the analysis of *VY* Canis Majoris by Herbig, 1970) whose properties can be interpreted in terms of disk-shaped nebulae surrounding them and through which they are seen at different angles (Poveda, 1965). One may perhaps hope that the accumulation of significant observations in this respect will finally help locate the critical stages where these extraneous factors play an important rôle as well as their respective importance giving a precious lead to the theorists.

References

Bodenheimer, P.: 1966, *Astrophys. J.* **144**, 709.
Bodenheimer, P.: 1968, *Astrophys. J.* **153**, 483.
Böhm-Vitense, E.: 1958, *Z. Astrophys.* **46**, 108.
Bok, B. J.: 1948, Harvard Observ. Monograph, No. 7.
Bok, B. J. and Reilly, E. F.: 1947, *Astrophys. J.* **105**, 255.
Cameron, A. G. W.: 1962, *Icarus* **1**, 13.
Cameron, A. G. W.: 1969, in *Low Luminosity Stars* (ed. by S. S. Kumar), Gordon and Breach, New York, p. 423.
Chandrasekhar, S.: 1961, *Hydrodynamic and Hydromagnetic Stability*, The Clarendon Press, Oxford, Ch. XIII.
Chandrasekhar, S. and Fermi, E.: 1953, *Astrophys. J.* **118**, 116.
Cheung, A. C., Rank, D. M., Townes, C. H., Thornton, D. D., and Welsh, W. J.: 1968, *Phys. Rev. Lett.* **21**, 1701.
Cheung, A. C., Rank, D. M., Townes, C. H., Knowless, S. H., and Sullivan III, W. T.: 1969, *Astrophys. J.* **157**, L13.
Disney, M. J., McNally, D., and Wright, A. E.: 1968, *Monthly Notices Roy. Astron. Soc.* **140**, 319; 1969, *Monthly Notices Roy. Astron. Soc.* **146**, 123.
Ebert, E.: 1957, *Z. Astrophys.* **42**, 263.
Field, G. B.: 1962, in *Interstellar Matter in Galaxies* (ed. by L. Woltjer), Benjamin, New York, p. 183.
Field, G. B.: 1965, *Astrophys. J.* **142**, 531.
Field, G. B.: 1969, *Monthly Notices Roy. Astron. Soc.* **144**, 411.
Field, G. B. and Saslaw, W. C.: 1965, *Astrophys. J.* **142**, 568.
Field, G. B., Goldsmith, D. W., and Habing, H. J.: 1969, *Astrophys. J.* **155**, L149.
Gaustad, J. E.: 1963, *Astrophys. J.* **138**, 1050.
Goldreich, P. and Lynden-Bell, D.: 1965, *Monthly Notices Roy. Astron. Soc.* **130**, 125.

Graham, E.: 1970, in *XVIe Colloque d'Astrophysique de Liège, Mem. Soc. Roy. Sci. Liège* **19**, 141.
Hartmann, W. K.: 1970a, in *XVIe Colloque d'Astrophysique de Liège, Mem. Soc. Roy. Sci. Liège* **19**, 49.
Hartmann, W. K.: 1970b, in *XVIe Colloque d'Astrophysique de Liège, Mem. Soc. Roy. Sci. Liège* **19**, 215.
Hattori, T., Nakano, T., and Hayashi, C.: 1969, *Progr. Theor. Phys. Osaka* **42**, 781.
Hayakawa, S., Nishimura, S., and Takayanagi, K.: 1961, *Publ. Astron. Soc. Japan* **13**, 184.
Hayashi, C.: 1961, *Publ. Astron. Soc. Japan* **13**, 450.
Hayashi, C.: 1965, *Publ. Astron. Soc. Japan* **17**, 177.
Hayashi, C.: 1970, in *XVIe Colloque d'Astrophysique de Liège, Mem. Soc. Roy. Sci. Liège* **19**, 127.
Hayashi, C. and Nakano, T.: 1965, *Progr. Theor. Phys. Osaka* **34**, 754.
Hayashi, C., Nakano, T., Narita, S., and Ohyama, N.: 1969, in *Low-Luminosity Stars* (ed. by S. S. Kumar), Gordon and Breach, New York, p. 401.
Heiles, C.: 1967, *Astrophys. J. Supp. Ser.* **15**, 97.
Heiles, C.: 1969, *Astrophys. J.* **157**, 123.
Herbig, G.: 1970, in *XVIe Colloque d'Astrophysique de Liège, Mem. Soc. Roy. Sci. Liège* **19**, 13.
Hoyle, F.: 1953, *Astrophys. J.* **118**, 513.
Hunter, C.: 1962, *Astrophys. J.* **136**, 594.
Hunter, C.: 1964, *Astrophys. J.* **139**, 570.
Hunter Jr., J. H.: 1966, *Monthly Notices Roy. Astron. Soc.* **133**, 181 and 239.
Hunter Jr., J. H.: 1967, in *XIVe Colloque d'Astrophysique de Liège, Mem. Soc. Roy. Sci. Liège* **15**, 307.
Hunter Jr., J. H.: 1969a, *Monthly Notices Roy. Astron. Soc.* **142**, 473.
Hunter Jr., J. H.: 1969b, cf. presentation of the paper by Dr. D. McNally, *Observatory* **89**, 44.
Jeans, J.: 1902, *Phil. Trans. Roy. Soc. London* **A199**, 49; cf. also: 1928, *Astronomy and Cosmogony*, Cambridge Univ. Press, Cambridge, p. 337.
Knowles, S. H., Mayer, C. H., Cheung, A. C., Rank, D. M., and Townes, C. H.: 1969, *Science* **163**, 1055.
Larimer, J. W.: 1967, *Geochim. Cosmochin. Acta* **31**, 1215.
Larson, R. B.: 1969, *Monthly Notices Roy. Astron. Soc.* **145**, 271.
Larson, R. B.: 1970, in *XVIe Colloque d'Astrophysique de Liège, Mem. Soc. Roy. Sci. Liège* **19**, 145.
Layzer, D.: 1963, *Astrophys. J.* **137**, 351.
Layzer, D.: 1964, *Ann. Rev. Astron. Astrophys.* **2**, 341.
Layzer, D.: 1967, in *XIVe Colloque d'Astrophysique de Liège, Mem. Soc. Roy. Sci. Liège* **15**, 325.
Ledoux, P.: 1951, *Ann. Astrophys.* **14**, 438.
Ledoux, P.: 1965, in *Stellar Structure* (ed. by L. H. Aller and D. B. McLaughlin), Univ. of Chicago Press, Chicago, p. 499.
Ledoux, P.: 1967, in *Highlights of Astronomy* (ed. by L. Perek), D. Reidel, Dordrecht, The Netherlands, p. 12.
Lerche, I. and Parker, E. N.: 1968, *Astrophys. J.* **154**, 515.
Lin, C. C., Mestel, L., and Shu, F. H.: 1965, *Astrophys. J.* **142**, 1431.
Litvak, M. M.: 1969, *Astrophys. J.* **156**, 471.
McCrea, W. H.: 1957, *Monthly Notices Roy. Astron. Soc.* **117**, 562.
McNally, D.: 1964, *Astrophys. J.* **140**, 1088.
McNally, D.: 1970, in *XVIe Colloque d'Astrophysique de Liège, Mem. Soc. Roy. Sci. Liège* **19**, 195.
Mebold, N.: 1969, *Beitr. Radioastronomie* **1**, 97.
Mestel, L.: 1965, *Quart. J. Roy. Astron. Soc.* **6**, 161 and 265.
Mestel, L.: 1970, in *XVIe Colloque d'Astrophysique de Liège, Mem. Soc. Roy. Sci. Liège* **19**, 167.
Mezger, P. G.: 1970, in *XVIe Colloque d'Astrophysique de Liège, Mem. Soc. Roy. Sci. Liège* **19**, 325.
Mezger, P. G. and Robinson, B. J.: 1968, *Nature* **220**, 1107.
Mezger, P. G., Altenhoff, W., Schraml, J., Burke, B. F., Reifenstein, E. C. III, and Wilson, T. L.: 1967, *Astrophys. J.* **150**, L157.
Nakano, T.: 1966, *Progr. Theor. Phys. Osaka* **36**, 515.
Nakano, T., Ohyama, N., and Hayashi, C.: 1968, *Progr. Theor. Phys. Osaka* **39**, 1448.
Nakano, T., Ohyama, N., and Hayashi, C.: 1970, *Progr. Theor. Phys. Osaka* **43**, 672.
Narita, S., Nakano, T., and Hayashi, C.: 1969, *Progr. Theor. Phys. Osaka* **41**, 856.
Ney, E. P. and Allen, D. A.: 1969, *Astrophys. J.* **155**, L193.
Oort, J. H.: 1954, *Bull. Astron. Inst. Neth.* **12**, 177.
Palmer, P., Zuckerman, B., Buhl, D., and Snyder, L. E.: 1969, *Astrophys. J.* **156**, L147.
Parker, E. N.: 1968, *Astrophys. J.* **154**, 875.

Penston, M. V.: 1966, *Roy. Observ. Bull.*, No. 117.

Penston, M. V.: 1969, *Monthly Notices Roy. Astron. Soc.* **144**, 159.

Pikelner, S.: 1967, *Astron. Zh.* **44**, 1915.

Poveda, A.: 1965, *Bol. Obs. Tonantzintla y Tacubaya* **4**, 15.

Reddish, V. C.: 1967, *Monthly Notices Roy. Astron. Soc.* **135**, 251.

Reddish, V. C.: 1970a, in *XVIe Colloque d'Astrophysique de Liège, Mem. Soc. Roy. Sci. Liège* **19**, 67.

Reddish, V. C.: 1970b, in *XVIe Colloque d'Astrophysique de Liège, Mem. Soc. Roy. Sci. Liège* **19**, 283.

Reddish, V. C. and Wickramasinghe, N. C.: 1969, *Monthly Notices Roy. Astron. Soc.* **143**, 189.

Savranov, V. F.: 1960, *Ann. Astrophys.* **23**, 979.

Schraml, J. and Mezger, P. G.: 1969, *Astrophys. J.* **156**, 269.

Sengbush von, K.: 1968, *Z. Astrophys.* **69**, 79.

Shklovsky, I. S.: 1967, cf. *Soviet-Bloc Research in Astronomy*, No. 178, 5.

Sim, M. E.: 1968, *Publ. Roy. Observ. Edinb.* **6**, No. 8.

Simon, R.: 1965, *Ann. Astrophys.* **28**, 40.

Simon, R.: 1967, in *XVIe Colloque d'Astrophysique de Liège, Mem. Soc. Roy. Sci. Liège* **15**, 155.

Simonson, S. C.: 1970, in *XVIe Colloque d'Astrophysique de Liège, Mem. Soc. Roy. Sci. Liège* **19**, 363.

Spitzer, L. Jr.: 1968, *Diffuse Matter in Space*, Interscience, New York.

Strittmatter, P.: 1966, *Monthly Notices Roy. Astron. Soc.* **132**, 359.

Verschuur, G. L.: 1969, *Astrophys. Lett.* **4**, 85.

Werner, M. W. and Salpeter, E. E.: 1970, in *XVIe Colloque d'Astrophysique de Liège, Mem. Soc. Roy. Sci. Liège* **19**, 113.

Wickramasinghe, N. C. and Reddish, V. C.: 1968, *Nature* **218**, 661.

Wright, A. E.: 1970, in *XVIe Colloque d'Astrophysique de Liège, Mem. Soc. Roy. de Liège* **19**, 75 and 95.

STELLAR EVOLUTION

A. G. W. CAMERON

Belfer Graduate School of Science, Yeshiva University, New York, N.Y., U.S.A.

and

Institute for Space Studies, Goddard Space Flight Center, NASA, New York, N.Y., U.S.A.

In these lectures I shall concentrate on those aspects of Stellar Evolution which appear relevant to the study of the structure and the evolution of the Galaxy. I shall not review the techniques used for the numerical computation of stellar evolution, but will assume that the interested reader can consult standard reference works for such topics. The lectures by Professor Ledoux discuss the formation of stars from the interstellar gas. Here we shall be concerned with the reverse process, in which stars lose mass back to the interstellar medium. We shall also be interested in late stages of evolution of stars in which they have properties especially interesting for the determination of galactic distances and we shall discuss the properties of some of the endpoints of stellar evolution.

1. The Earliest Stages of Stellar Evolution

We begin with the earliest stages of evolution of the star, in which it approaches the main sequence. There is a maximum radius which a stable star can possess. This arises from the virial theorem as applied to a star, and from its composition. The virial theorem states that half of the released gravitational potential energy in a star must be stored as internal thermal energy in order to have hydrostatic stability. In addition, if a star is formed from the interstellar medium, then not only are the main constituents the hydrogen and the helium, initially composed of neutral atoms, but the hydrogen is likely to be present in the form of molecules at some early stage in a collapse of an interstellar cloud. However, in order to come into hydrostatic equilibrium the star must have an interior temperature sufficiently high so that both the hydrogen and the helium are ionized throughout the bulk of the star. Energy is required to dissociate the hydrogen molecules and to ionize the hydrogen and helium, and the only source of energy available for this purpose is the released gravitational potential energy which arises from the collapse of the material. One therefore obtains a maximum possible stable radius by equating the released potential energy in a model to the sum of the internal thermal energy and the energy required for dissociation of the hydrogen molecules and ionization of the hydrogen and helium in the interior. For a body of one solar mass, this radius is some sixty times the present radius of the sun. To a good approximation, the maximum radius scales as the mass of a star. We call this radius the threshold stability radius of the star.

Figure 1 shows the evolutionary tracks in the HR diagram for a variety of stellar masses in the range 0.5 to one hundred solar masses as calculated by Ezer and Came-

L. N. Mavridis (ed.), Structure and Evolution of the Galaxy, 236–249. All Rights Reserved
Copyright © 1971 by D. Reidel Publishing Company, Dordrecht-Holland

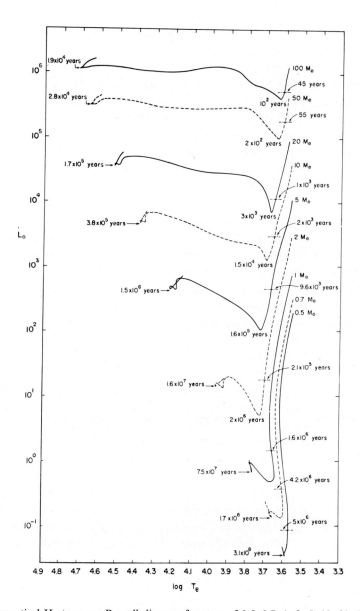

Fig. 1. Theoretical Hertzsprung-Russell diagram for stars of 0.5, 0.7, 1, 2, 5, 10, 20, 50, and 100 solar masses in the initial gravitational contraction stage. The calculations were continued until half the central hydrogen was exhausted. The dashed horizontal lines and corresponding times indicate the cessation of complete convection in the models. The final times for each track are the 'zero age', main sequence times.

ron (1967). These tracks start from the threshold of stability, proceed until the star has evolved onto the main sequence, and then until half of the central hydrogen has been depleted by hydrogen-burning processes. Near the threshold of stability all of the models are completely convective. The initial motion of the tracks in the HR diagram is principally downward at approximately constant surface temperature. Along most of vertical portion of these tracks, the models stay fully convective, but toward the bottom of the tracks the central regions become in radiative equilibrium. When about half of the mass is in the central radiative part, the track then turns toward the left and rises slightly as the star evolves toward the main sequence. The higher masses must traverse a fairly lengthy horizontal path before they approach closely to the main sequence, but the lower masses have only a very short horizontal path.

When these stars approach nuclear-burning temperatures at the center, thermonuclear reactions are ignited and the luminosity of the model decreases somewhat. This results from the fact that the energy source due to gravitational potential energy is widely distributed throughout the model, whereas the nuclear-burning energy source is highly concentrated toward the center. Under these circumstances, the energy arising from nuclear burning has considerably more matter to traverse in going from the region of production to the surface, and hence the star has an overall greater effective opacity, so that the total luminosity is decreased. When the energy generation at the center becomes fully equal to the luminosity of the model, then contraction ceases. These stars have now reached the main sequence. At this stage the more massive stars, which burn hydrogen into helium by the CNO reactions, will have developed a fairly large central convective core, owing to the extreme temperature sensitivity of the energy generation processes. As the central hydrogen burns into helium, the track of the star in the HR diagram departs from the position of the initial main sequence, rising vertically for these stars of lower mass and going toward the upper right for stars of larger mass. In each case, the increase in luminosity results from the fact that helium has a lower opacity than the hydrogen from which it was made, so that the radiation can emerge from the star somewhat more easily.

The vertical portions of these evolutionary tracks are often called Hayashi tracks, after C. Hayashi, who first pointed out that the surface boundary conditions of stellar models would not allow too low surface temperatures. One should not expect to find any stars near the tops of the Hayashi tracks. The threshold of stability was calculated on the assumption that the collapsing gas is unable to radiate any energy away into space before it forms a stable stellar model. This assumption is obviously wrong. Recent calculations by Hayashi of the hydrodynamics of a collapsing interstellar gas fragment show that the infalling material will be halted by a shock front, in which a great deal of energy can be radiated to space, so that the stellar model stabilizes on the Hayashi track only near the bottom of the vertical region. Such calculations are carried out with the assumption that the collapsing material possesses zero angular momentum. This too is an untenable assumption. The collapsing cloud fragment is expected to have so much angular momentum that it will flatten into a disc distribution of mass having dimensions comparable of those of the solar system. This disc distribution has recently been studied in a preliminary way by myself, and it is evident that no stellar body can be initially stabilized with spherical geometry at the center of the disc. Instead, the inner parts of the disc, out to several astronomical units, will be in thermally-driven convective equilibrium, as a result of which there will be a hydrodynamic dissipation of mass, with a net inflow of mass and a net outflow of angular momentum. Since this dissipation process necessarily takes considerably longer than hydrodynamic free fall times, there is again an opportunity for a great deal of radiation of the released gravitational potential energy into space, and again the star which is formed should probably be first recognizable near the bottom of a Hayashi track, although detailed calculations on this point have not yet been carried out.

2. Mass Loss

The tracks shown in Figure 1 are also unrealistic in another respect. They do not include effects due to mass loss. It is well known that the T Tauri stars are stars which have not yet reached the main sequence and which have rather large rates of mass loss to space, presumably by some kind of stellar wind process. Mrs. Ezer and I have recently started to study pre-main sequence stellar evolution with mass loss effects taken into account.

It is well known that the phenomenon of the solar wind arises from the high temperature of the solar corona, so that the coronal gases undergo a hydrodynamic expansion into space. What is not so well known are the details of the mechanisms responsible for heating the solar corona. The heating appears to result from a variety of forms of mechanical energy generated in the outer solar convection zone which lies just below the photosphere of the Sun. Part of this energy is emitted as acoustic waves, but this is probably mostly used up in heating the solar chromosphere. The corona itself can be heated by gravity waves, Alfvén waves, or various kinds of hydromagnetic waves. All of these are generated by mechanical motions, and some of them depend on the presence of magnetic fields. Indeed, the generation of ordinary hydrodynamic waves appears to be enhanced by the presence of magnetic fields. These various types of waves are dissipated in a variety of different ways in the corona.

It is evident that if we cannot understand the details of heating of the solar corona which is responsible for the generation of the solar wind, then we are unlikely to be able to make any reasonable predictions from fundamental theory about the temperature of the coronas of other stars and of the resulting rates of mass loss. For this reason it has been necessary to take a semiempirical approach to the problem. We have tried partially to parameterize the problem in the following way.

Presumably some fraction of the mechanical energy near the photosphere of the star will go into heating the corona and into subsequent mass loss. Therefore we would expect that the rate of mass loss should be proportional to the surface area of the star, R^2. We also have to take into account that the amount of mass which can be removed from a stellar surface per unit energy input is inversely proportional to the gravitational potential of this surface, GM/R. As a result of these two considerations, we write the rate of mass loss from a star in the following form:

$$- \mathrm{d}M/\mathrm{d}t = \alpha R^3/M.$$

In order to examine the function α, one must appeal to the observations of mass loss from stars. There are very few of these observations in existence. However, two massive red giant stars and the Sun have values of α in the vicinity of 3×10^{-14} solar masses per year, where both R and M in the above formula are expressed in solar units. This is encouraging, for it suggests that the main variations associated with the description of mass loss have already been taken into account. However, the mass loss deduced for various T Tauri stars yields values of α in the range $10^{-10} < \alpha < 10^{-8}$ solar masses per year. Thus it is evident that in general the rates of mass loss in the T Tauri stage are very much greater than observed either on the main sequence or in the post-main sequence evolutionary phases. Recently the rate of mass loss for a dwarf main sequence star which is also a flare star has been deduced, and the value of α for this star is intermediate between the Sun and T Tauri stars. This suggests the following empirical interpretation of this mass loss behavior. When stellar material is still contracting toward the main sequence, the interior magnetic fields probably have a higher magnetic energy density than is commonly encountered on the main sequence or after it. As a result of the larger magnetic fields, the heating of the stellar corona is greatly enhanced and mass loss rates are particularly high. When the star reaches the main sequence, the stronger magnetic fields probably die away to some residual level, with the rates of decay being faster if the surface convection zone of the star is rather shallow. The value of α which characterizes the later main sequence and the post-main sequence stars may still be dominated by magnetic effects in the coronal heating, but it is also possible that ordinary hydrodynamic effects have by that time become dominant, so that the value of α which fits the Sun and the two massive red giant stars may be a minimum which is achieved during stellar evolutionary history. This means that mass loss in the T Tauri stage, where the evolution time toward the main sequence is relatively short, may be approximately as important in terms of the total mass loss from a stellar body as that which occurs in the red giant phase, which lasts much longer because of nuclear burning, but where the mass loss is enhanced because of the larger radius of the star.

Mrs. Ezer has carried out some preliminary investigations of pre-main sequence stellar evolution using a mass loss rate $\alpha = 10^{-9}$ solar masses per year. This value of α was reduced to the present solar value when the star approached the main sequence. For this value of α in the T Tauri phase, it has been found that in order to end on the main sequence with one solar mass, the Sun must start at the threshold stability with a little more than 1.6 solar masses.

3. Nuclear Energy Processes

The major stages in stellar evolution depend strongly upon the details of the nuclear energy generation processes. There are only a small number of major stages of nuclear energy burning. These are as follows:

(1) H→He. Hydrogen-burning. Stars which are undergoing hydrogen-burning in their cores lie on the sequence or close to it. Stars which have evolved to the right of the main sequence generally are undergoing hydrogen-burning in shells surrounding the core. The more massive stars only have solely hydrogen-burning shells a small part of the distance toward the red giant branch, but stars of smaller mass generally have hydrogen-burning shells all the way into the red giant branch.

(2) He→C, O. Helium-burning. Massive stars tend to burn helium at an intermediate position on the way toward the red giant branch. When they get to the red giant branch, they will commonly have both hydrogen- and helium-burning shells. Stars of smaller mass have highly electron degenerate cores in the red giant branch, they leave the red giant branch when helium-burning is ignited in the core, and they appear to burn helium on the horizontal branch leading toward the red giant branch.

(3) C→(Mg). Carbon-burning. Massive stars may burn carbon in the red giant branch, but this point is uncertain. Stars of smaller mass, which have burned out helium in the center and have once again formed a degenerate core, may burn carbon, but it is quite likely that this burning will be an explosive ignition leading to a supernova event, as is discussed below. The products of carbon-burning are nuclei lying in the vicinity of magnesium.

(4) O→(Si). Oxygen-burning. Stellar evolution calculations have not yet reached the stage where the onset of oxygen-burning can be predicted with any degree of assurance. The principal products of oxygen-burning, following carbon-burning, are nuclei in the general vicinity of silicon.

(5) (Si)→Fe. Silicon-burning. This is a final complicated set of reactions which leads to the formation of the iron equilibrium peak. This stage is bound to be very short-lived in any stage of stellar evolution. It is not known whether it occurs in the ordinary evolution of any star.

(6) Neutrino energy losses. Neutrino-antineutrino pairs can probably be emitted by a variety of processes in a very hot stellar interior. At lower temperatures in non-degenerate matter, the pairs can be produced by the photoneutrino processes in which photons are converted to neutrino-antineutrino pairs by interaction with electrons. At higher temperatures in non-degenerate matter, the pairs are produced by the annihilation of electron-positron pairs. In hot degenerate matter the pairs can be produced by the conversion of plasmons to the neutrino-antineutrino pairs. These processes have a strong temperature dependence, although it is not as strong as the temperature dependence of ordinary thermonuclear reactions. In the absence of nuclear energy generation sources which might supply the energy which is lost by neutrino-antineutrino pair emission, the energy which is lost must be supplied by gravitational contraction, and this naturally tends toward the formation of highly degenerate cores in stars in advanced stages of evolution.

4. Evolution Beyond the Main Sequence

When hydrogen is exhausted in a stellar core, the core contracts in order to continue furnishing energy which can be emitted from the surface, and after some degree of contraction, the hydrogen surrounding the helium core ignites, and further evolution is governed by the progress of the hydrogen-burning shell which is thus formed. As the shell burning continues, the core, now enhanced in mass, continues to contract, and the envelope expands, thus leading the star toward the red giant branch. In a low mass star, such as the Sun, contraction leads to a highly degenerate core. In a massive star, contraction leads toward higher temperatures and the onset of helium-burning. However, in stars ranging in mass up to about ten solar masses, the ignition of the helium-burning reactions occurs at the tip of the red giant sequence.

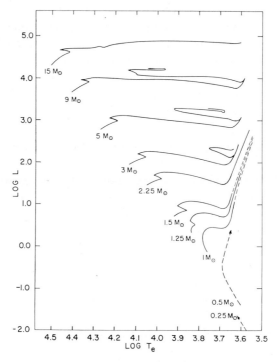

Fig. 2. Post main sequence evolution for stars of the indicated masses. After Iben (1967). Dashed lines are estimates.

Figure 2 shows the post-main sequence evolutionary tracks of stars ranging up to fifteen solar masses, taken from the work of Iben (1967). It may be seen that the stars of lower mass evolve toward the right and then the tracks turn steeply upward. This marks the onset of a deep convective envelope in these stars, and the stellar boundary condition pointed out by Hayashi prevents the surface temperature from falling too low. At the same time, the convection assists the energy flow out of the interior, so that the luminosity of the stars rises. The evolutionary tracks of the stars of larger mass are nearly horizontal, since they have a longer distance to go in the HR diagram before the Hayashi limitation would be reached. In fact, for these stars, helium-burning sets in the core before they have a chance to mount the Hayashi track, and therefore such stars stay at nearly constant luminosity during this entire phase.

As long as the energy released by the gravitational contraction of the degenerate core in a smaller mass star can be transported away from the center, the stellar core stays relatively cool and no ignition of helium reactions occurs. However, when the core releases energy at a faster rate than it can be transported away, the temperature rises and the helium reactions ignite. Because the equation of state at the center of the star is insensitive to the temperature, a thermonuclear runaway occurs which raises the temperature more and more rapidly until the electron degeneracy is lifted. Then the core expands rapidly and the helium reactions are damped out to a much lower level. At the same time, the envelope contracts towards smaller size, and the star swings away from the tip of the red giant branch onto the horizontal branch. In this phase we reach the end of the quantitative theory of the stellar evolution of stars of small mass. On the one hand, it is evident that mass loss plays an important role in the red giant phase, as can be judged from the R^3 dependence of the mass loss as indicated previously, and therefore the masses which stars will have on the horizontal branch are significantly less than the masses which they had on the main sequence. On the other hand, as Schwarzschild and Harm (1965) have shown, when a star on the horizontal branch has exhausted its central helium and has both a hydrogen-burning and a helium-burning shell source, then quite frequent instabilities can occur in which thermal runaways in the helium-burning shell take over, for varying lengths of time, the entire energy output of the star. Their calculations show that a number of these helium shell flashes will occur, and that eventually these shell flashes will lead to mixing of material between the region of the hydrogen-burning shell source and the helium-burning shell source, thus producing a variety of helium-burning reactions, as a result of which neutrons will be produced and quite large numbers of heavy elements may therefore also be produced as a consequence. Because of these various complications, no trustworthy calculated evolutionary track for the stars on the horizontal branch yet exists.

For the somewhat more massive stars shown in Figure 2, the calculations indicate that the onset of helium-burning in the core makes the model swing to the left in the HR diagram, with a further swing toward the right as the helium in the core becomes exhausted and the helium-burning shell is established. Needless to say, the evolutionary details beyond helium-burning in these more massive stars are quite untrustworthy, owing to neglect of mass loss and because of uncertainties regarding the neutrino-antineutrino emission and various instabilities which might be associated with the stars.

One feature related to stellar evolution which has been of considerable importance for galactic and extragalactic research is a question of stellar variability. There are a number of classes of variable for which approximate period-luminosity relations have been established. These variables lie in a special region of the HR diagram to the right of the main sequence, but to the left of the red giant branch. This instability strip is narrow for the horizontal branch stars which originally had about one solar mass, but is wider for massive stars. The instability appears largely to be associated with the existence of a helium ionization zone in the outer envelope of the stars and with the fact that a vibrational cycle leads through variations in opacity to the storage of energy which can be pumped into the vibrations, thus maintaining them. It is evident from Figure 2 that stars with a wide range of masses can be expected to traverse the instability strip. Since the period of a star is a function of its mean density, one might therefore expect that a wide variety of period-luminosity relations would exist, and that this would not be a very useful criterion for distance measurements. However, most stars initially traverse the instability strip very rapidly during their evolutionary history, so that very few such stars are likely to be found in the instability strip at any given time. On the other hand, stars in later stages of their evolutionary history may swing back and spend longer periods of time in the instability strip. This appears to be the case for stars initially of about one solar mass which have evolved onto the horizontal branch. These stars have a narrow mass range and give the RR Lyrae variable stars. Because of the narrow mass range involved in most old stars which are evolving off the main sequence, one thus gets a fairly good period luminosity relationship for the RR Lyrae variables. There are additional complications, as Christy (1966) has shown. Some RR Lyrae variables appear to be excited in their fundamental vibrational periods, whereas others are vibrating in an overtone. Christy has had an excellent degree of success in accounting for subtle details of this sort as a result of his hydrodynamic calculations.

As Iben has pointed out, there is also a relatively narrow mass range among the more massive stars in which quite large amounts of time are spent in the instability strip while core helium-burning occurs. Hence the Cepheid variables can be associated with a fairly narrow mass range, and one can understand that the empirical period-luminosity relation found for them has proved very useful in astronomy. However, one must also expect that Cepheid-type variables will occasionally be found in which the period-luminosity relation differs significantly from the usual one assumed. These stars will be scarce because they do not spend much time in the instability strip, but nevertheless they should be there and one should not be surprised if a few of the distance determinations based upon Cepheids should prove in error owing to this feature.

There are significant differences in the post-main sequence evolutionary tracks of Population I and Population II stars. Most Population II stars start with masses of about solar order, but after evolving off the main sequence, they turn up in the HR diagram at a much higher surface temperature than do the Population I stars. This is because the Population II stars start with masses of about solar order, but after evolving off the main sequence, they turn up in the HR diagram at a much higher surface temperature than do the Population I stars. This is because the Population II stars are deficient in heavier elements compared to Population I, and hence the surface opacities are lower and the Hayashi surface condition requires that the surface temperature should be higher. Relatively little work has been done on the more advanced stages of the Population II stars.

5. The Late Stages of Stellar Evolution

It is well known that a star of less than about 1.4 solar masses can be stabilized in its endpoint of evolution as a white dwarf star. When one takes mass loss into account, which has never been done quantitatively, it is evident that stars which started their evolutionary lifetimes with quite large masses may also end as white dwarfs, since they can have reduced their masses to less than the white dwarf limit during the course of their evolutionary history. There is a great deal of suggestive observational evidence concerning the late stages in the evolution of such stars, but there are very few relevant calculations.

Among the more suggestive of these calculations are some carried out by S. Vila, but unpublished. He constructed models in which there was a very large degenerate core surrounded by a hydrogen envelope containing about 0.1 solar masses. At such an advanced stage in evolution, one expects to have a hydrogen-burning shell source, but no helium-burning shell source. When he included the effects of neutrino-antineutrino energy losses from the interior, he found that the luminosity of the models was forced to become very high. With such a thin envelope, the radius of the models is not very large, and hence the models lie toward the left in the HR diagram. The opacity of the surface layers is very high, and the opacity increases rapidly below the surface. Vila found that he could not fit the equilibrium conditions precisely in these models, because the radiation pressure below the surface exceeded the radiation stress limit, at which there would be a net outward force on the matter exceeding that of gravity. This difficulty suggests that in such circumstances the outer hydrogen envelope of the star will be ejected by radiation pressure. One can make an estimate of the time required for this by examining the luminosity of the models and equating a fraction of the outgoing energy to the energy required to lift the envelope off the core. In this way, one can estimate that the time required to eject the hydrogen envelope will only be a few hundred years.

This is very suggestive of the planetary nebulae. Such nebulae consist of an ejected nebular shell containing about 0.1 solar masses, with a very hot, mostly degenerate remnant left at the center which is responsible for exciting emission lines in the nebula. A number of studies have been carried out of the evolution of these nuclei of planetary nebulae. Such studies have concentrated on the rapidity of cooling toward the ordinary white dwarf stage with and without neutrino-antineutrino pair emission. The results of these studies, when compared to observations of the relative numbers of such planetary nuclei observed in the Galaxy, support the existence of the basic interaction responsible for the neutrino-antineutrino emission.

Because the planetary nebula stage of evolution appears to be very brief, and because there are a large number of such objects visible in the Galaxy, it appears quite likely that a very large fraction of the single stars in the Galaxy complete their evolutionary histories in this way.

The details of stellar evolution can be greatly modified in the case of binary systems, particularly in the case of close binaries. When the more massive of a binary star pair finishes hydrogen-burning in the core and starts to expand into the red giant phase, mass will be transferred through the inner Lagrangian point from the more massive star to the less massive one. Some sample studies of this situation have been carried out by Kippenhahn and Weigert (1967). In their studies, the mass transfer appears to be remarkably complete. The more massive star is reduced to a white dwarf remnant having a mass of only about 0.3 solar masses. The distance between the binary pair increases dramatically, so that when the second star in its turn evolves to the red giant branch, it can expand very much more before a reverse mass transfer takes place. Although it is not possible to follow such details in an evolutionary calculation of this type, it is also very likely that a considerable amount of mass would be lost from the system through other Lagrangian points. Some of this mass may form rings around the entire system, as are sometimes observed in nature.

Most nova explosions appear to involve stars in binary systems. Although detailed hydrodynamic calculations of nova explosions have not yet been carried out, it appears likely that mass transfer onto the surface of a white dwarf star, as would happen in the case discussed above, leads to the establishment of a thin hydrogen envelope with temperature at the base sufficient to ignite a thermonuclear runaway in the hydrogen fuel. This thermonuclear runaway, occurring in degenerate material, generates sufficient energy to blow off the accumulated envelope at quite high velocities, so that the material is lost from the system. The evolution of the companion star into the red giant phase will continue, so that new material is transferred onto the new white dwarf surface, and additional nova explosions will follow in due course. It seems very likely that the component stars in binary systems will almost always each wind up as white dwarf remnants.

6. Supernova Explosions

The foregoing considerations only apply to stars which are sufficiently low in mass to begin with, or which are able to lose sufficient mass, so that they approach the endpoint of their evolutionary history with less mass than the white dwarf limit, about 1.4 solar masses. The behavior of a more massive star which exceeds this limit at the time it develops a highly degenerate core, or which reaches very high temperatures without developing a degenerate core, is much more controversial. Such stars probably undergo supernova explosions. The details of a supernova explosion are in considerable controversy at the present time.

The core of such a star may approach dynamical instability in two different ways, depending upon whether or not it becomes degenerate. If the star is sufficiently massive so that the core remains non-degenerate following the various stages of nuclear burning, then eventually the temperature will become sufficiently high so that a photonuclear decomposition of the iron equilibrium peak occurs, breaking it down into helium and nucleons. Such a decomposition absorbs energy, which must be supplied by a renewed period of gravitational contraction in the core, and the requirements for energy are sufficiently great that this contraction must in fact become a hydrodynamical implosion. On the other hand, a star of somewhat lower mass which forms a highly degenerate core following helium-burning may not survive the onset of carbon burning. In such a star, the carbon thermonuclear reactions are kindled by a rapid contraction of the core, which raises the temperature to about 10^9 °K despite the burden of neutrino-antineutrino energy losses, and the resulting thermonuclear runaway will lead to the completion of the carbon-burning reactions in the core in a time short compared to the hydrodynamic expansion time of the core. Indeed, oxygen-burning also is likely to occur explosively in such a core after being kindled by the carbon thermonuclear runaway. However, there are probably also some situations in which electron capture occurring on the nuclei in the core removes electrons which give the dominant contribution to the pressure, and the core may implode and carbon detonation may not occur until a sufficiently high density that the electron degeneracy is not lifted by the thermonuclear reactions at all. In that case the core will also implode toward nuclear densities.

In the case of an implosion, the compression of the core is expected to continue until nuclear densities greater than 10^{14} gm/cm^3, are reached. The core under such circumstances becomes extremely hot, and the number densities of the electrons, although higher than at previous stages in the evolution of the star, are considerably less than the number densities of neutrons. For several years it appeared promising that the supernova explosion would then proceed by means of an energy transfer from the hot core to the cooler descending envelope by means of the transport of neutrinos and antineutrinos out of the core, and that this energy transfer would then lead to the development of very high temperatures in the descending envelope, leading to a gigantic shock wave which would eject the descending material and also the material overlying it. Supernova theories based upon such a picture have been developed by Colgate and White (1966), by Arnett (1966, 1967), and by Schwartz (1967). However, more recently, Arnett (1969) has redone this problem with a better treatment of the physics, and he has found that the energy transfer process by means of neutrinos and antineutrinos is less efficient than previously thought, and it appears that rather little mass can be ejected by such means.

In contrast, Arnett finds that stars of less than about 15 solar masses should undergo the carbon detonation in their degenerate cores, and as a result of this a detonation wave will progress outwards through the core which will blow the entire structure apart. In this case, one gets an expanding supernova envelope, but no stellar remnant whatever is left behind.

There is some indirect evidence favoring such a carbon detonation process. The intermediate elements, from neon through the iron equilibrium peak, appear to be produced by the fast thermonuclear reactions accompanying the carbon detonation in abundances which agree rather well with those observed in the solar system materials. Thus explosions of this sort appear to constitute one of the major processes of the nucleosynthesis of the elements. On the other hand, some of the heavy elements observed in nature require formation in regions of extremely high neutron flux. Such a high neutron flux can probably only be obtained in the dense neutronized cores which are produced by the implosion processes which lead the stellar core to the range of nuclear densities. However, this component of the heavy elements constitutes only an extremely small fraction of the mass which is present in nature, and hence it is consistent to believe that the ejected material forming heavy elements by rapid neutron capture has come from a supernova explosion in which an implosion to nuclear densities has taken place, but in which only a small amount of mass has been ejected from the core by processes of neutrino-antineutrino energy transfer.

There is other indirect evidence for the occurrence of the full-scale implosion in nature. The recent identification of pulsars as rotating neutron stars indicates that some supernova explosions have involved implosions to nuclear densities. Since we do not know what the effective radiating lifetime of a pulsar is, nor the full extent of the solid angle into which the pulsar radiation is emitted, this knowledge is by itself insufficient to allow a determination of the fraction of supernova explosions likely to result in a neutron star. However, there is other indirect evidence which suggests something regarding this point.

Mrs. Charlotte Gordon has recently made considerable progress in understanding the emission features appearing in the spectra of Type I supernovae. She has found that the broad emission features correspond to emission lines of the common elements, such as magnesium, silicon, and iron, but the emission lines are forbidden lines which result when these elements are highly stripped of electrons and exist in a hot thin plasma at a temperature of order 10^6 K and relatively low density. These conditions are typical of the solar corona, and in fact it appears that Type I supernova spectra are essentially widened coronal spectra. Furthermore, these lines are emitted from the interior of the expanding supernova shell.

It seems to me that such a hot interior region in the expanding supernova shell can only exist at typical times of weeks or months following the supernova explosion if there is a strong energy input to this material which maintains its high temperature. The only continuing source of energy which is reasonable for this purpose is the energy emitted by a pulsar. If this is correct, then we would be able to identify the formation of neutron stars with the Type I supernova explosions. Apart from its own intrinsic astrophysical interest, this identification then implies that something like half of the supernova explosions are the core implosion type leading to the formation of neutron stars. Thus we can expect something like 10^8 neutron stars to have been formed during the history of the Galaxy.

Some recent results obtained by some colleagues and myself in New York on the properties of neutron stars may be of some interest. The maximum mass of a stable neutron star appears to be a little more than 2 solar masses, and a typical radius is about 13 km, not very sensitive to the mass. In the outer 2 km of the models, one can expect to find ordinary nuclei, although these become increasingly neutron rich at increasing depths below the surface. No neutrons are actually present in the outer 1.5 km of a 0.6 solar mass neutron star. When the nuclei disappear, the medium consists of neutrons, protons, and electrons, with each of the latter two constituents being present to the extent of one part in thirty of the neutrons. As the density increases inward toward the center, the relative numbers of protons and electrons increase relative to those of neutrons, so that eventually negative muons are also present. Towards the center, at greater than normal nuclear densities, a variety of hyperons may also become present, the first of these being the Σ^- hyperon. The outer shell which consists mainly of ions is expected to be crystalline and therefore very rigid. The inner regions may have superfluid properties. We have found that the simultaneous requirements of fitting the period and first derivative of the period of the pulsar in the Crab Nebula, and also providing for a rate of rotational energy loss of at least 10^{38} ergs/sec, requires that this neutron star should have a minimum mass of 0.4 solar masses.

Acknowledgements

This work has been supported in part by the National Science Foundation and the National Aeronautics and Space Administration.

References

Arnett, W. D.: 1966, *Can. J. Phys.* **44**, 2553.
Arnett, W. D.: 1967, *Can. J. Phys.* **45**, 1621.
Arnett, W. D.: 1969, *Astrophys. Space Sci.* **5**, 180.
Christy, R. F.: 1966, *Ann. Rev. Astron. Astrophys.* **4**, 353.
Colgate, S. A. and White, R.: 1966, *Astrophys. J.* **143**, 626.
Ezer, D. and Cameron, A. G. W.: 1967, *Can. J. Phys.* **45**, 3429.
Iben, I.: 1967, *Ann. Rev. Astron. Astrophys.* **5**, 571.
Kippenhahn, R. and Weigert, A.: 1967, *Z. Astrophys.* **65**, 251.
Schwartz, R. A.: 1967, *Ann. Phys. New York* **43**, 42.
Schwarzschild, M. and Harm, R.: 1965, *Astrophys. J.* **142**, 855.

EVOLUTION OF STELLAR CLUSTERS AND ASSOCIATIONS

P. BOUVIER

Observatoire de Genève, Sauverny, Switzerland

1. Introduction

Star clusters are usually defined as agglomerations of stars in space, whose average projected densities appear to exceed the density of background stars. Cluster members are identified either by parallax measurements or by common motion or also with the help of the color-luminosity diagram established from known members.

The *globular* clusters are distributed within a quasi-spherical subsystem of our Galaxy and seem to follow mostly elongated orbits around the galactic center (von Hoerner, 1955). They number about 120 on the whole and fairly complete data are available for 70 of them (Haffner, 1965). The linear diameter of such a cluster is generally of the order of 30–50 pc and the total number of its member stars lies somewhere between 5×10^4 and 5×10^7; on account of the crowding effect, star counts become impossible in the central region and must be replaced by surface brightness measurements. The photographic image of a globular cluster is generally circular, although some slight ellipticities occur in certain cases. It has sometimes proved useful to divide the globular clusters into a dozen classes according to their apparent degree of star concentration.

Galactic or *open* clusters are smaller objects of various apparent forms varying from nearly circular to very irregular. Average figures of linear diameters and masses for 128 open clusters are 6 to 10 pc and 1500 solar masses.

In contrast to the globulars, open clusters are narrowly concentrated toward the galactic plane and this, together with color-luminosity diagrams, evolutionary tracks of individual stars, and the type of variable stars occurring in a cluster, leads us to place the globular clusters among the oldest population of our Galaxy and most of the open clusters among the disk population. In other words, whereas the age of a globular cluster exceeds 10 galactic years (or 2×10^9 years), the spread of ages for open clusters, according to a sample of two dozen of these objects (Haffner, 1965), extend from 10^6 to 5×10^9 years.

Stellar *associations* are groups of stars having definite physical types (generally O-type or T Tauri stars); their average star density exceeds the background density for stars of that precise type, but is well below the overall background stellar density. Therefore, such systems are loosely bound. The members of O-associations are young bright stars which often disclose motions corresponding to a linear expansion of the system.

Stellar associations belong to the younger disk population. Their age is generally less than 10^7 years and, together with young galactic clusters and H$_{II}$ regions, they are useful in delineating the structure of spiral arms.

We shall not go further here into the problems of spatial distribution and external motions of clusters and associations, but rather turn to what is known about their structure and dynamical evolution.

2. Description of a Star Cluster; Relaxation Time and Mean Free Path

In considering a stellar cluster, we are essentially faced with the n-body problem of classical non-relativistic mechanics, and since we now have at our disposal fast and powerful computing machines, one line of attack will certainly pass through the numerical integration of the $6n$ first order equations describing the motion of anyone of the n members of the system in the resultant gravitational field due to all the other members. On the other hand, a star cluster can also be visualized as a mass of 'stellar gas' held together by its own gravitation and this would suggest taking advantage of gas kinetic considerations. We shall adopt this second point of view first, not so much on account of previous historical developments as because it demonstrates more readily several useful physical concepts.

The equilibrium state of a gas in a box is characterized by a definite pressure, due to the repeated impacts of molecules on the surrounding walls of the box and the inner interactions of the gas essentially consisting of collisions between the molecules. How far can we extend this classical gas picture to the 'gas of stars' making up a cluster? Confinement through inner gravitation forces is indeed very different to confining a gas by outer solid walls and we must estimate at first the importance of collisions or encounters within a self-gravitating system.

In an ordinary neutral gas, the particles (molecules or atoms) interact through short-range forces so that we may unambiguously define a mean collision time; such is not the case in a plasma or in a self-gravitating system which are both dominated by long-range interaction forces. If we define a smoothed-out gravitational potential at every point of the system and at every instant, we may try to describe the motion of a particular or test-star among the other members or field-stars of the cluster producing the smoothed-out potential; we shall thus obtain, from given initial conditions, a well determined 'regular' trajectory. But the smoothed-out potential does not take the granular structure of the system into account and when the test-star passes very near to a field star, it undergoes a binary encounter producing a deviation from the regular trajectory. We then call 'relaxation time' T_{rel} the time after which the position and velocity of the test-star are no longer correlated with the initial position and velocity.

To make this definition more precise, we may say that time T_{rel} for a given star has elapsed when the cumulated fluctuations of energy becomes comparable to the star's initial energy or alternately when the sum of squared deviations is of the order of unity. A crude argument shows that, in a system having stars of equal masses m, the relaxation time for a star of velocity v is proportional to v^3/nm^2 and this statement is only little modified in the more sophisticated calculations of T_{rel} (Chandrasekhar, 1942). Let us only recall here that the treatment is based exclusively on binary collision cross-sections; consequently, we meet the difficulty of having to cut off the impact parameters at a certain critical value which is usually chosen, somewhat arbitrarily, as the mean interstellar distance within the cluster. In a plasma, a natural cut-off is afforded by the Debye length.

Let us now denote by A the age of the system; then the ratio

$$\eta = T_{rel}/A$$

characterizes the relative importance of encounters. Wherever we have $\eta \ll 1$, the system is relaxed in the sense that it is dominated by the effect of stellar encounters; such is the case in the central region of globular clusters. If on the contrary $\eta \gg 1$, as in the outer part of any cluster and within elliptical galaxies, relaxation plays no role because the encounter effect has not yet influenced the regular trajectories in any appreciable amount.

A mean free path λ can formally be associated to the time T_{rel} in multiplying the latter, averaged over the cluster by the mean velocity of the test star. According to former considerations, λ is approximately proportional to the square of the star's kinetic energy and for stars of high energy, λ surpasses by far the dimensions of the system.

Therefore the fluid model of a gas, given in terms of pressure, is inadequate for star clusters and a description nearer to free particle physics, based on the Boltzmann equation, should be adopted.

3. Basic Equations

Denoting by $f_m(\mathbf{x}, \mathbf{v}, t)$ the distribution function of the position, velocity at time t of the stars of mass m and by $\phi(\mathbf{x}, t)$ the smoothed-out gravitational potential, we shall write the Boltzmann equation for these stars

$$\frac{\partial f_m}{\partial t} + \mathbf{v} \cdot \frac{\partial f_m}{\partial \mathbf{x}} - \frac{\partial \phi}{\partial \mathbf{x}} \cdot \frac{\partial f_m}{\partial \mathbf{v}} = \sum_{m'} \left(\frac{\partial_c f_m}{\partial t} \right)_{m'}. \tag{1}$$

The right hand side represents the variation per unit time of f_m, due to encounters of stars of mass m with stars of mass m' in the cluster. If the cluster is unrelaxed, as it could be in an early evolutionary phase, the right hand side of (1) vanishes; otherwise we will replace it by the Fokker-Planck expression

$$\left(\frac{\partial_c f_m}{\partial t} \right)_{m'} = - \frac{\partial}{\partial \mathbf{v}} \cdot (f_m \langle \Delta \mathbf{v} \rangle_{m'}) + \frac{1}{2} \frac{\partial}{\partial \mathbf{v}} \frac{\partial}{\partial \mathbf{v}} (f_m \langle \Delta \mathbf{v} \, \Delta \mathbf{v} \rangle_{m'}), \tag{2}$$

where $\langle \Delta \mathbf{v} \rangle_{m'}$ is the average velocity change per unit time of a star of mass m after collisions with stars of mass m'. Equation (2) is connected to the fact that for long-range forces such as gravitational attraction, the cumulative effect of distant encounter outweighs the influence of the very rare close encounters. The process is therefore akin to brownian movement.

Now the effect of collisions is twofold: there will be a mixing of the individual energies and angular momenta and some of the most energetic stars may escape from the cluster. Both effects tend to alter the structure of the system and cause its dynamical evolution.

For an isolated cluster, Φ is derived from the Poisson equation

$$\nabla^2 \Phi = 4\pi G \sum_m m \int f_m \, d^3 v \qquad (3)$$

considering, for simplicity, a discrete mass spectrum.

The integrodifferential system of Equations (1), (2), (3), nonlinear in f_m even without collisions, is too formidable to be solved in general. On the other hand, the mechanism of a quasi-continuous piling-up of energy for causing a star to escape, such as described by (2), does not work for an isolated cluster (Hénon, 1960), since a star having an energy near to that of escape spends most of its time in the outer region of the cluster and thus experiences no encounters. For the problem to bear a more definite meaning, we should include here the tidal force of the Galaxy; the simplest but rather unsatisfactory way to do so is to introduce a radius of stability r_s defining the point where the gravitational pulls of the cluster and the Galaxy balance each other. Any star which has sufficient energy to reach and go beyond the distance r_s from the cluster's center is lost to the cluster.

Let an open cluster of center C describe a circular orbit in the galactic plane, around the galactic center O; let a cluster star S lie on the line OC, at a distance r from C and with $OC = R$, M and M_g being the respective masses of the cluster and of the Galaxy reduced to a central mass-point. We can express the balance of forces in the frame of reference rotating around O with the angular velocity ω at point C:

$$-G\frac{M_g}{(R-r)^2} + G\frac{M}{r^2} + \omega^2(R-r) = 0$$

and since

$$\omega^2 = G(M_g/R^3)$$

then

$$r_s = (M/3M_g)^{1/3}. \qquad (4)$$

For globular clusters, we may consider an elliptic orbit of eccentricity e; the factor 3 in the denominator should then be multiplied by $1+e$ (King, 1962).

In several preliminary investigations a given distribution function f has been assumed, followed by an attempt to solve (3) in order to obtain the corresponding dynamical model. Conversely, assuming a given potential, one could attempt to derive f from (1). Most of the references to papers of that kind can be found in R. Michie's review of cluster dynamics (1964).

The problem is somewhat simplified by the assumption of spherical symmetry, which is justified for globulars and many open clusters. Among other simplifying assumptions, one often considers stars of equal masses for all the cluster, which is convenient but not very realistic as regards energy exchange and evaporation. One also assumes frequently an isotropic velocity distribution, which appears fairly adequate in the central part of a cluster, but much less in the outer part. When this last hypothesis is made, the very complicated expression (2) can then be converted into a relatively tractable form, which has been used by several authors.

4. Dynamical Cluster Models

Starting with a spherical cluster of equal masses and isotropic velocity distribution, Hénon (1961) computed the structure determined by the basic Equations (1) and (3), as well as the evolution of the cluster assumed to take place while keeping the model similar to itself (*modèle homologique*). During evolution, the central density immediately appeared to increase rapidly, producing a sharp cusp which compelled Hénon to start with an infinite central density. Further, the central region absorbed negative energy, which was accounted for by close binary star formation. The external radius increases while the total mass decreases linearly with time, at a rate of order 2 solar masses per million years. The density profile, when projected on the sky, agrees in a satisfactory way with that of the globular cluster 47 Tuc.

The same kind of agreement is claimed by Michie (1961), who started by adopting a distribution function of the form

$$f(E, J) = A \exp(-\alpha E - \beta J^2) Q(E) \tag{5}$$

in terms of stellar energy E and angular momentum J, A, α, β are constant parameters and we notice that (5) becomes a Maxwellian distribution at the center of the cluster, exhibiting a depopulation of stars possessing a high angular momentum and including a cut-off factor having its major effect at large distances from the center. Since account is taken of the velocity anisotropy, the expression (2) becomes very complicated and in order to avoid solving the Boltzmann Equation (1), Michie seeks to describe the cluster's evolution by a sequence of equilibrium states all characterized by the invariant form (5). The changes in the three parameters A, α, β are determined in first approximation by taking successive moments of (1). The evolution thus obtained reveals, like Hénon's model, a contraction of the central core and expansion of the outer region. The velocity anisotropy increases from the center outwards, in qualitative agreement with a former study of von Hoerner (1957) who regarded the outer zone of a globular cluster as populated by stars thrown out from a relaxed core, into orbits of high E and low J.

Neglecting velocity anisotropy, King (1966) computed a whole set of models using as distribution function the approximate steady-state solution (Michie, 1963a) of Equations (1), (2) which reduces to a Maxwellian distribution minus a constant,

$$f(r, v) = f_0 [\exp(-j^2 v^2) - \exp(-j^2 v_e^2)], \tag{6}$$

where j is constant, $v_e = -2\Phi(r)$, $\Phi(r_s) = 0$. In other words, the escape velocity v_e at distance r from the center allows any star to reach the radius of stability r_s given by (4). From (6) one calculates the mass density $\varrho(r)$ which reads

$$\varrho(r) = 4\pi m \int f(r, v) v^2 \, dv$$

for a cluster of equal stellar masses m, whence the projected density which is the quantity suited for comparison with observations.

Close examination of Palomar-Schmidt plates enabled King (1962) to follow out the projected density far enough to define a limiting radius r_t which appeared to agree satisfactorily with (4). The projected density of many globular clusters is well represented by the empirical formula

$$\varrho_p(r) = k \left\{ \left(1 + \frac{r^2}{r_c^2} \right)^{-1/2} - \left(1 + \frac{r_t^2}{r_c^2} \right)^{-1/2} \right\}^2 , \tag{7}$$

in which k is a number factor, r_c the 'core' radius and r_t the limiting radius, three parameters corresponding to the three quantities which fix the quasistationary state of the cluster, viz. its total mass, total energy and the tidal field.

Within the core, relaxation should be attained or at least well under way, while the outer layers are mostly under tidal influence. When $r \ll r_t$, expression (7) reduces to its first term which describes correctly the central region of the globular cluster 47 Tuc up to 5' from the center (Gascoigne and Burr, 1956) and also the main part of the open cluster Praesepe up to 60' from its center (Bouvier, 1961). With suitable values of k and r_c, we may represent the projected density in globular clusters of high or low concentration ratio as well as the distribution of both bright and faint stars in a given cluster.

The basic structural properties of clusters, i.e. the core and the outer halo or corona, have been lately discussed by Kholopov (1969) and in particular, for the Pleiades open cluster, Artyukhina (1969) has shown that the core extends to $1°$ from the center and the corona to $3°25$

All globular clusters in our Galaxy disclose a very similar structure which probably illustrates a trend toward a common final state; the regularizing mechanism seems to reside in the stellar encounters within the central core. Furthermore, according to their light distribution, elliptical galaxies also possess projected densities which can be made to fit models based on (6) and described by (7) with adequate k, r_c. This similarity in aspect cannot be induced here by collisions, because the mean relaxation time of ellipticals is larger than their age ($\eta \gg 1$), so the regularizing mechanism is to be found in the mixing of orbits of stars having different periods and moving in a time-dependent potential. Such a process must have occurred very quickly, in a time of the order of the mean period P of a star in the system, whereas the slow relaxation due to star-star encounters has a time scale T_{rel} much larger than P, by a factor proportional to $n/\log n$ where n is the total number of stars in the system. Consequently, the initial phase of dynamical evolution of a stellar system is entirely dominated by the orbital mixing. Moreover, the interaction of a star with the changing potential does not depend on the star's mass, in contrast to the encounter effect which leads to equipartition and therefore to mass segregation inside the system.

When encounters can be neglected, the basic dynamical problem is to obtain a stationary solution of Equations (1) and (3), for a vanishing right hand side of (1). Such a solution is a function of the integrals of motion E and J, i.e. the total energy and angular momentum, both per unit mass, and it could in principle be deduced from the observed star density. Assuming all stars to have equal masses, we would first transform the observed projected density into the space density

$$\varrho(r) = \frac{4\pi}{r} \, m \int\limits_{\Phi}^{0} dE \int\limits_{0}^{J_{max}} \frac{Jf(E, J)}{\sqrt{J_m^2 - J^2}} \, dJ \tag{8}$$

where $J_{max}^2 = 2r^2(E - \Phi)$, and then solve this double integral equation for $f(E, J)$. This problem has been dealt with independently by Veltmann (1961) and Bouvier (1962, 1966), who developed $f(E, J)$ in a power series in J^2, retaining a finite number of terms. They thus gained some insight into the various possible types of velocity anisotropy.

Now anisotropy entails a depletion of the higher angular momenta and therefore pulls down the density of stars in the outer regions of a cluster; the tidal part of the drop-off is correspondingly less severe and a larger tidal radius r_t will be inferred. In fact, r_t is too badly known to help us determine the degree of anisotropy; one may fit star counts with a model based on velocity isotropy but the same counts could also be fitted to models of rather marked anisotropy and with a larger tidal radius.

Insofar as we consider clusters of equal stellar masses, the rate of escape remains comparatively very small; the selective evaporation rate estimated by several authors under many simplifying assumptions, had shown a monotonous increase with decreasing mass of the escaping stars. As a consequence, the fainter end of the initial luminosity function will be depopulated, while the bright end also becomes depleted on account of the rapid evolution of the more massive stars. But, apart from the uncertainties in the calculated escape rates, comparison with observations is hampered by the probable lack of a universal initial luminosity function for star clusters (Michie, 1963b; Martinet 1966).

The essential cause of escape lies in energy gained by a star in close encounters with other cluster stars; a correct calculation must start with the expression giving the probability that a certain star suffers a close encounter at time t, which will change its velocity by a finite amount Δv. Following this method, Hénon calculated the rate of escape from an isolated cluster, first with equal masses (1965) then with an arbitrary mass distribution (1969) and adopting an initial isotropic velocity distribution corresponding to the polytropic model of index 5, in order to simplify the computation. The rate of escape is quite sensitive to the mass spectrum; it is multiplied by a factor of 30 or more when equal masses are replaced by unequal ones. This result had already been obtained in numerical experiments (Wielen, 1968) and we shall now turn to this alternative approach to cluster dynamics.

5. Numerical Experiments

The great advantage of the method of numerical experiments, initiated for stellar dynamics by von Hoerner (1960), resides in its being free from starting assumptions.

Given arbitrary initial conditions, we just have to solve numerically the $6n$ first order equations of motion pertaining to the n point-stars building the cluster: if x_{ik}, v_{ik} denote the respective position and velocity coordinates $(i=1, 2, ..., n; k=1, 2, 3)$ we then write

$$dx_{ik}/dt = v_{ik}, \tag{9}$$

$$\frac{dv_{ik}}{dt} = -G \sum_{j \neq i} \frac{m_j}{\sum_k (x_{ik} - x_{jk})^2}. \tag{10}$$

The expression in the right hand side of (10) consumes most of the computing time and the divergence occurring when $x_{ik} \to x_{jk}$ is in itself a problem. Even with the help of the fast computers now at our disposal, it becomes difficult, on account of prohibitive computing time and limited memory capacities, to deal with a very large number of stars; the method is therefore limited to multiple stars and small clusters. Von Hoerner had begun with 16, then 25 stars, van Albada worked on a few dozen (1967, 1968), Wielen 100 stars (1968), and Aarseth (1968) managed to cope with the 250-body problem, which represents an upper limit at the present time.

Numerical integrations show that the cluster exhibits a central density cusp after a finite time, as well as an extended halo. This agrees with results obtained by models described in Section 4 as well as with a recent study of von Hoerner (1968) on the high central densities in stellar systems, assuming local virial equilibrium and restrictive properties of energy exchange. Later on, the cluster seems to approach some sort of final state, but the numerical experiments performed with different initial conditions give no precise information as to whether the overall properties tend to become increasingly similar or not (Wielen, 1967).

In such computations, it is advisable to choose, for each star, a variable integration time-step.

Now errors tend to accumulate so that one is never completely assured to find again the initial conditions when time is reversed (Miller, 1964); the accuracy with which the constants of motion do indeed remain constant is thus to be as high as possible.

A comparative study of eleven numerical integrations of the same gravitational 25-body problem showed good agreement of the runs during the initial phase, but when encounters became important, the experiments could no longer be reproduced in detail (Lecar, 1968).

The initial evolutionary phase, dominated by the dynamical mixing of orbits, can also be studied by means of numerical experiments (Hénon, 1964). As stated before, the analytical problem would consist in trying to solve the simultaneous Equations (1) without right hand side and (3), which is a hopeless task in general. But we can avoid solving (1) in the very peculiar case where the initial distribution function $f(\mathbf{x}, \mathbf{v}, 0)$ is equal to a constant positive value C inside and vanishes outside a finite phase domain D_0.

We still have $f = C$ in the domain D_t occupied by the system at any later time and $f = 0$ outside D_t (Liouville's theorem of statistical mechanics).

But the state calculated by merely expressing the conservation of the fine-grained phase density f has no physical reality; we may call it a virtual final state. Indeed, as evolution proceeds, the phase domain D_t develops a filamentary structure and the filaments become always longer and thinner so that f finally loses its meaning. The real final state of orbital mixing has to be described in terms of a coarse-grained phase density $\bar{f}$, which is the average of f over cells of finite size. The virtual final state, which in case of a spherical system corresponds to a polytropic model of index 1.5, may nevertheless be useful as a reference state (Bouvier and Janin, 1970a).

6. Complementary Remarks on Stellar Clusters

The evaporation of stars from a cluster, which is certainly favoured by the tidal action of the Galaxy (Wielen, 1968; Hayli, 1970), entails a loss of mass and of energy for the cluster. On the other hand, the cumulative influence of passing clouds of interstellar matter, which is undoubtedly of importance for open clusters since they remain in the disk region, brings over an increase of energy and a tendency of these clusters to expand and gradually disrupt. The effect was examined some time ago by Spitzer (1958) and could partly explain the relatively low occurrence of very old open clusters.

Recently, Bouvier and Janin (1970b) performed some numerical experiments on the disruption of a small stellar cluster by passing interstellar clouds; the effect is sensitive to the initial state of the cluster and a more complete treatment of the problem should include both the continuous action of the galactic field and the stochastic influence of the clouds.

For the galactic clusters, it has proved strikingly illustrative to construct a composite color-luminosity diagram by fitting the respective main sequences to the so-called 'zero age main sequence', which cannot however be defined in a completely unambiguous manner since the physical conditions prevailing at the birth of the cluster, in particular the chemical composition of the protocluster gas cloud, should be different from one cluster to another.

The theory of stellar evolution tells us that, in general, a main-sequence star after having exhausted its hydrogen fuel within the central core, evolves to the right of the main sequence toward the red giant state. The larger the star's mass, the earlier this happens, so we expect the upper main sequence of any cluster diagram to become depleted in course of time. The present age determination will then follow from the absolute magnitude at the turn-off point for the particular cluster considered.

But such a determination is most sensitive to the metal content and cannot claim to be accurate to better than a few billion years (Iben and Rood, 1970).

The oldest open clusters, like M67 and NGC 188, having respectively around five and ten billion years, lack all their upper main sequence but possess well-developed giant branches; their color-luminosity diagram resembles that of globular clusters.

For the latter objects, we have no direct observational evidence about their ages, but they contain no observable interstellar gas and are highly stable and therefore

long-lived stellar systems. According to their space distribution and kinematic properties, they belong to the oldest of the stellar populations and their age may go up to
20 billion years. Stellar evolution may account for the red color of the brightest stars
within globular clusters of our Galaxy.

In comparing color-luminosity diagrams with theoretical evolutionary tracks, one
implicitly assumes that all the stars of the same cluster are coeval; this is justified only
if the age spread is noticeably less than the average age of all the members. Such is the
case for the great majority of clusters except for the younger one where a large spread
of ages is present and where the rate of star formation, still under way, seems even to
increase with time (Iben and Talbot, 1966).

We had noticed earlier (Section 4) the similar appearance revealed by all globular
clusters; this does not prevent them from differing widely in chemical composition.
Some, like M15, M5 or M92 are made of anomalously metal-poor stars while others,
as NGC 6356 or 47 Tuc hardly differ in metal content from typical rather young open
clusters such as Hyades and Pleiades.

The metal content Z of a globular cluster appears to be generally correlated to the
structure of its horizontal branch (Faulkner, 1966) but not always in the same way
(Sandage and Wildey, 1967; Rood and Iben, 1968). One more parameter at least, in
addition to Z, must vary from one cluster to another; according to Castellani et al.
(1970), globular clusters form a two-parametric family of objects.

Moreover, globular clusters have been observed in extragalactic systems like M31
(Andromeda) and the Magellanic Clouds, where two types of globular clusters occur,
the red and the blue. We first have the same kind of globular clusters as those of our
Galaxy, but with ages which do not seem to exceed 5×10^9 years and which correspond
to the youngest of the globular clusters of the Galaxy (Bok, 1969); some other globular
clusters form a sub-group of intermediate-age clusters, and are only 10^9 years old.
Secondly, a number of globular clusters in the Large Magellanic Cloud appear much
bluer in integrated light than those of the Galaxy and their color-luminosity array,
after having been matched to the Hyades main sequence with allowance for differences
in chemical composition and evolutionary effects, resemble those of very young galactic clusters such as h and χ Persei. Thus, the birth of globular clusters in the Magellanic
Clouds presumably took place considerably later than in our Galaxy.

These facts, together with the indications of the theory of stellar evolution, led
Kholopov (1966) to postulate that shortly after cluster formation, all color-luminosity
diagrams of globular clusters were similar to those of young open clusters today and
only with evolution did they develop into the presently observed diagrams, typical of
the globular clusters in our Galaxy. In contrast to the Magellanic Clouds, clusters
formed today in the Galaxy will no longer be globular.

7. Stellar Associations

The O-associations are spherical and since, as mentioned in the introduction, their
densities are well below the necessary limit for stability against tidal disruption,

Ambarzumian (1955) argues that the O-associations are expanding systems of positive total energy. The expansion motions have been effectively found by Blaauw (1964) and others; the total positive energy is to be understood after deducting the negative binding energy of the numerous multiple stars generally present in associations. According to Ambarzumian, associations contain a nucleus formed by one or more open clusters and in case they are very young, multiple stars of Trapezium type and sometimes star chains. If linear expansion prevails, which means that the member stars have left a common center of formation simultaneously, each of them with a particular velocity remaining constant in time, and if r is the distance gone through from the center during time t, by a star of velocity v, constant with respect to matter remaining at rest in the medium of formation, then the ratio v/r, called the rate of expansion, is characteristic for the association and is equal to the inverse of the time elapsed since the stars left their origin, i.e. of the kinematical age of the system. O-associations are generally found because O stars occur nearly always in groups in the Galaxy and in extragalactic systems, as for instance in the central section of the Larger Magellanic Cloud.

The numbers of stars of an O-association may be of the order of several hundreds and the diameters of the groups range from 30 to 200 pc (Haffner, 1965). With increasing size however, the boundaries become increasingly vague because the association gradually merges into the general stellar field; furthermore several of the larger associations appear to be divided into subgroups with different evolutionary stages. The properties of O-associations have been discussed by Blaauw in his (1964) review of this topic. As a by-product of O-associations, we shall note here the so-called 'runaway' stars; they possess velocities of up to 200 km/sec and their mass distribution differs from that of ordinary stars by a high relative frequency of large masses. Their origin, for the few known cases, in the O-associations is well established. Blaauw (1961) proposed to account for the process causing the runaway stars, which are always single, by assuming that they were former companions of massive stars which underwent rapid disintegration. Another cause of runaway stars could also reside in the dynamical behavior of collapsing stellar clusters (Poveda et al., 1967) or perhaps also in the instability of multiple systems (Worrall, 1967).

T-associations are likewise groups of T Tauri stars, irregular variables with emission lines; such associations, which are also unstable aggregates of an age less than 10^7 years, are connected with strong clouds of obscuring matter and sometimes with bright galactic nebulae (Herbig, 1962).

The study of associations leads to the idea that all the massive stars of flat subsystems of the Galaxy, as well as the T Tauri stars, are born in groups; this cannot be ascertained for every class of stars; data are too sparse for small masses. Further, the existence of subgroups within many associations indicates that star formation develops in irregular fashion through complexes of interstellar clouds. We have emphasized the youth of associations; let it be said that after 10^7 years or so, O stars will have evolved and about simultaneously, the O-association will have dispersed in the general stellar field.

References

Aarseth, S.: 1968, *Bull. Astron. Paris* (3), **3**, 105.
Albada, T. S. van: 1967, *Bull. Astron. Paris* (3), **2**, 59.
Albada, T. S. van: 1968, *Bull. Astron. Inst. Neth.* **20**, 47.
Ambarzumian, V. A.: 1955, *Observatory* **75**, 72.
Artyukhina, N. M.: 1969, *Soviet Astron.* **12**, 987.
Blaauw, A.: 1961, *Bull. Astron. Inst. Neth.* **15**, 265.
Blaauw, A.: 1964, *Ann. Rev. Astron. Astrophys.* **2**, 213.
Bok, B. J.: 1969, *J. Roy. Astron. Soc. Canada* **63**, 105.
Bouvier, P.: 1961, *Publ. Observ. Genève*, No. 61.
Bouvier, P.: 1962, *Publ. Observ. Genève*, No. 63.
Bouvier, P.: 1966, in *The Theory of Orbits in the Solar System and in Stellar Systems*, IAU Symposium
    No. 25 (ed. by G. Contopoulos), Academic Press, London and New York, p. 57.
Bouvier, P. and Janin, G.: 1970a, *Astron. Astrophys.* **5**, 127.
Bouvier, P. and Janin, G.: 1970b, *Astron. Astrophys.*, in press.
Castellani, V., Giannone, P., and Renzini, A.: 1970, internal report, Univ. of Rome.
Chandrasekhar, S.: 1942, *Principles of Stellar Dynamics*, Univ. of Chicago Press, Chicago, Ch. II.
Faulkner, J.: 1966, *Astrophys. J.* **144**, 978.
Gascoigne, S. C. B. and Burr, E. J.: 1956, *Monthly Notices Roy. Astron. Soc.* **116**, 570.
Haffner, H.: 1965, in *Landolt-Börnstein Zahlenwerte und Funktionen aus Naturwissenschaften und
    Technik, Gruppe VI: Astronomie-Astrophysik und Weltraumforschung, Band I: Astronomie und
    Astrophysik* (ed. by H. H. Voigt), Springer-Verlag, Berlin, Heidelberg, New York, p. 582.
Hayli, A.: 1970, *Astron. Astrophys.* **7**, 17.
Hénon, M.: 1960, *Ann. Astrophys.* **23**, 668.
Hénon, M.: 1961, *Ann. Astrophys.* **24**, 369.
Hénon, M.: 1964, *Ann. Astrophys.* **27**, 83.
Hénon, M.: 1965, *Ann. Astrophys.* **28**, 68.
Hénon, M.: 1969, *Astron. Astrophys.* **2**, 151.
Herbig, G. H.: 1962, *Adv. Astron. Astrophys.* **1**, 47.
Hoerner, S. von: 1955, *Z. Astrophys.* **35**, 255.
Hoerner, S. von: 1957, *Astrophys. J.* **125**, 451.
Hoerner, S. von: 1960, *Z. Astrophys.* **50**, 184.
Hoerner, S. von: 1963, *Z. Astrophys.* **57**, 47.
Hoerner, S. von: 1968, *Bull. Astron. Paris* (3), **3**, 147.
Iben, I. and Rood, T. R.: 1970, *Astrophys. J.* **159**, 605.
Iben, Jr., I. and Talbot, R. J.: 1966, *Astrophys. J.* **144**, 968.
Kholopov, N.: 1966, *Soviet Astron.* **9**, 928.
Kholopov, N.: 1969, *Soviet Astron.* **12**, 625 and 978.
King, I. R.: 1962, *Astron. J.* **67**, 471.
King, I. R.: 1965, *Astron. J.* **70**, 376.
King, I. R.: 1966, *Astron. J.* **71**, 64.
Lecar, M.: 1968, *Bull. Astron. Paris* (3), **3**, 91.
Martinet, L.: 1966, *Publ. Observ. Genève*, No. 72.
Michie, R.: 1961, *Astrophys. J.* **133**, 781.
Michie, R.: 1963a, *Monthly Notices Roy. Astron. Soc.* **125**, 127.
Michie, R.: 1963b, *Monthly Notices Roy. Astron. Soc.* **126**, 499.
Michie, R.: 1964, *Ann. Rev. Astron. Astrophys.* **2**, 49.
Miller, R. H.: 1964, *Astrophys. J.* **140**, 250.
Poveda, A., Allen, C., and Riuz, J.: 1967, *Bol. Observ. Tonantzintla Tacubaya* **4**, 86.
Rood, T. R. and Iben, I.: 1968, *Astrophys. J.* **154**, 215.
Sandage, A. and Wildey, R.: 1967, *Astrophys. J.* **150**, 469.
Spitzer, L.: 1958, *Astrophys. J.* **127**, 1.
Veltmann, U. L.: 1961, *Publ. Tartu Astron. Observ.* **33**, 387.
Veltmann, U. L.: 1966, *Astronomia*, Moscow 1968, Ch. 1.
Wielen, R.: 1967, *Veröffentl. Astron. Rechen-Inst. Heidelberg*, No. 19.
Wielen, R.: 1968, *Bull. Astron. Paris* (3), **3**, 127.
Worrall, G.: 1967, in *XIVe Colloque d'Astrophysique de Liège, Mem. Soc. Roy. Sci. Liège* **15**, 365.

THE EVOLUTION OF SPIRAL STRUCTURE

E. M. BURBIDGE

Dept. of Physics, University of California at San Diego, La Jolla, Calif., U.S.A.

1. Introduction

In attempting to study the evolution of spiral structure in galaxies, one may concentrate either on our own Galaxy, or on external galaxies. In our own, only a small part of the total time scale can be spanned and it is best therefore to consider other galaxies as well and try to order them in a sensible morphological sequence, in the hope that this will represent an evolutionary sequence. But one should not forget the cautionary tale of stellar evolution: in early studies the Hertzsprung-Russell diagram was interpreted as a sequence in which stars began their existence as red giants, then moved on to the upper end of the main sequence and evolved down it. An understanding of stellar evolution depended on the realization of two crucial factors in the problem: the role played by the masses of stars, and the development of a chemical inhomogeneity through the exhaustion of the nuclear fuel, hydrogen, in the central core. In galactic evolution, I fear we may be lacking some vital information or understanding that is as basic and fundamental as these were to stellar evolution.

With these reservations in mind, however, we can try to understand the development and maintenance of spiral structure in our own and other galaxies, within the frame work of present-day knowledge. In studying our own Galaxy the advantages are the large scale and the fact that three-dimensional velocities, at least for the nearer stars, can be obtained, but these advantages are offset by the problems caused by the obscuring dust for discerning the large-scale structure, and by the difficulties in determining distances. In other galaxies, the spiral structure can be clearly traced out, but the scale is small (in the Virgo Cluster one second of arc corresponds to 75 pc). Further, only one component of velocity – the line-of-sight component – can ever be measured, and this is hard to measure with precision.

One simplifying assumption that can be made is that magnetic fields are apparently not important in affecting the large-scale spiral structure in our Galaxy (since measurements of magnetic field strength concur in giving a low value, only a few microgauss, for an organized field). Thus, gravitational forces will dominate in our Galaxy, and it can be assumed as a working hypothesis that the same situation will hold in other galaxies, although the possibility that stronger magnetic fields might exist, for example in barred spiral galaxies, should not be ignored.

2. Relevant Measurements

Other lecturers at this course are covering the types of measurement relevant to the study of spiral structure in our Galaxy – radioastronomical measures of positions and velocities by means of the 21-cm line of neutral hydrogen, distances and motions of stars, and so on. I shall speak about the measurement of velocities by optical methods in external galaxies, by the Doppler shift in spectrum lines.

First, we should consider what kind of velocities are to be measured. The mean velocity of a galaxy as a whole relative to the center of our Galaxy – its systemic velocity – is of interest only as a means of deriving the distance of the object from Hubble's expansion law. What we are interested in are the internal velocities, of both stars and gas, within a galaxy. The principal velocity field in a spiral galaxy is its rotation; the circular or rotational velocity in a normal spiral galaxy can be assumed to have axial symmetry and the plot of this circular velocity as a function of distance from the center of the galaxy is the *rotation curve*.

In addition to the circular velocity field, there may be noncircular velocities, particularly in the central regions. We shall consider these in detail later; it is clear that the study of noncircular motions in the gaseous component is of importance in considering the origin and evolution of the spiral arms.

Very little work has been done on rotational velocities in the stellar component of galaxies. These must be measured by means of absorption lines, which are much more difficult to measure accurately than emission lines, because they come from an integrated population of stars of differing spectral types and they are usually rather diffuse. By contrast, emission lines produced in the ionized gas component can be measured with good accuracy, depending on the instruments available, and on the intrinsic velocity dispersion in the gas. Also, emission lines can be measured even in the outer parts of a galaxy, wherever there is ionized gas, whereas it is feasible to measure absorption lines only where there is a strong continuum background spectrum, i.e. where the star density is not too low.

All the observations I shall be talking about, therefore, refer to the ionized gas component, but since the gas is concentrated in the spiral arms, this is what we need (although it would eventually be interesting to be able to study motions in the smooth stellar spiral arms seen in some galaxies, e.g. NGC 4826 shown in Figure 1).

We now consider the instruments to be used for the measurements. In the first place, it is sometimes convenient to determine the location of H II regions in a galaxy before taking slit spectrograms, in order to be able to choose good orientations and locations for the slit. This can be done by means of photographs taken through interference filters isolating a narrow wavelength region around the wavelength of Hα with the appropriate redshift (see e.g. Hodge, 1969). For nearby galaxies of large angular diameter, like M31 and M33, individual H II regions can be set on the spectrograph slit; their line-of-sight velocities and distances from the center as seen projected on the plane of the sky must then be corrected to true distances from the center and circular velocities in the plane of the galaxy (for the latter, the preliminary assumption is made that only circular velocities are present). Non-circular velocities will have to be deduced from scatter or systematic discrepancies when the results from many H II regions are combined into a single rotation curve.

A convenient method for galaxies of smaller angular diameter, e.g. those in the

Fig. 1.   NGC 4826 (120-inch telescope, Lick Observatory).

nearby groups outside the Local Group, and in the Virgo Cluster, as well as for more distant galaxies, is to use a fast spectrograph with a long slit. The slit can be set along the major axis of a galaxy, or in any other orientation, and velocities along the whole diameter of the galaxy can be recorded on one single spectrogram. This method is economical in observing time, and is very useful for obtaining rotation curves. My husband, K. H. Prendergast, and I used it for a series of studies of rotations and mass distributions in galaxies, published in the *Astrophysical Journal* between 1959 and 1965 (see e.g. Burbidge and Burbidge, 1970).

A new and promising instrumental method has been developed and applied to a few galaxies of large angular diameter by Courtès and his collaborators, at the Haute Provence Observatory in France (Courtès, 1960, 1964). The instrument used is a Fabry-Pérot interferometer; interference rings are projected on to the image of the galaxy being studied and are photographed with a reducing camera. Emission lines from HII regions appear as knots in which differences in velocity are measured by the changes in radius of the interference rings. Thus velocities can be measured all over the area, wherever ionized gas is present, whereas in the long-slit method velocities over the whole galaxy have to be obtained by taking many spectrograms, each one being a cut across the galaxy in a chosen orientation. Also, since higher dispersions than the traditional nebular spectrographic dispersions can be used, velocities can be measured with greater precision.

That concludes what I want to say about observational techniques. We turn now to the results of such observations and their bearing on theories of spiral structure.

3. Spiral Arms and Differential Rotation

I want to talk now about an old problem that has been solved by the new theories of C. C. Lin and others concerning the nature of spiral arms. This old problem is what may be called 'the winding-up of spiral arms'. This problem came about because all studies of rotations of galaxies, by the methods just described, showed that galaxies do not rotate like solid bodies, or wheels, but are in differential rotation. Now we must specify what is meant by a spiral arm – do we refer to what Lin has called the 'grand design' of spiral structure, or are we, for example in our Galaxy, merely talking about short segments and separate arcs and loops forming a roughly spiral pattern? If we are talking about the 'grand design', i.e. the traditional picture of a spiral galaxy, with two main spiral arms that trail with respect to the rotation, making 1 or $1\frac{1}{2}$ turns about the galaxy, then there definitely used to be a dilemma over the persistence of these arms. On the other hand, no such dilemma was obvious if one was merely talking about short nonpersistent arcs and loops, because any density fluctuation will tend to be drawn out into a spiral-like structure by differential rotation. This used to be demonstrated in lectures by von Weizsäcker, who used to draw the outline of an animal in a differentially rotating disk and show how, after a few turns, the outline would be drawn out into a good-looking spiral, bearing no resemblance to its original form!

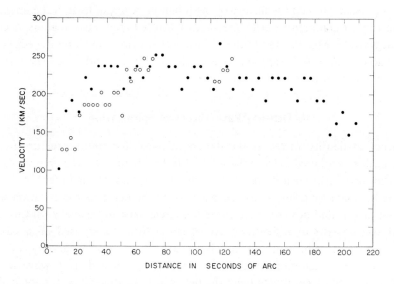

Fig. 2.   Rotational velocities measured along major axis of NGC 5055, uncorrected for orientation of plane of galaxy. Filled and open circles are from opposite sides of galaxy. Cross: central velocity determined from stellar component (Burbidge *et al.*, 1960a).

The 'winding-up problem' was very well exemplified by NGC 5055 (Burbidge *et al.*, 1960a). This is an Sc galaxy in which the rotation curve, shown in Figure 2, has been determined by measures in the ionized gas alone – the material constituting the spiral arms – and from this observed curve, the period of rotation varies from about 10^7 years in the inner part to about 2×10^8 years in the outer part, and spiral structure would be wound up out of all recognition in the presumed age of the galaxy, $\approx 10^{10}$ years.

I believe it was the emphasis laid on this problem by Oort at a symposium on interstellar gas in galaxies (Oort, 1962) which first led C. C. Lin, who was attending the symposium, to turn his attention to the problem, and, with his colleagues, to develop the density-wave theory of spiral structure which is described in the next section. While this is a first approach and there are other workers attacking the problem from different viewpoints, and while, as we shall see, there are still many unanswered problems, the 'winding-up problem' has been conquered.

The main observational features that need to be accounted for are as follows:

(1) The spiral arms we see in the usual blue-ultraviolet photographs of galaxies are regions of H II, dust, OB stars; other tracers of spiral arms, besides H I, are long-period Cepheids, certain red stars evolved from massive main-sequence stars, and in fact any star which can be shown to be young.

(2) Spiral structure – the 'grand design' of Lin – does extend over most of the galaxy; the arms do not make many turns.

(3) There are usually two arms, and they *trail* (the old controversy over leading or trailing spiral arms was resolved for the normal spiral galaxies in favor of trailing arms).

(4) Dust usually lies on the insides of the luminous arms, at least in the inner parts, but sometimes the luminous arms are sandwiched between dusty regions, according to recent work by Mrs B. Lynds; the brightest H II regions lie along the edges of the dusty regions and, being strung out like slender strings of beads, they show that star formation occurs in restricted regions only.

4. Density-Wave Theory of Spiral Arms

The theory of density waves as the dynamical basis for spiral structure in highly flattened galaxies originated from the work of B. Lindblad (followed up by P. O. Lindblad). During a time when it was fashionable to 'explain' the maintenance of spiral structure by magnetic fields, Lindblad persisted in the belief that gravitation was the dominant factor, and now we have come full-circle back to this view. In this theory, a spiral arm consists of a density wave of spiral form superposed on a basic axisymmetric mass distribution, and we have to ask whether such a density wave, once formed, is able to maintain itself, and, if so, how such a density wave might be formed in the first place. By studying the orbits of individual stars, Lindblad showed that a spiral gravitational field introduced as a perturbation in an axisymmetric gravitation field will tend to produce a redistribution of density which tends to maintain the imposed density wave, i.e. a rotating thin disk will be unstable against the formation of such a density wave.

In his work, Lindblad considered only the stars – point masses – and not the gas; he calculated the orbits of stars as systems of epicycles, and he noted that it is possible for a resonance to exist between the density waves and the epicyclic motion of the stars. Consider spiral structure consisting of two arms, and imagine a star moving in an epicycle about a center which is itself carried around the galaxy by the general rotation. At the instant when the center of the epicycle is crossing one arm, let a star be at a particular point in its epicycle. Then if the star has returned to that same point in the interval taken by the general rotation to move the epicyclic center around to the other spiral arm, a resonance will be set up. In fact, for two-armed spirals with a particular form of the rotation curve, resonance can extend over a substantial part of the disk and in the inner part of our Galaxy the rotation curve does indeed have approximately this form.

The opposition to Lindblad's theory came because he advocated spiral arms that lead, whereas an increasing body of evidence was showing that spiral arms actually in general trail. Also, he placed too much emphasis on the occurrence of exact gravitational resonance over finite portions of the galactic disk. What was new in the theory introduced by Lin, Shu, and others was that they concentrated on the gaseous interstellar medium, studied by means of its collective behavior, rather than on orbits of stars, and thus removed the emphasis from the resonances. The theory thus became one in which a somewhat inconspicuous density wave in the stellar distribution produces a spiral gravitational field as a fluctuation upon the smooth field, and the gas, reacting strongly to the presence of such small fluctuations, collects at these places and acts as a very efficient tracer for the gravitational potential troughs.

Accounts of this work, giving references, have been published by Lin and Shu (1967), Lin (1967), Prendergast (1967), and further papers appear in the Proceedings of IAU Symposium No. 38 held in September 1969, in Basel. A spiral pattern, in a galaxy in which $(\tilde{\omega}, \theta)$ are polar coordinates in the plane, may be described by a function

$$F(\tilde{\omega}, \theta, t) = A(\tilde{\omega}) \exp \{i [\omega t - m\theta + \Phi(\tilde{\omega})]\}$$

where t is the time, m is the number of arms (2 in cases of interest to us), $A(\tilde{\omega})$ is a slowly varying function of $\tilde{\omega}$, $\Phi(\tilde{\omega})$ is a slowly varying monotonic function multiplied by a large parameter, and ω gives the angular velocity at which the pattern rotates, Ω_p, by

$$\Omega_p = \omega/m.$$

For a given model of the distribution of mass in a galaxy, the pattern speed can be calculated, and Lin and Shu (1967) derived

$$\Omega_p = 11 \text{ km/sec per kpc}$$

using the Schmidt (1965) mass model for the Galaxy. Lin et al. (1969) have recently given a pattern speed of 13.5 km/sec per kpc. This, then, describes a system of two trailing arms which are density waves propagating around the center at an angular velocity Ω_p in the direction of the general galactic rotation of stars.

Lin and his colleagues emphasize that this is only a first-order theory, dealing with small density fluctuations in the form of spirals of small pitch angle. Several people have argued that we have no theory for the initiation of the density fluctuation, nor for the formation of the steeply-inclined arms often found in the outer parts of galaxies, particularly barred spirals; papers, references, and discussions on these aspects, e.g. by Toomre and Kalnajs, will be found in the Proceedings of the Basel IAU Symposium No. 38.

One feature of the density-wave theory is that it can explain the absence of regular spiral arms of ionized hydrogen in the inner 4 kpc of our Galaxy. This will be the distance at which the Lindblad resonance sets in; inside here, the solution to the equations giving a rotating pattern breaks down.

We shall return to a consideration of what may initiate spiral structure, and the place of barred spirals in such a scheme, later.

5. Observations Relating to Evolution of Spiral Arms

We consider three types of observational test that can be applied to the Lin-Shu theory. One concerns the place of birth of stars in our Galaxy. Another considers the spiral structure in external galaxies as delineated by stars of different ages. The third deals with small departures from a smooth rotation curve.

5.1. PLACE OF ORIGIN OF YOUNG STARS

For a pattern rotation $\Omega_p = 11$ km/sec per kpc, as worked out by Lin and Shu (1967), the pattern speed at the location of our sun, 10 kpc from the center of the Galaxy, will be 110 km/sec. Galactic rotation at this place is 250 km/sec, so the speed of separation of the density wave from a particular location is 140 km/sec. If the spiral arm is inclined at about $5°$, or $\frac{1}{10}$ radian, to the circular, then we can calculate how far a star, say 10^7 years old, will have moved in a radial direction from its place of origin. This will be

$$r = \frac{1}{10} \cdot \frac{140 \times 10^5 \times 10^7 \times 3 \times 10^7}{3 \times 10^{18}} \text{ pc} = 140 \text{ pc}.$$

This is a small distance, so that such young stars will still effectively define the spiral arm where they originated. In order to have the possibility of considering stars that have moved right out of the spiral arm where they originated, Strömgren (1967) considered stars with ages of some 10^8 years. Turning the problem around, he took stars in the present solar neighborhood, with known distances, space motions, and ages, and asked where would be the location of their birth-places. For B8-B9 stars on which accurate narrow-band photometry had been done, yielding good ages and distances, and with accurate proper motions and radial velocities, he used tables of plane galactic orbits computed by Contopoulos and Strömgren to calculate the distances of the places of origin from the galactic center. For the majority, the place of origin was obviously the local arm, with $9.4 < R < 11.3$ kpc, but some 10–15% had $12.2 < R < 13.6$ kpc, and clearly formed in a spiral arm some 2–3 kpc further out, i.e. the Perseus arm.

Next, taking older stars (5×10^8 yrs), with space velocities such that they could have come from the Perseus arm, he considered what pattern speed in the Lin-Shu theory would give a place of location actually in the Perseus arm. If he had not assumed that star formation occurred in a spiral arm pattern that rotates as a rigid body, but rather in a big cloud complex moving with circular velocity, these stars would have come (in the usual diagram, anticenter region upwards, galactic longitude increasing clockwise) from a region inside and to the left of the present-day location of the outer spiral arm. With a pattern speed $\Omega_p = 20$ km/sec per kpc, the place of origin coincided with the computed past position of the Perseus arm. But note that a pattern speed *greater* than that computed by Lin *et al.* (1969) was required to obtain agreement.

5.2. OLDER SPIRAL STRUCTURE IN GALAXIES

A further way to consider observationally the evolution of spiral arms is to try to delineate the older spiral structure in galaxies, remnants of the passage of the density wave prior to its current position given now by the gas, dust, and youngest stars. Such older spiral structure would be expected to be smoother and broader, and to consist of A type stars and the red giants evolved from them, which have diffused through random motions out of an initially narrow arm, and to be displaced from the present spiral structure. This is just what is shown by some beautiful composite photographs of the Sc galaxy M51 and its SO companion NGC 5195, by Zwicky (1955). He took photographs through different filters, and superposed negatives with positives in different combinations. A smooth yellow-red spiral structure, much more regular than the usual spiral arms, was revealed in M51, and the general tendency was for this to lie on the insides of the presently bright spiral arms. Filter photographs shown by Morgan and Osterbrock (1969) also show smooth yellow arms in the Sc galaxy NGC 628, as compared with the very knotty arms in the light of Hα coming from the H$\scriptstyle\rm II$ regions.

5.3. 'BUMPS' ON ROTATION CURVES

Dr. Kerr has talked about the rotation curve of our Galaxy, from radioastronomical measures, and we have seen small bumps on it coinciding with places of crossing the spiral arms. In other galaxies, the studies of rotation of the ionized gas, by G. Burbidge, Prendergast, and myself, have often revealed quite large bumps or departures from a smooth rotation curve, often coinciding with the crossing of a spiral arm. But these bumps are larger, in km/sec, than the bumps in our Galaxy referred to by Dr. Kerr, and a dynamical study of them has not been made. Until recently, the scatter in measurements of rotation velocity in most external galaxies has been too great to permit a search for the small effects we are discussing.

Carranza *et al.* (1968), using Courtès's Fabry-Pérot interferometer, have, however, done a high-precision study of M33 (in the Local Group). They found a velocity discontinuity of 15 km/sec, i.e. just comparable with the effect in our Galaxy, between the ionized gas in the spiral arms and the general field of lower-density ionized gas between the arms. The arms are rotating faster than the disk by this amount, as in our Galaxy.

To relate the density-wave theory to observations in this way is a very exciting new field of research. We need more high-precision studies of velocity fields in galaxies, and we need also to take account theoretically of the larger 'bumps' already known to exist in some galaxies, e.g. NGC 2903 and NGC 4258.

Finally, we should not forget that, for all its successes, the first-order density-wave theory may not be the whole story. For example, Dixon (1967) made an observational study of the local spiral arm (gas and young stars), and he concluded that non-gravitational forces are important; they tend to oppose the gravitational forces and cause the interstellar gases to orbit at less than the gravitational circular velocity, and they tend to force the gases away from the mean galactic plane. This needs to be looked into in more detail.

6. General Considerations of Galactic Evolution

My husband is talking about the origin and evolution of galaxies, so all I want to do now is to give a brief discussion of the Hubble morphological classification in relation to the evolution of spiral structure. We would like to know, for example, whether spiral structure appears at a particular evolutionary stage in all galaxies, whether Sc galaxies may be at an earlier or later evolutionary stage than Sb or Sa, and so on.

We need, therefore, to arrange the galaxies in some classification scheme that makes sense as a time sequence. As Shapley (1950) pointed out, the Hubble classification taken in the direction Irr$\rightarrow$Sc$\rightarrow$Sb$\rightarrow$Sa$\rightarrow$SO, E does this in so far as it gives a sequence of decreasing amount of uncondensed matter (gas and dust) and young stars. Morgan (1958) has extended this argument, classifying galaxies according to the relative brightness of the central to the outer parts, and his scheme fits well with spectral classification according to the strength of emission lines coming from the H$_{II}$ regions and the earliness of the integrated stellar spectral types (see Morgan and Osterbrock, 1969). The spiral arms have maximum visibility at type Sc where they are also most highly inclined to the circular; the inclination decreases on going toward the galaxies with highly amorphous nuclei. The luminosities and numbers of H$_{II}$ regions decrease on going from Sc to Sb. There is a general sequence of increasing central concentration, decreasing amount of gas, and increasing mean luminosity per unit area.

These considerations, however, do not by any means prove that the Hubble sequence is an evolutionary sequence in the sense that the red giant branches in color-magnitude diagrams of star clusters represent approximately an evolutionary sequence of stellar structure. Rather, the Hubble sequence may be like the main sequence in the

Hertzsprung-Russell diagram, which is a locus determined by initial conditions – mass and chemical composition.

Holmberg (1964) has argued that one morphological type of galaxy cannot evolve into another, and, under certain assumptions, has demonstrated this by an argument relating density with evolutionary stage as shown by integrated color index for a large number of galaxies. First, to obtain a bigger sample than these with actual mass determinations, he noted that there exists a good correlation between the mass-to-light ratio (in solar units), M/L, and the color index, C. Then, measuring C and L for a large number of galaxies, he read M/L off his curve and obtained M, from which, by measuring the dimensions, he derived the density, D.

By plotting $\log D$, in solar masses per pc^3, against C, Holmberg then demonstrated that the galaxies fall into distinct groups, with a clear separation according to the Hubble type (see Figure 3). He argued that the mass and volume are constant during a galaxy's lifetime. He said that the only effect of aging will be a steady increase in the color index, as all the gas and dust become used up so that no new young stars are formed, and the stellar population consists of stars ever lower down the main sequence, and hence redder, because of stellar evolution which eats its way down the main sequence. Thus, the direction of evolution of a galaxy in Holmberg's diagram would be parallel to the horizontal axis, and this would preclude evolution of one type into another.

A number of assumptions, about whose validity we are uncertain, are implicit in this interesting argument, however. In particular, do the mass and volume of galaxies really remain constant? I shall return to this point later. We have so little theoretical knowledge of galactic evolution at present that it is very difficult to set out a valid theoretical framework for the observational facts.

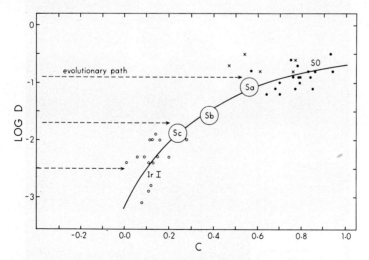

Fig. 3.   Holmberg's (1964) plot of log mass density D (solar masses per pc^3) against C. Open circles: type Irregular I; filled circles: type SO; crosses: type Irregular II. Evolutionary tracks corresponding to a time-independent mass density are indicated by dashed lines.

7. Theoretical Simulation of Galactic Structure

Large fast computers with big memories have made possible the numerical simulation of galactic structure. Miller and Prendergast (1968) have devised a very clever numerical 'game' in which they can represent a galaxy by 10^5 mass points and follow its evolution under gravitational forces alone. They have done this in two dimensions, which will give a good approximation for highly flattened galaxies, but they see no real difficulty with the new generation of automatic computers becoming available, in extending this to three dimensions. Prendergast and Miller presented their results in the form of a movie film at the Basel symposium on galactic structure (IAU Symposium No. 38).

In this scheme, each of the 10^5 points is considered to move in a certain number of integral steps around a path which represents a harmonic oscillation; the force field governing its motion is provided by the sum of all the remaining points. By doing the computations with integral steps, the accumulation of rounding-off errors is avoided; such rounding-off errors in other computational schemes for representing galaxies have led to a kind of 'numerical relaxation' which gives false results, because

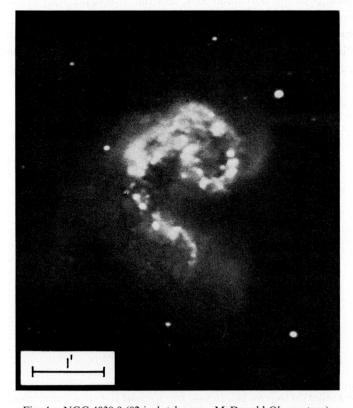

Fig. 4.   NGC 4038-9 (82-inch telescope, McDonald Observatory).

real galaxies have very long relaxation times. The new scheme is also completely reversible in time.

To make the 10^5 masses represent a real galaxy as well as possible, Prendergast and Miller started the points as gas clouds. Collisions between gas clouds are inelastic, whereas encounters among stars are elastic. As time passes, regions of higher than average density form and the gas clouds are allowed to form stars at a rate proportional to the square of the density (the exponent 2 and the constant of proportionality can, of course, be varied). The 'galaxy' is set in motion by giving it a circular velocity field such that for each mass point, the centrifugal and gravitational forces are very nearly balanced.

The results, shown in the movie film, were spectacular. After a short initial period during which the numerical effects of the initial conditions are eliminated, one sees the beginnings of spiral structure; a two-armed trailing system develops in which the arms are first quite irregular and then settle down into a regular form. It is always the gas which delineates the spiral arms; the stars, drifting out of the arms, form a substratum with rather large random velocities. During the early stages, one sees large knots and blobs forming far out on the spiral arms, and you will remember that this sort of structure is found in a large portion of the 'peculiar' or 'interacting' galaxies shown in the atlases prepared by Voroncov-Vel'jaminov (1959) and Arp (1966). At one stage, the structure looked very reminiscent of the curious looped arms in the peculiar double galaxy NGC 4038-9 (shown in Figure 4). Later on, when the amount of uncondensed gas is much reduced and the spiral arms are regular, ring structures of stars can form, and we know that these are often seen, particularly in Sa galaxies.

8. Barred Spirals – Angular Momentum – Asymmetries

I have said nothing so far about the barred spiral galaxies. An obvious question is: do barred spirals evolve into normal spirals, or vice versa, or are the two types always distinct? The next question to ask is: what is the role of angular momentum in the evolution of spiral types? Do the barred spirals have more or less angular momentum than normal spirals?

8.1. BARRED SPIRALS AND ANGULAR MOMENTUM

Some years ago, Burbidge et al. (1960b) studied the rotation of the SBc galaxy NGC 7479. It has large dimensions and an open spiral structure (steeply inclined arms); the bar is 18 kpc long. Since the bar has ionized gas along its whole length, the rotational velocity could be determined all along the bar, and it was found to give a straight-line rotation curve, i.e. the bar rotates with constant angular velocity. This is what one hoped to find, because a real bar can only persist for any appreciable length of time if it rotates like a solid body. We estimated the mass of the bar as $2 \times 10^{10} \, M_\odot$ by assuming it to be a prolate spheroid of uniform density rotating end-over-end, which is stable configuration. The angular momentum of the bar alone does indeed turn out to be larger than in normal Sc galaxies that have well-determined rotation curves and density distributions. We can see this as follows: The angular momentum is

$$J = \int mr^2 \omega$$

summed over the whole object; the angular velocity ω is of the same order of magnitude as in the outer parts of normal spirals, r is large, the density which gives the mass elements m is uniform, and the mass of the bar is the same order of magnitude as the masses of normal Sc galaxies.

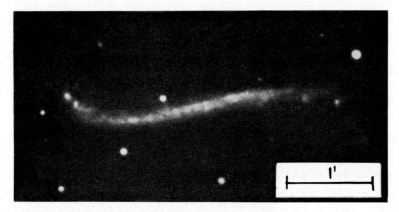

Fig. 5.   MCG 12-7-28 (82-inch telescope, McDonald Observatory).

It should, however, be pointed out that J is not easy to evaluate accurately in normal spiral galaxies, for the following reasons. The rotation curve is usually least well determined in the outermost parts of a galaxy where the light intensity is lowest, and therefore the mass distribution here is subject to the greatest uncertainty, yet the outermost parts contribute substantially to the total angular momentum, despite their low density, because of the large r.

Qualitatively, at least, we can say that long bars or bar-like structures, combined with angular velocities comparable with those found in the outer parts of normal spirals, will tend to give a large angular momentum. In this connection, Figure 5 shows an interesting galaxy of a form called by W. W. Morgan 'integral-sign' galaxies. This is MCG 12-7-28 (numbering in the Morphological Catalogue of Galaxies by Voroncov-Vel'jaminov), and it was studied by Burbidge *et al.* (1967). The object is long; if its redshift ($+3300$ km/sec) can be used to give its distance d by the Hubble relation, then

$$d = \frac{3300}{H} \times 10^6 \text{ pc}$$

(H is the Hubble constant), and its overall dimension which is 3.5 arc minutes gives a length $l = 34$ kpc for $H = 100$ km/sec per Mpc or $l = 45$ kpc for $H = 75$ km/sec per Mpc.

MCG 12-7-28 has a rotation but we cannot determine its orientation to the line of sight. If ξ is the unknown angle of inclination, the mass M is

$$M = 5 \times 10^{10} \, \mathrm{cosec}^2 \, \xi M_\odot.$$

It is clear, then, that this object, with a mass of this order of magnitude and a large dimension, will have a larger angular momentum.

We should like to study the rotations of many barred spiral galaxies, but of the 'classical' objects in this class, most do not have ionized gas well distributed along the bar, and for most, in addition, it is difficult or even impossible to make a good estimate of the spatial orientation, because the barred spirals do not have axial symmetry. We could only estimate the orientation in NGC 7479 by drawing an elliptical contour around the outermost faint extent of the spiral arms, and assuming that this outline was circular in the plane of the galaxy. The ratio of the minor to major axis then gave the cosine of the angle of inclination.

Four barred spirals in which we were able to study rotations in the central regions, however, all proved to have fast rotations. These are NGC 613 (SBc), NGC 1097 (SBb), NGC 1365 (SBc), and NGC 5383 (SBb). These nuclei obviously cannot be in solid-body rotation with the rest of the bar. The nuclei all have interesting structure; for example, NGC 1097 has an inner ring-like single spiral arm, first noted by Sersic (1958). The connection of these fast-rotating inner regions with the bars is puzzling and would provide an interesting topic for research.

8.2. GAS FLOW IN BARRED SPIRALS; EVOLUTION OF BAR

The gas in the bars of barred spiral galaxies will be circulating around the bar, in the same sense as the rotation of the whole galaxy; this was shown mathematically by Prendergast (1962). He studied the flow pattern of gas, in a rotating frame of reference. It turns out that the gas circulates around the bar, and then it can leave the bar altogether from the ends of the bar. This gas, emerging from the ends, gets

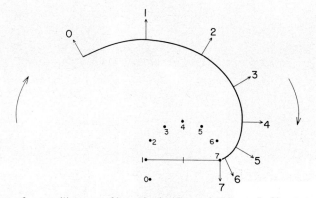

Fig. 6.   Formation of one trailing arm of barred spiral by gas leaving end of bar tangentially to circle traced out by end of bar. Line 1–7 represents position of bar, which rotates clockwise as rigid body. Right-hand end of bar has been successively at points 0, 1, 2, ..., 7; matter which left end of bar at 0, 1, 2, ..., now lies along spiral arm with velocities given by arrows.

'left behind' because it experiences a lesser gravitational attraction; thus steeply inclined arms can be formed as in Figure 6, and maybe we can take this as the basic mechanism of formation of spiral arms in barred spirals.

There will be various consequences of this type of gas circulation:

(1) Consideration of Figure 6, showing the development of spiral arms, shows that gas on the minor axis of the galaxy as seen by an observer can actually have an outward velocity component in the line-of-sight. This helps to explain some of the velocity measurements, e.g. in NGC 5383 (Burbidge *et al.* 1962a).

(2) It ought to be possible to detect the gas flow in the form of a velocity gradient at right angles to the bar, or as velocities measured along the length of the bar. There are no observations on the former; de Vaucouleurs *et al.* (1968) found streaming along the bar of NGC 4027, but this is an irregular 'non-classical' barred spiral, like one of the early stages of Miller and Prendergast's numerical simulations and it would be good to look for flow in a more regular-type bar.

(3) Gas coming out from the center is fed with the necessary angular momentum and kinetic energy by the gravitational torque of the bar, and then the gas, together with its angular momentum and energy, is lost from the end of the bar. Thus the direction of evolution will be such that the bar becomes more dense, more stubby, rotates more rapidly, and becomes more tightly bound (see a series of papers by

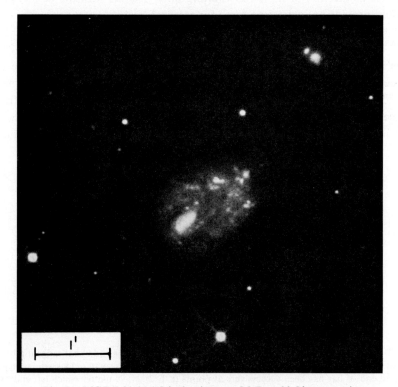

Fig. 7.   MCG 4-31-14 (82-inch telescope, McDonald Observatory).

Freeman (1965, 1966) for detailed studies). The sense of evolution, if this is correct, is from barred spirals to normal spirals.

8.3. ASYMMETRIES

The barred spirals do not have axial symmetry, although they have symmetry about one diameter. We have seen how such an object may evolve into a normal spiral with axial symmetry. Any irregular galaxy possessing rotation can be expected to evolve toward axial symmetry, so this will be the sense of evolution unless something else (e.g. an explosive event in the center) disturbs this scheme. It is of interest, therefore, to study very asymmetrical objects and look at the velocity field in their ionized gas component. One interesting asymmetrical galaxy is MCG 4-31-14 (shown in Figure 7). It looks like just one half of a normal Sc galaxy! A very preliminary study of it shows a redshift of $+2560$ km/sec and a diameter of 10–13 kpc (according to a Hubble constant of 100 or 75 km/sec per Mpc). There is a velocity gradient with a range of about 250 km/sec, which seems to be steepest at the end *away* from the 'nucleus', but this needs more observational work. If it is confirmed, it could mean that the apparent nucleus is not really the region of highest density, but merely a region of greatest recent activity in star formation. I think pathological cases like this may throw light on the processes of development of spiral structure and symmetry.

9. Outflow from Centers of Galaxies

Despite the successes of the first-order theory of density waves for explaining the maintenance of spiral structure, there remains the question, already alluded to, of the initial establishment of a spiral pattern. I have described a way in which the outer arms of barred spirals could be formed, but we should look also at what goes on in the central regions of galaxies. There has been great interest recently in the occurrence of violent events in nuclei, as the means of producing the radio galaxies and Seyfert galaxies, and further, on a much smaller scale, we know there is activity in a number of well-studied galaxies because gas is observed to be flowing outward from their nuclei. Examples are: our Galaxy, M31, M51, NGC 253, and NGC 1365.

In our Galaxy, the evidence comes from radio-astronomical studies of 21-cm radiation from the neutral gas, and so is outside the scope of my lectures. I shall say a few words about the other four galaxies.

9.1. M31

In Section 2 I spoke about the circular or rotational velocity field in the ionized gas of normal spiral galaxies, and a way of measuring this by setting the slit of a long-slit nebular spectrograph along the major axis of the galaxy. If the slit is set instead along the minor axis of the galaxy, then if circular velocity parallel to the equatorial plane of the galaxy is the only motion present, the velocity field of the gas on the minor axis will be everywhere perpendicular to the line of sight and no Doppler displacements of the spectrum lines should be seen (except for the motion of the galaxy as

278 E.M.BURBIDGE

a whole with respect to the observer). If any Doppler shifts are seen along the minor axis, they must be produced by non-circular motions, inward or outward from the center.

Münch (1960, 1962) found that velocities measured in the [OII] λλ3726, 3729 doublet, on spectra taken along the minor axis of M31 in its central region, do in fact show such differential Doppler displacements. Since we know which is the near side of M31, from the distribution of dust, Münch was able to conclude that these velocities denote a radial *outflow* from the center amounting to about 60 km/sec.

Now let us consider what would happen to gas that has been ejected from the nucleus *out* of the plane but not directly along the axis of rotation. Suppose such gas (as is likely) has little or no angular momentum. It will eventually fall back in to the plane at some distance from the center, where, having little angular momentum, it will be moving at very much less than the local circular velocity. Some new observations that can perhaps be interpreted in this way have been made by Mrs. Rubin and W. K. Ford (1970). They made a new determination of the rotation curve of M31, using emission lines from the ionized gas. At less than 20' or 4 kpc from the center, there are no HII regions and the rotation of the ionized gas component was not previously known in this region. Rubin and Ford found the [NII] λ6583 emission line was visible, produced in low-density diffuse ionized gas within the central 20', and were able to fill in the gap. They found that the velocity first rises very steeply to about 220 km/sec in the first few minutes of arc, and then drops as sharply to 50 km/sec (or possibly less) at 7'–10' from the center (see Figure 8). This behavior is rather similar to the old observations of the stellar component by Babcock (1939); he obtained a curve that rose steeply and dropped to near zero about 10' from the center, and his observations were never confirmed nor was a satisfactory interpretation given. For the stars, eccentric and steeply inclined galactic orbits are probably the explanation, and now for the gas it seems to me that one could make the hypothesis just outlined, i.e., matter is ejected from the nucleus out of the plane, and falls back

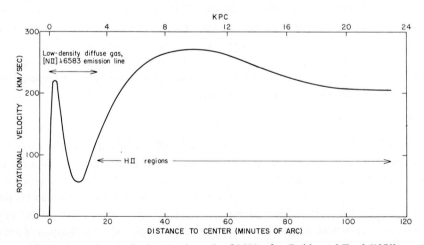

Fig. 8. Rotation curve along major axis of M31, after Rubin and Ford (1970).

Fig. 9.   M51 (120-inch telescope, Lick Observatory).

some 7′–10′ from the center where it creates a zone of slowly-rotating gas. It must eventually fall back in to the center since it will not be supported against gravity by centrifugal force.

9.2. M51

This spiral galaxy is oriented more nearly face-on than M31 (Figure 9). Burbidge and Burbidge (1964a) studied the velocity field of its ionized gas with a long-slit spectrograph and found that in several directions between the major and minor axis, the ionized gas shows velocities which cannot be explained as projections of the circular rotational velocity into the various position angles of the observations. These peculiar velocities are unsymmetrical and the largest effect is found about 20° to the minor

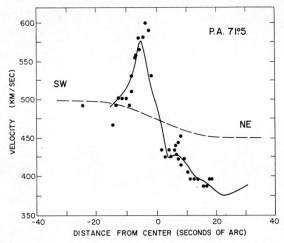

Fig. 10. Velocities measured in ionized gas in position angle about 20° to minor axis of M51. Dashed curve is rotation curve projected into this position angle.

axis, on the south-west side of the center of M51 (Figure 10). An unambiguous interpretation of the measurements could not be made, but we suggested one possible interpretation: a stream of gas emerges in two opposite directions, flowing outward from the center above the equatorial plane on one side and below it (as seen by the observer) on the opposite side; on the underside the outward-flowing gas is not seen close to the center but becomes visible after it has fallen back into the equatorial plane. In contrast to the observations in M31 by Rubin and Ford, where velocities on the major axis were observed to be *less* than the expected local circular velocity, in M51 the observations are near the minor axis and the measured velocities, containing a considerable component of outflow velocity from the center, are *larger* than the rotation curve projected into this position angle.

9.3. NGC 1365 AND NGC 253

NGC 1365 is a large barred spiral galaxy, with a bright nuclear region some 3 kpc in diameter in which there is plenty of ionized gas, made evident by strong emission lines in its spectrum. The velocity difference measured between opposite sides of the nuclear region is large, some 800 km/sec. Burbidge *et al.* (1962b) measured the velocity gradient in five different position angles across the nuclear region and concluded that these gradients were not consistent with circular rotational velocities seen in different projections, although one cannot determine the spatial orientation of this barred spiral with certainty. Clearly some kind of large-scale streaming motions are present, and the steeply-inclined spiral dust arms that connect with the nucleus support this impression. But there is not sufficient information to construct an unambiguous model for the streaming motions. The most plausible model is that there is an outflow from the center, which is not symmetrical about the rotation axis but is mainly directed in two opposite streams.

In the large Sc spiral NGC 253, the orientation to the plane of the sky can be determined and therefore it is easier to make a geometrical interpretation of the line-of-sight velocities. NGC 253, like M51 and NGC 1365, has much ionized gas in its central region, and non-circular motions are clearly present. Mlle. Demoulin and I (1969) have measured velocities in several position angles across the nucleus, and we have concluded that gas is being ejected out of the nucleus, probably at an angle between the equatorial plane and the rotation axis. To preserve symmetry and make a flow pattern analogous to that apparently existing in the neutral gas in the center of our Galaxy, one may make the hypothesis that an oppositely-directed stream emerges on the underside of the equatorial plane; dust in the plane would prevent this being seen. Here again, as in M51, the ejection is seen not far from the minor axis; in NGC 253 it is seen as excess *approach* velocities on the near side of the center, while in M51 it is seen as excess *recession* velocities on the far side of the center.

To sum up, one may conclude that non-circular velocities in the gas component in the central regions of galaxies may be fairly common, and when the observations can be interpreted with a fair degree of unambiguity, they suggest outflow, probably out of the equatorial plane, and possibly in two opposite directions. The possibility that this outflow initiates the formation of spiral arms seems to be a promising line of investigation.

10. Inflow of Gas into Galaxies?

In the past, the accretion of intergalactic gas by galaxies has sometimes been considered. However, all attempts to detect intergalactic gas have so far led to null results, and in the absence of any positive evidence for diffuse material lying between the galaxies, discussion of accretion has dropped out of fashion.

There is one set of observations, however, for which the infall of gas from outside has been tentatively invoked as a possible explanation. The Leiden radio astronomers have found high-velocity clouds of neutral hydrogen in high galactic latitudes, in regions of galactic longitude where they might be interpreted as gas falling into our Galaxy from outside. Since the distances of these clouds are still unknown, it is not possible at present to do more than suggest possible interpretations of the observations,

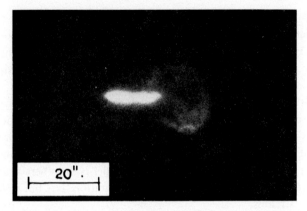

Fig. 11.   Mayall's Object (120-inch telescope, Lick Observatory).

and since they are made by means of 21-cm radiation, they do not fall within my subject matter; I mention them only for completeness' sake. However, if the high-velocity clouds should prove to be infalling material, the increase in mass of our Galaxy would amount to an increase of 5% in the gas in the disk in one revolution. It might then not be safe to take the mass of a galaxy as effectively constant over 10^{10} years, as in the argument presented in Section 6.

Let me finish by showing in Figure 11 one of my favorite astronomical pictures – a real gas ring and not one of Dr. Cameron's smoke rings projected on the screen! This is Mayall's Object; a peculiar galaxy at a redshift of about 10000 km/sec with a rotating ring of gas connected to the main body of the galaxy. I have no idea what the origin or dynamical state of this system is; the observations merely give line-of-sight velocities of the gas along the main body and across the ring (Burbidge and Burbidge, 1964b) which show that the ring and main body are connected and the ring is rotating. Objects like this show that there is a great deal we do not yet understand about galactic formation and evolution.

References

Arp, H. C.: 1966, *Astrophys. J. Suppl. Ser.* **14**, 1.
Babcock, H. W.: 1939, *Lick Observ. Bull.* **19**, 41.
Burbidge, E. M. and Burbidge, G. R.: 1964a, *Astrophys. J.* **140**, 1445.
Burbidge, E. M. and Burbidge, G. R.: 1964b, *Astrophys. J.* **140**, 1617.
Burbidge, E. M. and Burbidge, G. R.: 1970, in *Stars and Stellar Systems* **9**, Univ. of Chicago Press, Chicago, in press.
Burbidge, G. R. and Prendergast, K. H.: 1960, *Astrophys. J.* **131**, 243.
Burbidge, E. M., Burbidge, G. R., and Prendergast, K. H.: 1960a, *Astrophys. J.* **131**, 282.
Burbidge, E. M., Burbidge, G. R., and Prendergast, K. H.: 1960b, *Astrophys. J.* **132**, 654.
Burbidge, E. M., Burbidge, G. R., and Prendergast, K. H.: 1962a, *Astrophys. J.* **136**, 704.
Burbidge, E. M., Burbidge, G. R., and Prendergast, K. H.: 1962b, *Astrophys. J.* **136**, 119.
Burbidge, E. M., Burbidge, G. R., and Shelton, J. W.: 1967, *Astrophys. J.* **150**, 783; catalogue number identified in *Astrophys. J.* **153** (1968) 703.
Carranza, G., Courtès, G., Georgelin, Y., Monnet, G., and Pourcelot, A.: 1968, *Ann. Astrophys.* **31**, 63.

Courtès, G.: 1960, *Ann. Astrophys.* **23**, 115.
Courtès, G.: 1964, *Astron. J.* **69**, 325.
Demoulin, M. H. and Burbidge, E. M.: 1969, *Astrophys. J.* **159**, 799.
Dixon, M. E.: 1967, *Monthly Notices Roy. Astron. Soc.* **137**, 337.
Freeman, K.: 1965, *Monthly Notices Roy. Astron. Soc.* **130**, 183.
Freeman, K.: 1966, *Monthly Notices Roy. Astron. Soc.* **134**, 1.
Hodge, P. W.: 1969, *Astrophys. J. Suppl. Ser.* **18**, 73.
Holmberg, E.: 1964, *Uppsala Astron. Observ. Medd.* No. 148.
Lin, C. C.: 1967, *Ann. Rev. Astron. Astrophys.* **5**, 453.
Lin, C. C. and Shu, F. H.: 1967, in *Radio Astronomy and the Galactic System*, IAU Symposium No. 31 (ed. by H. van Woerden), Academic Press, London, p. 313.
Lin, C. C., Yuan, C., and Shu, F. H.: 1969, *Astrophys. J.* **155**, 721.
Miller, R. H. and Prendergast, K. H.: 1968, *Astrophys. J.* **151**, 699.
Morgan, W. W.: 1958, *Publ. Astron. Soc. Pacific* **70**, 364.
Morgan, W. W. and Osterbrock, D. E.: 1969, *Astron. J.* **74**, 515.
Münch, G.: 1960, *Astrophys. J.* **131**, 250.
Münch, G.: 1962, in *Problems of Extragalactic Research*, IAU Symposium No. 15 (ed. by G. C. McVittie), Academic Press, London, p. 119.
Oort, J. H.: 1962, in *Interstellar Matter in Galaxies* (ed. by L. Woltjer), Benjamin, New York, p. 234.
Prendergast, K. H.: 1962, in *Interstellar Matter in Galaxies* (ed. by L. Woltjer), Benjamin, New York, p. 217.
Prendergast, K. H.: 1967, in *Radio Astronomy and the Galactic System*, IAU Symposium No. 31 (ed. by H. van Woerden), Academic Press, p. 303.
Rubin, V. C. and Ford, Jr., W. K.: 1970, in *The Spiral Structure of Our Galaxy*, IAU Symposium No. 38 (ed. by W. Becker and G. Contopoulos), D. Reidel, Dordrecht, The Netherlands, p. 61.
Schmidt, M.: 1965, in *Galactic Structure* (ed. by A. Blaauw and M. Schmidt), Univ. of Chicago Press, Chicago, p. 513.
Sersic, J. L.: 1958, *Observatory* **78**, 123.
Shapley, H.: 1950, *Publ. Observ. Univ. Michigan*, **10**, 79.
Strömgren, B.: 1967, in *Radio Astronomy and the Galactic System*, IAU Symposium No. 31 (ed. by H. van Woerden), Academic Press, London, p. 323.
de Vaucouleurs, G., de Vaucouleurs, A., and Freeman, K.: 1968, *Monthly Notices Roy. Astron. Soc.* **139**, 425.
Voroncov-Vel'jaminov, B. A.: 1959, *Atlas of Interacting Galaxies*, Moscow.
Zwicky, F.: 1955, *Publ. Astron. Soc. Pacific* **67**, 232.

THE EVOLUTION OF THE GALAXY AS A WHOLE

G. BURBIDGE

University of California, San Diego, Calif., U.S.A.

Nearly all of the lectures here are concerned with the physics of the Galaxy as it is now. To describe its evolution we must ask how it was formed, how it has evolved to its present configuration, and what it can be expected to contain and how it will look in perhaps another 10^{10} years. These are the same questions that have been asked, and fairly satisfactorily answered, for stars over the last thirty years. But the corresponding problem, as it relates to galaxies, is very far from solution.

There are several reasons for this. In the first place, we still know very little about the detailed structure of galaxies. Second, we have not found a way of unambiguously determining their ages. Third, we do not understand how they were formed, and this problem is intimately tied up with cosmology and, in particular, in whether or not the universe has expanded from an initial state of very high density. These uncertainties will be discussed in what follows.

1. The Cosmological Background

There are four observational approaches to cosmology which have been heavily pursued in recent years. We consider them in turn.

1.1. Determination of the deceleration parameter q_0

The classical approach to observational cosmology is to measure the redshift-apparent magnitude relation for ever fainter galaxies. This method, first in the hands of Hubble and Humason, then of Humason, Mayall, and Sandage, and in recent years, of Sandage and his colleagues, led to the discovery that the universe is expanding, and in recent years the rate of the expansion (the Hubble constant) has been determined with increasing precision. However, the attempts to measure the deceleration parameter q_0 have not been very successful. This is essentially because it is necessary to measure the departure from linearity of the redshift-apparent magnitude relation, and this only becomes appreciable at redshifts which are close to the limit so far achieved, that is at $z \simeq 0.25$. Attempts to derive a value for q_0 have frequently been reported to give positive values with large uncertainties. While it has often been stated that this result rules out the steady-state cosmology in which $q_0 = -1$, an objective evaluation of the data hardly warrants this conclusion. Only with the measurement of far more objects with redshifts in the range 0.25–0.5 can this question be settled.

1.2. Counts of radio sources

For more than a decade attempts have been made to test cosmological models by

L. N. Mavridis (ed.), Structure and Evolution of the Galaxy, 284–293. All Rights Reserved
Copyright © 1971 by D. Reidel Publishing Company, Dordrecht - Holland

investigating the slope of the $\log N - \log S$ curve for radio sources. A similar attempt made with optical galaxies was abandoned by Hubble in the late 1930's.

It is well known that the slope of the $\log N - \log S$ curve in a Euclidean universe with sources at a constant density is -1.5 and that, in all cosmological models, the effects of the curvature of space tend to flatten this curve to values less than -1.5 at the faint end. Now the observed slope for counts of radio sources at comparatively low frequencies, from the 3C and 4C radio source counts, and the Parkes results tended to be much steeper, about -1.8. From this Ryle and his colleagues concluded that there must be an excess of faint sources, and, since they assume that the faint sources are the most distant, this implies the frequency of radio sources varies with epoch. This is in contradiction to the simple steady-state cosmology. However, as one goes to fainter flux levels, the counts tend to flatten off to values considerably below the Euclidean slope, and Hoyle has pointed out that this part of the curve is entirely compatible with the predictions of the steady-state theory. Thus, he has contended that it is at the large values of S where the greater uncertainties arise.

For the sample of radio sources for which redshifts are available, it is clear that the slope of the $\log N - \log S$ curve is largely due to the luminosity function for sources, and not to distance-volume effects. The basic uncertainty of the method is still simply due to the fact that we are counting large numbers of sources about which very little is known. For these reasons I believe that this cosmological test has not given conclusive results.

1.3. THE MICROWAVE BACKGROUND RADIATION

One of the few predictions made in astronomy in modern times, which can be directly tested, is that by Gamow and his colleagues that, if the universe began in a big bang, there would have resulted a black-body radiation field which, at this epoch, would have a temperature of a few degrees Kelvin. The detection of radiation of cosmic origin (first by Penzias and Wilson and later by others) over a wavelength range from about 20 cm to ~ 1 cm with a spectrum corresponding to the Rayleigh-Jeans part of a black-body curve with a temperature of about $3°$ K, is therefore preliminary evidence of this big bang origin. If, on the other hand, we do not live in an evolving universe, but in a steady-state universe, this background radiation must be due to the integrated effect of discrete sources, and it is not likely to have a black-body form. A fundamental question is therefore whether this radiation really has a black-body form. To be sure of this it is necessary to make measurements near the peak of the curve and on the exponentially falling side. This is where the trouble begins. For there are conflicting data near the peak of the curve. Direct observations using rockets and, more recently, balloons show that a large flux of radiation much greater than that expected at the peak of a $3°$ black-body curve is apparently present. These results are in conflict with the indirect measurements or upper limits on the background radiation of the Galaxy, which were obtained by investigating the state of excitation of the CN, CH, and CH^+ molecules in the interstellar gas, except for the unlikely possibility that the radiation measured directly is due to lines which happen to fall between the wavelengths where limits are set by the interstellar molecules. At present we can assume one of three possibilities:

(1) That the direct measurements of millimeter wave radiation at the top or above the Earth's atmosphere are not correct and that there is no *truly cosmic* radiation at this power level.

(2) That it is cosmic in origin and is line or continuum radiation superposed on a black-body spectrum which originated in a primeval fireball.

(3) That the primeval fireball origin theory is not correct and that the background radiation is due to the integrated effect of large numbers of discrete sources.

Only if we accept (1) or (2) can we still argue that the existence of this radiation is evidence for an evolving universe. The acceptance of (3) enables us to argue that we live in a steady-state universe.

2. The Formation of the Galaxy

From the discussion of cosmology in the previous section, it is clear that we do not know whether our Galaxy, and other galaxies, were formed in an early period, when the density was very much higher in an evolving universe, or whether it was formed in conditions similar to those found at present, in a steady-state universe.

We consider these possibilities in turn. Most attention has been paid to the problem of galaxy formation in an evolving universe. The idea is that density fluctuations present initially will lead to gravitational instability and the formation of protogalaxies. Recent work has been done involving the microwave background radiation. However, the problem which has been known since Lifshitz's work in 1946 is that it is not possible to make density fluctuations grow in an expanding universe unless very large fluctuations are artificially put into the problem at the beginning, i.e., we must put in fluctuations which are cosmological in origin; there is no physical justification for them. Given that such fluctuations are present, it is possible to understand how galaxies can be formed though, at present, we have no theory which will explain their observed masses, rotations, and clustering tendencies. Given this type of galaxy formation, the evolution will be one of increasing condensation. Such a collapsing protogalaxy was studied by Eggen, Lynden-Bell, and Sandage, who investigated the correlations between the orbits of stars and their gross chemical compositions. They found that the stars with the lowest metal abundances (largest UV excesses) are invariably moving in highly elliptical orbits. Stars with low metal abundances have low angular momenta. On the other hand, stars with normal compositions have circular orbits and large angular momenta.

If these results are interpreted in terms of a collapsing protogalaxy, it is argued that the oldest stars with low metal abundances were formed out of material collapsing radially toward the plane. The timescale for collapse is a few times 10^8 years. It is important to realize that, while many have interpreted these results as evidence for collapse of a protogalaxy, the results can equally be interpreted in terms of ejection from the center of matter which forms the halo population of stars.

What is the alternative approach to galaxy formation? It is that proposed by Ambartsumian in the framework of general relativity with 'delayed core' models, and by Hoyle and Narlikar, and McCrea in the steady-state cosmology. In these theories it is argued that the initial state is one of matter in a very dense form, or that matter is created in regions of highest densities, the nuclei of galaxies. In these models galaxies are continuously being formed, so that the existence of young galaxies can be understood. In this theory also it is very difficult to understand the angular momentum of galaxies.

While both theories are in a rudimentary form, it is important to realize that, when one comes to explain the observations, neither is strongly preferred over the other.

3. The Chemical Evolution of the Galaxy

The origin of the elements remains one of the problems of greatest importance. It is now generally accepted that, even if the universe did originate in a big bang, nuclear physics shows that, starting from baryons and fermions, only D, He^3, He^4, and Li^7 could have been built in such an epoch. The remainder of the elements at the abundance level at which we see them in the solar system and in stars in our Galaxy and in other galaxies must have been built elsewhere. And, of course, if the universe is in steady state, all of the elements must have been built in stars of one kind or another. We now understand the physical conditions which are required to build all of the elements in stars, and if nucleosynthesis had proceeded steadily since the Galaxy was first formed we would expect that age and metal/hydrogen ratio would be smoothly correlated. Since the development of the detailed theory of nucleosynthesis (Burbidge, Burbidge, Fowler, and Hoyle, 1957) observational evidence has been obtained which shows that the metal abundance has not been a smoothly increasing function of time. Old galactic cluster stars have normal solar system abundances. The evidence now suggests that at a comparatively early time in the history of the Galaxy a large part of the total enrichment in heavy elements occurred. This is the case, even though we have direct evidence that nucleosynthesis is taking place in stars at present.

There are two possibilities for early nucleosynthesis. If numbers of very massive stars were formed, $\geqslant 10^5 M_\odot$, nucleosynthesis in little big bangs (Wagoner, Fowler, and Hoyle) may have been responsible. The other very interesting development has come through investigation of processes which may go on in fairly massive ($\sim 20 \, M_\odot$) stars. Arnett, Truran, and Clayton have recently shown that much of nucleosynthesis of the abundant elements between atomic weights ~ 20 and 60 may occur in explosive oxygen burning and other reactions in such stars. If much of the stellar nucleosynthesis took place through this mechanism we must argue that the relative number of comparatively massive stars formed at this epoch was very much greater than at present. Almost all of the qualitative arguments concerned with the rates of star formation indicate that star formation is likely to have been more important near the centers of galaxies where the densities are higher. It is not clear why more massive stars were much more plentiful.

4. The Helium Problem

A special problem in the nucleosynthesis theory is to understand the origin of the helium. It has been known for more than a decade that, if the helium/hydrogen ratio seen in young stars and gaseous nebulae (~ 25–30% by mass) is representative of the Galaxy as a whole, the helium cannot have been synthesized by hydrogen burning in which all of the energy released was radiated by stars without there having been a phase in the life of the Galaxy when it was far more luminous than it is today. Thus, we can argue either that there was such a phase, or that the bulk of the helium was made before the Galaxy formed, or that the average helium abundance is much lower than that given by the young stars and gaseous nebulae.

The development of interest in the primeval fireball model of the universe led to calculations of the helium production that is to be expected in such a cosmological model. For a number of simple cases it is found that He/H abundance ratio arising in a big bang is ~ 25–30%. This led many to the view that perhaps this was the origin of the helium in the Galaxy, and that this ratio is universal. There are a number of other indirect arguments which also suggest that this rather high value of the He/H ratio is representative of the Galaxy. On the other hand, there is direct evidence from analyses of the atmospheres of some B stars that very low He/H ratios are present, while the other elements have normal abundances. Since it is exceedingly difficult to destroy helium, evidence that there are stars with very low He/H ratios would mean that probably helium was not made in a primeval fireball. This does not mean that the universe may not have passed through this phase, since we now know of ways in which very little helium need have been synthesized in a big bang.

If the evidence just referred to is correct, the most plausible explanation is that the high helium/hydrogen abundance ratio is due to early nucleosynthesis in the Galaxy involving very massive stars. Differences in the He/H ratio between different stars must then be due to lack of complete mixing of the material processed in the early nucleosynthesis with the remaining gas. After several generations of stars have evolved, it is likely that there will only be rather small pockets in which the mixing has not been effective.

It is also of interest to point out that the conversion of hydrogen to helium in $\sim 25\%$ of the mass of galaxies would lead to a release of energy of about 3×10^{-12} erg/cm^3 throughout the universe. This energy released in the early phases of galaxies is adequate to explain the microwave background radiation (assuming that it comes from galactic nuclei) and a universal cosmic ray flux.

5. Activity in the Galaxy and Galactic Nuclei in General

We do not really know how our Galaxy formed and evolved. However, one fact of great interest that we have become aware of in the last few years is that many, if not all, galactic nuclei, including our own, are exceedingly active. They are pouring out energy in many forms, through non-thermal processes which must ultimately have a gravitational origin, and the importance of the nuclei from the point of view of galactic evolution must be very great. What sort of evidence for this activity is now available?

5.1. The galactic center

The galactic center shows in its non-thermal radio source and its infrared nucleus that it is a miniature version of a Seyfert nucleus, radiating at a power level of 10^{-4} of the nucleus of NGC 4151, for example. The recent observations in the far infrared by Low and his colleagues and by Neugebauer and Becklin show that the central core from which much of the far infrared radiation is emitted is an object with a size of about 0.7 pc. It appears most likely that the flux in the frequency range between 10^{10} and 10^{14} Hz is synchrotron radiation. The steeply increasing flux shortward of 10^{14} Hz reaches a maximum at about 3×10^{12} Hz and then falls steeply. This form of the spectrum is likely to be due to synchrotron self-absorption, and as Low and his colleagues pointed out, if the source were a single coherent object, this would require that the magnetic field be extremely large, in excess to 10^{20} G. One is therefore forced to conclude that the source is composite, and it is easily shown that, provided a number of very small components are present, the magnetic fields associated with each of them are likely to be comparatively modest. The shape of the spectrum which is observed is then due to the superposition of radiation from a cloud of small sources all contained within a dimension of about a parsec. The electrons responsible for the emission have comparatively low energies ($\sim 10^8$ eV) and they must be injected into each region where they radiate without being able to escape from the immediate vicinity of the source.

5.2. Gravitational radiation

Weber has now been able to detect pulses of gravitational radiation which appear to be very frequent and powerful. While there is still some doubt about the reality of the pulses, if they are really cosmic, their most likely place of origin is the galactic center. If this is the case, then gravitational energy is being emitted in pulses as a rate of about 1 per week or more, and the power radiated is $\sim 10^{48}$–10^{50} erg/sec. The fact that Weber sees separate events suggests that gravitational energy release is not due to the collapse of a single massive coherent object, but that many separate objects are involved. Thus, again the evidence points to a source which is composed of many separate pieces.

The power radiated is the equivalent of about 10–1000 $M_\odot$/year. Thus one can argue

(1) that the frequency of events seen at present is far higher than the average over the life of the Galaxy;

(2) that the Galaxy was much more massive in the past and is losing its mass through this mechanism;

(3) that this is a manifestation of continuous creation in the nucleus of the Galaxy.

5.3. Mass loss from Seyfert nuclei and QSOs

Over the last decade many studies of Seyfert galaxies have suggested that mass is being ejected from the nuclei at a very high rate. The frequency of occurrence of the Seyfert phenomenon suggests that the excitation and ejection process goes on for at least $\sim 10^8$ years. Recently Anderson and Kraft identified sharp absorption lines in the spectrum of the nucleus of NGC 4151 which are due to expanding shells of gas. They estimated that the rate of mass loss is $\sim 10-10^3$ $M_\odot$/year. Thus they concluded that the bulk of the mass of such a galaxy would be ejected in $\sim 10^8$ years so that matter must in some way continue to appear in the nuclear region. Cromwell and Weymann have found evidence that this phenomenon may be sporadic, variations of the absorption line strengths having taken place in time scales of the order of a year. This has led them to reduce the estimated mass-loss rate by about three powers of 10. Evidence has also been found for very large mass ejection from NGC 1275. Here the ejection rate can be crudely estimated to be of order $\geqslant 10$ $M_\odot$/year, based on the large amount of ionized gas seen moving out at velocities $\geqslant 3000$ km/sec relative to the nucleus.

Apart from the magnitude of their redshifts, the spectra of Seyfert nuclei and QSOs are very similar. Narrow absorption lines are often seen in the QSOs with large redshifts. The majority of absorption-line redshifts, but not all, are less than the emission redshifts. Whether or not the emission redshifts have a cosmological or an intrinsic origin, the difference between absorption and emission is attributable either to the ejection of gaseous shells, as in the case of NGC 4151, or it may be gravitational in origin. The data presently available do not support the view that the absorption is due to intergalactic gas. If gaseous shells are being ejected the mass loss rate must be very high.

5.4. Radiation fields in QSOs and galactic nuclei

The evidence from the direct measurement by interferometric methods, or from scintillation techniques, shows that exceedingly small structures are present in nuclei. Direct measurement of sizes of the order of a few thousandths of a second of arc means that upper limits on size $\sim 0.1-1$ pc are being found for objects at distances $\gtrsim 50$ Mpc. Variability in flux suggests objects with sizes $\gtrsim 10^{17}-10^{15}$ cm are present. The problem raised by the Compton effect in QSOs is marginally present even in galactic nuclei at comparatively small redshifts $\gtrsim 0.1$, if one considers the optical flux, and becomes more severe for the infrared objects *if* the nucleus is a single coherent source. However, the problem is alleviated if the multiple source model required to explain the infrared source in the nucleus of our Galaxy is used. In this case the magnetic fields in the subunits are large enough to dominate the radiation fields in these regions, and electrons radiate only in the vicinity of the objects which generate them.

5.5. Ejection of condensed plasmoids from nuclei

The presence of small radio source components far from the nuclei of the parent galaxy or QSO is most likely to be explained by the ejection of coherent objects which

have enough mass to hold themselves together, and from which particles are continuously generated. The suggestion that such small components are contained and
compressed by a surrounding intergalactic medium does not appear to work for
exceedingly small objects. The very small-scale structures which appear to make up
the optical jet in M87 also suggest ejection of such objects.

5.6. POSITIVE TOTAL ENERGY IN GROUPS AND CLUSTERS OF GALAXIES

The original suggestion of Ambartsumian and the early data suggesting that most
physical groups and clusters of galaxies appear to have positive total energy have been
amply confirmed by the study of many more new groups. Of particular interest are the
chains of galaxies which have large velocity dispersions or which contain one member
with a highly discrepant redshift. The most plausible interpretation is that the object
is moving with a velocity of many thousands of km/sec with respect to the other
galaxies in the group. While the presence of enough hidden mass to stabilize the
system may plausibly be invoked in a few cases, this is very unlikely when we come to
the small groups. The magnitude of the kinetic energy of the objects being ejected in
such systems is of order $10^{51}(M/M_\odot)$ ergs, where M is the mass of the galaxy being
ejected. If Ambartsumian is correct, these systems must be ejected from nuclei which
have very high densities.

From all of these data we can conclude that the rate of mass loss from regions of
high density in the nuclei are very high. All of these phenomena associated with the
nuclei of Seyfert galaxies, strong radio galaxies, and QSOs have been detected because
very high radio, optical, or infrared luminosities are involved, or because very large
velocities are seen, etc. It was the realization that only a small fraction of all galaxies
show such strong effects which led us originally to think in terms of comparatively
short-lived explosive events which are rare. However, there are now so many indications of nuclear activity in apparently normal galaxies that it appears that one should
now think more in terms of continuous activity taking place in all galaxies all of the
time, though of course the rate of release of energy is usually at a very low level compared with the highly conspicuous objects.

5.7. POSSIBLE INTERPRETATIONS

This circumstantial evidence all points to the conclusion that the nuclei of many, and
perhaps all, galaxies are made up of a very dense core containing many small subunits
which are able to generate relativistic particles with high efficiency, and that this core is
able to throw out very large amounts of matter, much of it in the form of coherent
objects.

The very large energies known to be present in the radio sources have shown us for
a long time that large masses must be present in such cores. These masses are of order
$(E/c^2)f^{-1}$ where E is the energy present in the form of relativistic particles and magnetic flux in a radio source and $f(<1)$ is the efficiency with which energy can be conver-
ted into this form. For $E \simeq 10^{61}$ erg and $f = 10^{-1}$–10^{-2}, $M \simeq 10^{8}$–10^{10} $M_\odot$. Given that
such large masses are present in volumes with sizes of order 1 pc or less, a number of

theories have been proposed to explain how gravitational energy can be released.

While that problem has not been solved, the basic problem is to understand how the dense cores were formed in the first place. It is through this that we are brought back to the problem of galaxy formation discussed earlier. On the conventional picture of galaxy formation the sequence of events is a complex one. The universe expands from a singularity, the mean density falls, but in some way large density fluctuations persist and grow, so that successively higher and higher density phases develop in the galaxies. Their infall leads eventually to catastrophe in which implosion is reversed, generating outflow of matter and radiation, which is fairly continuous. In view of the observational evidence, a much simpler proposal is that galaxies formed, not from diffuse clouds but from highly condensed objects, as Ambartsumian proposed long ago, and, moreover, that not all galaxy formation took place nearly 10^{10} years ago.

What rates of mass loss from nuclei are we talking about? Given that galaxies were formed with masses in the range that we measure today, average masses of $\sim 3 \times 10^{10}$ $M_\odot$ for luminous spirals, and $\sim 4 \times 10^{11}$ $M_\odot$ for bright ellipticals, do the mass loss rates give cause for alarm, as far as conventional ideas are concerned? It appears that the mass loss rates obtained from observation range from ~ 0.1 to 10^3 $M_\odot$/years. If such loss rates are maintained over 10^{10} years, galaxies are far from being conservative systems. However, can it be established that mass comparable with the total galactic mass present now is ejected in a galactic lifetime? The uncertainties are very great.

The instantaneous mass loss rates are themselves uncertain though they often may be underestimated rather than overestimated. We cannot determine directly the duration of the mass loss phase, and we are never sure that the mass loss that we see is all that is taking place. The timescale may be comparatively short, in which case the mass loss may still be only a small fraction of the total mass. For time scales $\sim 10^6$–10^8 years the total mass ejected may be $< 10^9$ $M_\odot$. If some galaxies are comparatively young, then it might be argued that the mass loss occurs early in their history and tapers off. But it must still be treated as a serious possibility that masses of order $\sim 10^{10}$–10^{11} $M_\odot$ can be ejected in the life of a galaxy.

What are the conceivable explanations if we do accept the point of view that very large masses are ejected? Is it possible that the inflow of mass from an intergalactic medium replenishes the system? The infalling clouds of neutral hydrogen may be evidence for this. The mass accreted in this way is difficult to estimate but it may amount to ~ 1 $M_\odot$ per year. The rate at which it flows in will be determined by the local density of the intergalactic gas, and the velocity and mass of the galaxy relative to this gas. However, it must be remembered that there is no evidence for the presence of any diffuse intergalactic medium other than these 21-cm data, and possible alternative explanations for these observations have been proposed.

Another possibility is that galaxies, when they formed, had much greater masses than they have today, and they have steadily shrunk due to mass ejection. Such an evolution will have important effects in galactic dynamics, but apart from arguments associated with our Galaxy, there is little chance of testing this idea. If such an evolution takes place it would mean that, if we live in an evolving universe, galaxies observed at great redshifts would be more massive and presumably more luminous than the nearby galaxies. However, over the range of redshifts investigated for galaxies (up to $z \simeq 0.25$) the very tight relationship between apparent brightness and redshift for the brightest ellipticals argues against this.

The other alternative is to suppose that galaxies or galactic nuclei are indeed feeding matter into the universe as Jeans proposed long ago in another connection. As was mentioned earlier, within the framework of general relativistic cosmology this only appears to be possible in delayed core models. Creation of matter is required in a steady-state theory, and both McCrea and Hoyle and Narlikar have suggested that the nuclei of galaxies are places where creation may occur.

6. Conclusion

From what has been discussed here it can be concluded that we are a long way from understanding the evolution of the Galaxy. We cannot understand its early history until we are able to deduce what kind of a universe we live in. Also, the comparatively recent discoveries associated with galactic nuclei, including the nucleus of our Galaxy, suggest that nuclei are more important in the evolution of galaxies than we had thought before.

Acknowledgements

Extragalactic research at UCSD is supported by the National Science Foundation and by NASA under grant NGL–05–005–004.

SUMMARY AND DESIDERATA

A. BLAAUW

Kapteyn Astronomical Laboratory, Groningen, The Netherlands*

1. Introduction

A great variety of subjects has been discussed in the lectures of this course, yet they represent only part of the many aspects of modern galactic research. They have, however, emphasized those fields upon which nowadays most of the attention is focussed. For those among the participants – for instance the students of physics – who might wish to become more acquainted with present trends in galactic research the following references to recent or forthcoming literature may be useful.

(a) As an introductory text: D. Mihalas, *Galactic Astronomy*, Freeman and Company, San Francisco, 1968;

(b) IAU Symposium No. 31, *Radio Astronomy and the Galactic System* (ed. by H. van Woerden), Academic Press, London, 1967;

(c) The IAU Transactions of the Prague Assembly, 1967, especially the reports of Commissions 33 and 40;

(d) Vol. VII of the compendium *Stars and Stellar Systems, Nebulae and Interstellar Matter* (ed. by B. M. Middlehurst and L. H. Aller), University of Chicago Press, Chicago, 1968;

(e) IAU Symposium No. 38 on *The Spiral Structure of the Galaxy*, D. Reidel, Dordrecht, 1970; IAU Symposium No. 39 on *Interstellar Gas Dynamics*, D. Reidel, Dordrecht, 1970;

(f) The forthcoming report on the Liège Symposium (1969) on Prestellar Evolution.

2. The Stellar Populations

As a reminder of some of the basic properties of the Galaxy, we summarize in Table I the population scheme as it was adopted in 1957. Although it should be updated in several respects now, we copy the data from the original scheme (see Blaauw, 1965) since the main features remain unchanged. Note that the bulk of the galactic mass is represented by the Disk Population and the Intermediate Population II, containing about two thirds of the total mass. A small fraction only is in the Extreme Population I and in the Older Population I which contain together about 10% of the total mass.

Yet, most of the items dealt with in this course refer mainly to this 10%. This is due to two circumstances: the recent progress in the understanding of the spiral structure, and the observational and theoretical work concerning the process of star formation. Both these topics belong to the top division of Table I: the Extreme Population I.

These special circumstances should not hide the fact that, masswise, the problem of the overall galactic evolution is one concerning mostly the Disk Population and the Intermediate Population II. We shall therefore return to these later.

* Now: European Southern Observatory, Hamburg-Bergedorf, German Federal Republic.

TABLE I

Principal galactic population components

Population component	Total mass ($10^9 \odot$)	Age (10^9 yrs)	Examples
Population I: 'Extreme'	2	< 0.1	Gas, Associations, Supergiants, Youngest clusters, Youngest spiral population
Population I: 'Older'	5	0.1–1.5	A stars, Strong-line stars, Young spiral population
Disk population	} 47	1.5–8	Planetary Nebulae, Weak-line stars
Population II: 'Intermediate'		8–10	High-velocity stars ($|Z| > 30$ mk/sec), Long period variables
Population II: 'Halo'	16	~ 10	Subdwarfs, Globular clusters, RR Lyrae variables period > 0.4 day

3. The Density-Wave Theory of Spiral Structure

Although references have been made in some of the lectures to this newly developed theory of spiral structure it appears useful to briefly summarize a few points. The Lin-Shu theory seems to go a long way to explaining why a spiral pattern in the distribution of gas and young stars, once it exists, can survive for time intervals of the order of 10^9 years (that is: several or more revolutions of the Galaxy at the Sun's distance from the centre). It seems to have solved the problem of the maintenance of spiral-like features which previously appeared to meet a seemingly unsurmountable difficulty in the distortion effect of differential galactic rotation. For a recent article, dealing with the main points of the theory and with the comparison with observations, we refer to Lin *et al.* (1969). Note, however, that no satisfactory explanation exists yet for the mechanism giving rise to the spiral pattern.

The Lin-Shu theory envisages both two-arm and multiple arm patterns, but emphasizes the two-arm case because this appears to be of particular significance for the application to the galactic situation. Multiple arm patterns, because of the shape of the circular velocity curve $\theta_c(R)$ as a function of distance from the galactic centre, would have to be confined to a shorter range in R. The main spiral features observed in neutral hydrogen and ranging from about $R = 4$ kpc to $R = 15$ kpc appear to fit very well into a two-arm pattern envisaged by theory. Choosing the spacing between the arms so as to provide coincidence of the theoretical pattern with the Sagittarius and the Perseus arm, the inclination of the arms which then follows from theory to be about $6°$ agrees well with observation. Note that in this pattern the Orion arm is attached only a secondary role, as an interarm feature, or a 'spur'.

The important point to notice is, that in this theory, the spiral arms represent the location of a wave of excess density in the galactic plane (or, rather, in the flat Population I), and that this wave progresses through the system. It moves in the direction of galactic rotation, with (approximately) constant angular speed, Ω_p (pattern speed), which, in the case of the Galaxy near the solar distance R_0 from the centre, is approximately half the angular speed of the flat population. In fact, the pattern speed is given by $\Omega_p = \Omega - \kappa/2$ where Ω is the angular velocity for circular motions and κ is the 'epicycle frequency'. This latter measures the frequency of oscillation around the circular velocity in the imaginary case where one gives a star moving in circular orbit a slight perturbation perpendicular to the circular motion resulting in epicyclic motion. κ depends on R: for large R, where the field is approximately newtonian, and, hence, stellar orbits nearly Keplerian, the epicyclic oscillation has virtually the period of circular motion. For smaller R, where the field deviates more and more from the Keplerian due to the deviation from the case of central mass, the epicyclic period becomes smaller and smaller as compared to the circular period. The quantity $\Omega - \kappa/2$ for the Galaxy happens to be approximately constant over a large range of R, a fact already noted by Lindblad who was the first to postulate the existence of density waves in the Galaxy.

In the region $R \sim R_0$, we have $\Omega = 25$ km/sec per kpc and Ω_p about 13 km/sec per kpc, hence for $R_0 = 10$ kpc there is a relative velocity of about 120 km/sec with which particles in circular orbit move with respect to the pattern. Their response to the potential field of the pattern, as the Lin-Shu theory has shown, is such that then this field can maintain itself. The total density excess in the density wave is about 10% over that of the neighbouring field, and of these 10% about half are due to the stars and half to gas. As, however, the total gas density is considerably smaller than that of the stars, the 5% mass excess due to the gas represents a considerable excess of the gas density with respect to the interarm gas density (factors up to 10 are quoted), whereas for the stars the excess with respect to the interarm density should be of the order of 5 to 10%. It should be noted that only that part of the stellar population which has low or moderate velocity dispersion, smaller than about 35 km/sec, responds to, and helps maintain, the density waves.

The importance of the Lin-Shu theory is, that we now have a theoretical framework within which we can discuss a variety of phenomena. One of these is the process of star formation. If we visualize this as due to a contraction process in the interstellar medium, we now notice that this process must be governed to a considerable extent by the regular passage of the gas, moving in approximately circular orbits, through the density wave pattern. The duration of the passage through the pattern must be of the order of several times 10^7 years. During this period, or as a consequence of several such passages, the contraction toward cloud complexes or individual clouds may take place which subsequently may ignite the star formation. The nature of such processes, involving a shock wave through the gas, is now the subject of detailed study.

4. Comparison with Observed Features

The density wave theory, in its more detailed form, leads to predictions concerning the velocity and density field which should allow comparison with observations. Part of these may be obtained by the study of extragalactic systems. Some can be obtained only from the study of the Galaxy.

A. STELLAR POPULATION

Whereas, so far, the spiral pattern in the gas, and to some extent that in the stars, has been determined via the galactic rotation curve $\theta(R)$, it clearly now becomes more urgent than ever to obtain a direct determination independent of the adopted velocity pattern. Only after having obtained this may we hope to be able to study in an un-biassed way the velocity pattern associated with the spiral structure. The only direct way applicable to the majority of the stars to be considered is the photometric one. How severe the observational requirements are may be judged by the fact, that for a spiral arm at about 5000 pc, i.e. at only half the distance of the galactic centre, an accuracy in distance of 500 pc – half the adopted width of the arm – corresponds to an inaccuracy in distance modulus of 0.2 magnitude. This implies errors below this limit in the correction for interstellar absorption, in the system of absolute magnitudes used, and in the transition from the luminosity criterion used to the absolute magnitude. All three requirements are, at this moment, very hard to meet, especially the first one concerning the interstellar absorption; see the lectures by Drs. Elsässer and Golay. In terms of desiderata this means:

(a) Accurate determination of the interstellar reddening curve for the particular directions studied, including the red and infrared wavelengths.

(b) Further work on the fundamental distance determinations which form the cornerstone for our distance scale, especially on the Hyades and on the trigonometric parallaxes and to some extent also on the Scorpio-Centaurus cluster.

(c) Further work on the choice and calibration of luminosity criteria; it is recommended that the very difficult calibration problem always should be foremost in the minds of those who embark upon the development of photometric or spectroscopic luminosity criteria.

(d) Whereas, so far, in determining the stellar distribution pattern we have had in mind especially the early type supergiants, it would seem to be important that special attention be given now to the M2-M4 giants. As it was pointed out in the lecture of Dr. Mavridis, results obtained by him, Blanco and Westerlund indicate that stars of these spectral classes seem to follow the distribution of the early type spiral population rather than a distribution with concentration toward the galactic centre as it is found for the later type M stars. The study of these objects offers several great advantages: the detection and further photometry in the infrared reduce considerably the difficulties presented by the interstellar extinction, and the stars appear to show a more uniform space distribution, less closely correlated with that of the interstellar clouds, and therefore less sensitive to local irregularities in the interstellar reddening law. It is not clear whether we are dealing here with a spiral indicator of recent stellar formation, or whether we observe the general response to the density wave formation of stars of moderate peculiar velocity. The predicted 5 to 10% density increase for the general stellar field would seem to be very hard to detect; yet the effort seems worthwhile.

B. THE DARK MATTER

An important feature in the patterns of many spiral nebulae is the apparent shift of the dark lanes with respect to the luminous main spiral arms: in the inner parts of the spiral structure the dust lies at the inside, (the trailing side) of the arms. This feature fits well into the density wave theory if we assume that the dark lanes mark the regions of highest gas concentration whereas the luminous parts represent the, mostly, very young and luminous population which was formed due to the passage of the gas through the waves and therefore show up with a certain time lag. Exact location of the dark matter with respect to the gaseous and young star spiral structure in the Galaxy is therefore desirable; recent work in this field was described by Elsässer; increased efforts to extend this to large distances, although very difficult, would seem to be justified in directions of strategic importance.

C. THE INTERSTELLAR GAS

Whereas no direct determination of the distance of HI is possible, one might consider to use instead the location of HII regions through the distances of the exciting stars. Evidence from large-scale surveys in the Galaxy based on radial velocity studies indicates good coincidence, but as soon as we are interested in finding the differential effects referred to in the preceding section, the method fails.

D. SPECIAL STUDIES OF DIFFERENTIAL LOCATION

Bok has recently reported on an observational program for determining the position of a variety of objects associated with the Carina arm (see the report of the Basel symposium). The basic idea is the following: we observe a large section of the Carina arm almost tangentially: the line of sight at longitude 285° virtually coincides with the arm and only at distances exceeding 7 kpc or within 4 kpc deviates from this. Hence if we follow the population of the arm as a function of longitude from $l = 275°$ to $l = 300°$ at distances within the range mentioned, then we should be able to observe the relative position of trailing and leading peaks of the arm population. The approach is favoured by the fact that the line of sight from the sun to the Carina arm first passes through an interarm region of relatively low interstellar obscuration. The results being obtained seem to warrant application to other similarly chosen directions. These studies include radio measures of HI and HII, and of molecular lines.

E. THE VELOCITY PATTERN

The basic role of the velocity curve $\theta(R)$ for the computation of the HI spiral pattern has been explained in Dr. Kerr's lectures. This curve is, however, derived from the 21 cm measures themselves on the assumption of circular velocities around the galactic centre. As Kerr has pointed out there are strong indications of systematic deviation from an axisymmetric circular velocity pattern, and a specially important question is, whether systematic outward or inward motions exist.

Clearly, independent determinations of the deviations from circular velocity are of paramount importance; only after these have been made can we proceed to a satisfactory comparison of the observed and the theoretical detailed velocity pattern. A few lines of approach are suggested.

(a) *The expansion component*

Since all measures, including those of 21 cm velocities, are referred to the Sun, an obvious determination would be that of the velocity of the Sun with respect to the galactic centre. Radio measures of sources of 21 cm or other interstellar radiation at the galactic centre are not suitable because of the turbulent motions in the central regions. Suitable stellar objects, in principle, are those which have a spherical or spheroidal distribution around the galactic centre; we may expect these to possess no systematic expansion or contraction and to be little affected by non axial symmetric density fluctuations in the flat subsystem of the Galaxy. Such objects are the globular clusters, the RR Lyrae variables with periods exceeding 0.4 day, the subdwarfs and, to a lesser extent, the planetary nebulae. A weighted mean of the velocity of these systems with respect to the Sun in the direction away from the galactic centre – the so-called U component, counted positive in the direction $l = 180°$ – is found to be $U = +9$ km/s± 5 (p.e.). Since for most stars as well as for HI, values between $U = +5$ and $U = +12$ km/sec are found, there is no proof of a systematic expansion or contraction of either of these kinds of objects. On the other hand, Kerr's proposed expansion velocity for matter near the Sun of $+7$ km/sec falls entirely within the limits of uncertainty. Clearly, these limits should be reduced, and this should not be impossible in principle by observing large numbers of radial velocities for Halo and Disk population objects at intermediate latitude around the galactic longitude of the galactic centre.

(b) *The circular velocity near the sun*

Denoting the velocity component with respect to the Sun in the direction of galactic rotation ($l = 90°$) by V, the problem is, to find the velocity V_c of an object with circular velocity with respect to the Sun, in order that, for instance, the velocity of HI with respect to the Sun can be reduced to velocity with respect to circular velocity. There is no direct way to determine V_c. One method to be explored is to predict the velocity with respect to circular velocity for certain kinds of stars and to compare these with their velocity relative to the Sun; this should produce V_c. This prediction can be made on the basis of relatively simple dynamical considerations based on the assumption of a steady state axisymmetric stellar system. The velocity sought for, V_*, then is determined by the velocity dispersion σ_π of the class of stars considered and the gradient of the stars' space density as a function of R:

$$V_*(2\theta_c - V_*) = \sigma_\pi^2(\partial \log v/\partial R).$$

Application of this formula to the common types of stars like the G and K giants and dwarfs leads to the prediction that V_c is between -7 and -14 km/sec. This implies that for H I the velocity relative to the circular velocity, V_H is between 0 and -7 km/sec. This would indicate a slight lag of the rotational motion of H I. Since the question of this lag is an important one for both the dynamics of gas and that of the young stars formed out of it, V_* should be determined much better. In such a determination one should try to avoid the use of the common G and K giants and dwarfs since these, belonging mostly to Population I, may not be free of local streaming effects. A better approach would be, to select old Disk Population stars at latitudes around, say, $\pm 20°$, and observe their velocity distribution for fields near $l = 90°$. This would seem feasible with the present objective prism radial velocity techniques. In fact, by applying this technique to stars of various brightness a very valuable check may be obtained on the velocity curve $\theta(R)$ up to several kiloparsecs from the Sun.

F. THE ORION ARM

The Lin-Shu theory does not consider the Orion arm to belong to the main two-arm density wave pattern, but rather as a secondary wave pattern. Since we are in a better position to study detailed properties of this feature than we can for the main spiral arms, the theoretical predictions for such a secondary feature deserve to be looked into in detail: we would wish to compare it with such observational quantities as the local stellar density pattern for young and youngest stars (which are known to show interesting differential effects) and the velocity field of gas and of stars.

5. Statistics of Double Stars and of Clusters

The lectures of Ledoux and Cameron have described progress in the understanding of the process of prestellar evolution. An important desideratum which here comes to my mind concerns the problem of the formation of double stars. This is particularly urgent in the case of the massive stars, those of 5 solar masses and larger.

We know that, for this mass range, so many binary systems occur, that it seems to be the rule rather than the exception that the star formation results in a binary or a multiple system. Recent work by T. S. van Albada and myself, still unpublished (see also, however, Van Albada, 1968) suggests that for binaries of masses in the range indicated, we should consider formation through two different processes: that of the narrow double star (separations of the order of 1 AU. and less) and that of the wide doubles (separations of the order of 10^2 to 10^4 AU). In the latter case, capture in the early protostellar stage within the confinements of the very small protocluster or protoassociation (dimension $\leqslant 0.1$ parsec) may well be the dominating process (see Van Albada, *l.c.*). Further observational work on this problem may lead to an estimate of the volume within which the star formation took place and of the number of stars formed in the protocluster.

The formation of the close doubles, on the contrary, probably is a fairly common phenomenon somewhere in the contraction process from globule (?) to stars. Little understanding of it has been reached so far, yet we must expect the process of this formation (fission?), or even the mere existence of close companions, to affect seriously the evolution tracks towards the main sequence, and hence the interpretation of the colour-magnitude diagrams of the very young clusters.

In the study of the small, very young clusters as well as in that of clusters with large numbers of stars and higher ages, the technique of numerical experiments, as Bouvier has shown in his lecture, now has acquired a place of its own next to the analytical and the observational approaches. Whereas at this moment the limit for such experiments seems to be around 200 stars per cluster, it is clear that for the application to open clusters in general numbers of one or two orders of magnitude larger should be treated. One of the very interesting problems in this domain is the explanation of the statistics of open clusters of different ages: whereas one would expect the number of clusters to increase steeply with advancing earliest spectral type because of the rapidly increasing ages, there is a fairly sharp decline around the age of about 10^9 yrs. This has been explained in a qualitative way by Spitzer (1958), a more refined calculation would, however, be interesting and in particular it would be of great value if from the statistics of presently observed clusters an estimate could be made of their rate of formation in the past.

6. Work on the Intermediate Age Stars

It was stated in the introduction that the bulk of the mass of the Galaxy is in the Disk and the Intermediate Halo Population. The large variety of space distributions and kinematical properties of the various categories of stars in these populations is the result of the past evolution which the Galaxy has undergone during the 'collapse' from the nearly spherical to nearly flat shape. Masswise, the term 'evolution of the Galaxy' should apply first of all to this past evolution.

It is tempting to assume that the various subsystems, like the metal-rich globular clusters, the RR Lyrae stars, and the planetary nebulae, which have different degrees of flattening and of concentration towards the centre, represent different stages in this 'collapse'. It would clearly be of the greatest interest if, by assigning ages to these various subgroups, we could identify certain present degrees of flattening and of central concentration with certain epochs of star formation in the Galaxy. (The degrees of flattening and of central concentration may have changed in the course of the further galactic evolution, therefore the present values for these parameters cannot without more be identified with those at the epoch of the star formation; they contain, however, a clue to tracing them).

In this approach, we need more complete knowledge of the various population components. In certain special cases, much knowledge is forthcoming, for instance for the various types of variables found in the Groningen-Palomar survey (Plaut, 1968) which reaches well beyond the galactic centre, or for the planetary nebulae. The fact that we are not, for the study of halo and disk, seriously hampered by the interstellar absorption helps a great deal in this work. On the other hand, what we shall need is knowledge of the evolutionary stages and estimates of the ages of the objects concerned. Here, we are only at a very beginning of what is to be acquired. For instance, we do not yet see clearly what stage is represented by the planetary nebulae, a problem referred to in Cameron's lecture. In the general approach sketched here, the, in a sense, peculiar objects like the planetary nebulae and the variable stars must naturally be considered as representing a whole populating subgroup of the corresponding age, and the extrapolation from these peculiar objects to the subgroup, with an estimate of its total mass will be the next, large and difficult, step.

A more modest, but nevertheless very useful program consists of determining the space distribution and velocity properties for the disk population up to distances of several kiloparsecs from the sun. Such studies can be based on surveys at high and intermediate latitudes. If we aim at studying, for instance, the late type giants of low metal content, distances as indicated correspond to stars of about 12th magnitude. Such studies are within present possibilities for intermediate band photometry, for radial velocity determinations and proper motion work. The main problem is still, the choice of suitable photometric criteria for the selection of a homogeneous subgroup of stars, and the development of such criteria forms one of our main desiderata.

The project just described is closely related to an important problem that has been with us for many years: the accurate determination of the field of force perpendicular to the galactic plane. There are two principal reasons why we should solve this problem: the accurate determination of the total mass density in the solar neighbourhood, and providing a solid anchor point for the construction of mass models of the Galaxy. At short distances, up to a few hundred parsec from the galactic plane, the force perpendicular to the plane is determined mainly by the local mass density, but at larger distances, the component of the force due to the general mass distribution in the Galaxy becomes more and more important. This problem requires basic observational data of much the same kind as that described in the previous paragraph: identification of suitable population components up to a kiloparsec at least in the direction of the galactic pole and radial velocities of the stars selected.

Ideally, one should not only obtain this force field for the column perpendicular to the plane at the sun, but also at somewhat larger and smaller distances from the galactic centre. An interesting application of this would be the calculation of the 'flight time' of stars ejected with intermediate velocities at the time of formation in, for instance, the Perseus arm, and observed at their descent, or at a subsequent passage, through the galactic plane, near the sun. For stars with initial velocity components perpendicular to the plane of 20 km/sec, this semi-oscillation time around the plane is estimated to be about 80 million years, and it increases strongly with increasing velocity. Hence, identification of stars at such passages should provide a useful, independent check on the determination of stellar ages. The maximum heights above the plane reached by such stars during their flight are of the order of 500 pc.

References

Albada, T. S. van: 1968, *Bull. Astron. Inst. Neth.* **20**, 57.
Blaauw, A.: 1965, 'The Concept of Stellar Populations', Chapter 20 in *Galactic Structure* (ed. by A. Blaauw and M. Schmidt), University of Chicago Press, Chicago.
Lin, C. C., Yuan, C., and Shu, Frank H.: 1969, *Astrophys. J.* **155**, 721.
Plaut, L.: 1968, *Bull. Astron. Inst. Neth. Suppl. Series* **2**, No. 6.
Spitzer, L.: 1958, *Astrophys. J.* **127**, 17.

INDEX OF NAMES

(Numbers in italics refer to the page on which the reference is listed.)

INDEX OF SUBJECTS

ASTROPHYSICS AND SPACE SCIENCE LIBRARY

Edited by

J. E. Blamont, R. L. F. Boyd, L. Goldberg, C. de Jager, Z. Kopal, G. H. Ludwig, R. Lüst,
B. M. McCormac, H. E. Newell, L. I. Sedov, Z. Švestka, and W. de Graaff

1. C. de Jager (ed.), *The Solar Spectrum. Proceedings of the Symposium held at the University of Utrecht, 26–31 August, 1963*. 1965, XIV + 417 pp.

2. J. Ortner and H. Maseland (eds.), *Introduction to Solar Terrestrial Relations. Proceedings of the Summer School in Space Physics held in Alpbach, Austria, July 15–August 10, 1963 and Organized by the European Preparatory Commission for Space Research*. 1965, IX + 506 pp.

3. C. C. Chang and S. S. Huang (eds.), *Proceedings of the Plasma Space Science Symposium, Held at the Catholic University of America, Washington, D.C., June 11–14, 1963*. 1965, IX + 377 pp.

4. Zdeněk Kopal, *An Introduction to the Study of the Moon*. 1966, XII + 464 pp.

5. Billy M. McCormac (ed.), *Radiation Trapped in the Earth's Magnetic Field. Proceedings of the Advanced Study Institute, Held at the Chr. Michelsen Institute, Bergen, Norway, August 16–September 3, 1965*. 1966, XII + 901 pp.

6. A. B. Underhill, *The Early Type Stars*. 1966, XIII + 282 pp.

7. Jean Kovalevsky, *Introduction to Celestial Mechanics*. 1967, VIII + 427 pp.

8. Zdeněk Kopal and Constantine L. Goudas (eds.), *Measure of the Moon. Proceedings of the Second International Conference on Selenodesy and Lunar Topography held in the University of Manchester, England, May 30–June 4, 1966*. 1967, XVIII + 479 pp.

9. J. G. Emming (ed.), *Electromagnetic Radiation in Space. Proceedings of the Third ESRO Summer School in Space Physics, held in Alpbach, Austria, from 19 July to 13 August, 1965*. 1968, VIII + 307 pp.

10. R. L. Carovillano, John F. McClay, and Henry R. Radoski (eds.), *Physics of the Magnetosphere. Based upon the Proceedings of the Conference held at Boston College, June 19–28, 1967*. 1968 X + 686 pp.

11. Syun-Ichi Akasofu, *Polar and Magnetospheric Substorms*. 1968, XVIII + 280 pp.

12. Peter M. Millman (ed.), *Meteorite Research. Proceedings of a Symposium on Meteorite Research held in Vienna, Austria, 7–13 August, 1968*. 1969, XV + 941 pp.

13. Margherita Hack (ed.), *Mass Loss from Stars. Proceedings of the Second Trieste Colloquium on Astrophysics, 12–17 September, 1968*. 1969, XII + 345 pp.

14. N. D'Angelo (ed.), *Low-Frequency Waves and Irregularities in the Ionosphere. Proceedings of the 2nd ESRIN-ESLAB Symposium, held in Frascati, Italy, 23–27 September, 1968*. 1969, VII + 218 pp.

15. G. A. Partel (ed.), *Space Engineering. Proceedings of the Second International Conference on Space Engineering, held at the Fondazione Giorgio Cini, Isola di San Giorgio, Venice, Italy, May 7–10, 1969*. 1970, XI + 728 pp.

p.t.o.

16. S. Fred Singer (ed.), *Manned Laboratories in Space. Second International Orbital Laboratory Symposium*. 1969, XIII + 133 pp.

17. B. M. McCormac (ed.), *Particles and Fields in the Magnetosphere. Symposium Organized by the Summer Advanced Study Institute, held at the University of California, Santa Barbara, Calif., August 4–15, 1969*. 1970, XI + 450 pp.

18. Jean-Claude Pecker, *Experimental Astronomy*. 1970, X + 105 pp.

19. V. Manno and D. E. Page (eds.), *Intercorrelated Satellite Observations related to Solar Events. Proceedings of the Third ESLAB/ESRIN Symposium held in Noordwijk, The Netherlands, September 16–19, 1969*. 1970, XVI + 627 pp.

20. L. Mansinha, D. E. Smylie and A. E. Beck, *Earthquake Displacement Fields and the Rotation of the Earth. A NATO Advanced Study Institute. Conference Organized by the Department of University of Western Ontario, London, Canada, 22 June – 28 June, 1969*. 1970, XI + 308 pp.

21. Jean-Claude Pecker, *Space Observatories*. 1970, XI + 120 pp.

In preparation:

22. L. N. Mavridis (ed.), *Structure and Evolution of the Galaxy. Proceedings of the NATO Advanced Study Institute, held in Athens*, September 8–19, 1969.

23. A. Muller (ed.), *Magellanic Clouds*

24. B. M. McCormac (ed.), *The Radiating Atmosphere*.

SOLE DISTRIBUTORS FOR U.S.A. AND CANADA:

SPRINGER-VERLAG NEW YORK, INC., 175 Fifth Ave., New York, N.Y. 10011